ESSAYS ON EXPLANATION AND UNDERSTANDING

SYNTHESE LIBRARY

MONOGRAPHS ON EPISTEMOLOGY

LOGIC, METHODOLOGY, PHILOSOPHY OF SCIENCE,

SOCIOLOGY OF SCIENCE AND OF KNOWLEDGE,

AND ON THE MATHEMATICAL METHODS OF

SOCIAL AND BEHAVIORAL SCIENCES

VOLUME 72

ESSAYS ON EXPLANATION AND UNDERSTANDING

Studies in the Foundations of Humanities and Social Sciences

Edited by

JUHA MANNINEN

Department of Philosophy, University of Helsinki, Finland

and

RAIMO TUOMELA

Department of Philosophy, University of Helsinki, Finland

D. REIDEL PUBLISHING COMPANY

DORDRECHT-HOLLAND / BOSTON-U.S.A.

Library of Congress Cataloging in Publication Data

Main entry under title:

Essays on explanation and understanding.

(Synthese library ; 72)
Includes bibliographical references and index.
1. Hermeneutics—Addresses, essays, lectures. 2. Intention (Logic)—Addresses, essays, lectures. 3. Comprehension—Adresses, essays, lectures. 4. Causation—Addresses, essays, lectures. 5. Wright, Georg Henrik von, 1916–
I. Manninen, Juha. II. Tuomela, Raimo.
BD241.E84 121 75-29214

ISBN 90 277 0592 5

Published by D. Reidel Publishing Company,
P.O. Box 17, Dordrecht, Holland

Sold and distributed in the U.S.A., Canada, and Mexico
by D. Reidel Publishing Company, Inc.
Lincoln Building, 160 Old Derby Street, Hingham,
Mass. 02043, U.S.A.

Printed in The Netherlands by D. Reidel, Dordrecht

TABLE OF CONTENTS

PART IV / REPLIES TO COMMENTATORS. SECOND THOUGHTS ON EXPLANATION AND UNDERSTANDING

PREFACE

The present anthology consists of contributions to the philosophy of the humanities and the social sciences by European and American scholars. These writers represent not only the analytical tradition but also hermeneutic philosophy and Marxism.

Of the papers included in this volume those by Professors Bubner, Hintikka, Kron, Tuomela, Kenny, Stoutland, Martin, and Niiniluoto were presented at an international colloquium on explanation and understanding held in Helsinki, Finland, on January 24–26, 1974. In addition to these papers, all the others, except Professor Winch's, were written specially for this anthology and have not been published before.

Most of the papers are related to the work of Professor Georg Henrik von Wright in this philosophical field. The anthology includes an essay by Professor von Wright himself, together with his extensive replies to his commentators. In these replies he elaborates and, to some extent, revises his earlier views. On the whole, the anthology reflects the on-going confrontation of the analytical and continental trends. It produces new philosophical and methodological ideas and results concerning the theory of human action, intentionality, understanding, causality, and determinism.

Basil Blackwell Publishing House has kindly permitted the reprinting of Professor Winch's paper, originally published in *Metaphilosophy* **4** (1973), pp. 63–75. We gratefully acknowledge the financial support by the University of Helsinki, the Philosophical Society of Finland, and the Finnish Cultural Foundation (*Suomen Kulttuurirahasto*), which made possible the Helsinki colloquium.

May 1975

JUHA MANNINEN
RAIMO TUOMELA

PART I

APPROACHES TO TELEOLOGY, INTENTIONALITY, AND HISTORICAL UNDERSTANDING

MANFRED RIEDEL

CAUSAL AND HISTORICAL EXPLANATION

The Problem of Teleology in Analytical and Dialectical Philosophies of History

In the discussion between the analytical philosophy of science and dialectics there has recently been a remarkable change. At the same time as the latest debate about method was staged in German sociology on inverted positions, with essentially ideological arguments and culminating in the misleading confrontation of a double positivism,[1] a change of argumentation was made within analytical philosophy, largely hidden from the participants of the so-called positivism debate. It is focused on the *analytical philosophy of action*, or, more precisely: the interest in the concept of action and in the logic of practical argumentation has been growing in the analytical philosophy of science as a result of the realization that the foundations of social and historical sciences were so far insufficiently clarified. The change originated thus in the *analytical philosophy of history*. The debate upon the methodical basis of the historical sciences[2] dissolved the traditional positivistic frame of mind in analytical philosophy and put into motion the well worn fronts in the discussion of the philosophy of science. Analytical philosophy of history – as has been shown by the long controversy about the suggestion from Hempel/Oppenheim for a comprehensive theory of explanation (Covering Law-Model)[3] – is only possible under restriction of the analytical program.

According to the Covering Law Model an individual event E can be deductively nomologically explained by a certain number of other similar events $E_l, \ldots, E_m$, called basis, and one or more general laws $L_l, \ldots, L_n$, – or in a weaker version – can be derived from a *given* basis $E_l, \ldots, E_m$ with the probability p. The Covering Law Model, which in my opinion has been more appropriately called the Subsumption Theory of Explanation in the discussion led by G. H. von Wright,[4] is not, however, as comprehensive as the analytical philosophers would like to think. As a theory of scientific explanation not only does it remain incomplete, but it also simplifies in an inadmissible way the structure and purposes of scientific activity in the area of the social and historical sciences. There are essentially two reasons for this, which have already been developed by P. Gar-

Manninen and Tuomela (eds.), Essays on Explanation and Understanding, 3–25. *All Rights Reserved.*

diner with sufficient clarity. Gardiner's examination was oriented towards the historical sciences: it presented the question about the nature of historical explanations. From the point of research practice ("what the historian is doing") Gardiner can prove that the Subsumption Theory of Explanation is not sufficient enough for either (a) the aim of scientific activity or (b) the complex means of language the historian must use in the pursuit of these aims. According to Gardiner, to whom Hempel's characterization of historical explanation as merely an explanation sketch appeared vague and undefined,[5] the historian has to deal with *action*, occurring in a particular place at a particular time and under particular conditions. He will not be able to explain the actions by subsuming them under general laws like natural events. His aim is much more "to talk about what happened on particular occasions in all its variety, all its richness, and his terminology is adopted to this object."[6] To reach this aim it is necessary besides the (natural science) Subsumption Theory of Explanation (called by Gardiner "explanation in terms of 'causes' and 'effects'"), to have an explanation in terms of intentions and plans. In this case – and that is the usual case of his methodical activity – the historian explains in a language expressing what it is reasonable to do under given circumstances to obtain the particular ends desired.[7]

Whereas Gardiner psychologically abridged this method – apparently in succession to the methodical dualism of Dilthey and Rickert – and introduced it solely as a supplement to the Subsumption Theory, W. Dray completely rejected the positivistic conception of the Subsumption Theory in his criticism of C. G. Hempel. The contradiction between the ambitious pretensions of the Subsumption Theory and its limited explanatory value for the historical sciences, which the analytical philosophers have admitted themselves, has its origin in the simple fact that historical explanation has a methodological structure unaccommodating to the logic of subsuming events under general laws. What interests the historian is not so much laws or general characteristics of a group of events, but individual events in an individual situation. His question is not: "What causes *y*'s?; he asks, What is the cause of this *y*? – and he asks this about an *y* in a determinate situation."[8] Although Dray emphasizes much more than Gardiner the methodological independence of historical explanation, he is unable to adequately explain where its uniqueness lies. Explanation of a historical event for Dray also means to speak in terms of intentions and plans – the historian does not have to explain natural phenomena, but has to explain human action. The problem of historical explanation is therefore the problem of how is it possible to explain action. In his answer Dray approximates the hermeneutic theory of understanding Droysen and Dilthey attempted to develop in the 19th century as a method for the cultural sciences (*Geisteswissenschaften*). To explain an action means to make the intention of the person carrying out the action understandable. The understanding of intentional action is more than a heuristic device. The historian, as Dray expresses it in the lightly emphatic tone of the hermeneutic tradition, "must *revive*, *re-enact*, *rethink*, *re-experience* the hopes, fears, plans, desires, views, intentions etc., of those he seeks to understand. To explain action in terms of the covering law model would be to achieve, at most, an external kind of understanding."[9] But Dray is unable to show plausibly in what way this understanding is to be methodically substantiated. He repeatedly emphasizes that historical events cannot be explained through empathy with the psychic conscious acts of monological individuals (personal action), but only through consideration of the actions of groups (social action), here also adhering to the factual intentions of the participants as the only possible explanatory principle. What Dray has sorted out among the various ways of acting is that kind of action Max Weber called purposeful action (*zweckrationales Handeln*). To explain an act means nothing else than to show that in a given situation it was reasonable (in the sense of purposeful) to execute this action.

I

If it is still worthwhile in philosophy to discover problems and methodically advance their formulation, then it belongs to the significant achievements of von Wright's examination that it pointed out the unsolved questions of substantiation in the analytical philosophy of history in an examination of them which opened up new points of relevance in wake of a far-reaching reorientation in analytical philosophy. According to von Wright, the methodical peculiarity of explanation in the social and historical sciences can only be adequately reconstructed when – instead of seeking refuge in the usual ideas about the role of empathy in understanding – the implied teleological elements are taken into consideration. What Gardiner and Dray called 'explanation through a motive' or 'rational explanation', could find – according to von Wright's supposition – their substantiation in a particular kind of teleology which the analytical philosophers, orientated towards natural science and positivism, stubbornly, but in vain attempted to avoid. In his assertion von Wright feels encouraged by the rise of the analytical philosophy of action (Anscombe, Taylor),[10] which, combined with the advance of cybernetics, reawakened the temporarily obsolete discussion about the use of teleology or the legitimacy of teleological conceptualization in the philosophy of science.

To avoid confusion with the traditional metaphysical interpretation, von Wright distinguishes two groups of teleological terms. The first group, belonging primarily to the language of biology, consists of the basic terms function, purposefulness, and organic unity, the second group consists of the terms aim and intentionality usually found in the social sciences. The language of cybernetics and its explanatory model, which has spread to the other sciences, uses only quasiteleological concepts void of intentionality. A minimal terminological basis is the distinction drawn between behavior (biological and cybernetical sciences) and action: "Among things to which intentionality is attributed actions occupy a prominent place."[11]

The position von Wright wins from here can be characterized as a somewhat ambitious systematic attempt to bring the *quaestio facti* in the debate surrounding the analytical philosophy of history back to the basic arguments. His suggestions receive additional weight in so far as he would like to reformulate the problems of the volatile history of the reaction of natural sciences and positivism against Hegelian philosophy, the counter-reaction of the cultural sciences against the natural sciences and the debate ensuing from it in regard to the relationship between understanding and explanation. Up until now these problems have been either left behind as unsolved or shoved off on to one of the various 'schools' to be administered. The methodically relevant suggestion, briefly stated, leads to consideration of historical explanation as a special kind of explanation, designated as teleological and differentiated from causal explanations. The introduction of this suggestion is supported essentially by a *historical* and *systematic* argument. The first is *methodical-historic*, and it recapitulates a piece of modern scientific development. In the formulation of conditions which an explanation has to fulfil in order to be considered scientifically respectable the Aristotelian and

Galilean tradition stand opposite one another. Both are historically associated with the opposition between causal and teleological thinking on the one hand, explanation and understanding on the other. The Galilean tradition in science corresponds to the efforts made (especially in modern times) to explain and predict phenomena, the Aristotelian tradition to the parallel efforts to make facts teleologically and finalistically understandable. The two main directions in the philosophy of scientific method determined, in the 19th century, the relationship between the natural and human sciences, and also the contemporary opposing positions of dialectical-hermeneutical and analytical philosophies of science.[12]

The second argument is *methodological* and relies on reflections from the analytical philosophy of action oriented towards Aristotle. The methodical peculiarity of the teleological explanation according to von Wright can be most adequately compared with the theory of practical inference – by Aristotle obviously only weakly developed – that (according to Anscombe) can be followingly reconstructed: the major premise of the syllogism is the end *Z*, a thing desired or a purpose of the action: the minor premise relates some action to this thing, roughly as a means to the end; the conclusion consists of the application of the means to the achievement of the end. The affirmation of the premises in a practical inference leads to the corresponding action according to the same rule through which an assertion of the premises in a theoretical inference necessarily leads to an assertion of the conclusion. Von Wright has given the following schema:

(PI) *A* intends to bring about *p*.
A considers that he cannot bring about *p* unless he does *a*.
Therefore *A* sets about to do *a*.

The schema of practical inference (which can be formulated in various ways) contains all the methodical steps in a reverse order necessary for an explanation of actions; it is, as von Wright expresses it, the schema of teleological explanation "turned upside down".[13] The starting point of a teleological explanation of action is that someone sets himself to do something or, more commonly, that someone does something. In answer to the usual question "Why?", the answer often is simply: "In order to bring about *p*". It is taken for granted that the agent considers the behavior which we are trying to explain causally relevant to the bringing about of *p and* that the bringing about of *p* is what he is aiming at or intending with his behavior. It is thereby irrelevant for the steps of methodical procedure as well as for the validity of the suggested explanation whether or not the agent is mistaken in thinking the action causally related to the end in view: "What the agent *thinks* is the only relevant question here".[14]

We can pass over the further argumentation in the debate between causal and intentional positions and restrict ourselves to a discussion of the basic methodological principles connected with the teleological model of explanation. According to von Wright the so-called practical syllogism does not represent a form of demonstration, but a different form of reasoning, which he calls practical reasoning. Practical reasoning – so the central argument goes – is of great importance to the explanation and understanding of action. "It is a tenet of the present work that the practical syllogism provides the sciences of man with something long missing from their methodology: an explanation model in its own right which is a definite alternative to the subsumption-theoretic covering law model."[15]

In connection with this surprising thesis in the fundamental discussion of the analytical philosophy of science, von Wright refers not only to Aristotle, but also to Hegel and the non-scientific school of humanistic (anthropological) Marxism influenced by Hegel. According to the historical argument, Hegel is considered as the great reviver of the Aristotelian tradition in the philosophy of methodology. The systematic argument is confirmed through Hegel's teleology. "Neglect of practical reasoning, however", von Wright says in a note, "has not been as complete as Miss Anscombe seems to think. Hegel's doctrine of what he occasionally also calls 'Schluss des Handelns' is interestingly similar to the idea of a practical syllogism as treated in the present work".[16] In fact the chapter dealing with teleology in Hegel's *Science of Logic* is built along the lines of von Wright's suggestion for reconstruction (a) the first premise being the subject's aiming at an end (*subjektiver Zweck*); (b) the second premise is the contemplated means to the end (*Mittel*); (c) the conclusion consists of the objectivation of the aim in action (*ausgeführter Zweck*). An indication of this is given in the following quotation from the text: "the end joins itself with objectivity (*Objektivität*) through a means and in this objectivity is joined with itself... the means is therefore the *formal* middle term of a *formal* syllogism; it is an *external element* (*Äußerliches*) against the *extreme* of the subjective end as well as also against the extreme of the objective end".[17] Is this kind of teleology, as we have introduced and explained it, suitable to fill the methodological gaps in the humanistic sciences? Is it a way out of the dilemma between the analytical and dialectical-hermeneutical philosophies of science opening up here, a dilemma von Wright correctly considers unsatisfactory? And finally, how secure is the foundation of the idea of practical reasoning when it is based on Aristotle and Hegel?

II

Under these aspects it would be helpful if we explained what teleology means according to Hegel. So that the problem is not unnecessarily isolated I will deal with it in conjunction with a theory of historical explanation, an interpretation convincingly made by von Wright. What comes into question here is the relationship between Hegel's teleology and Hegel's philosophy of history. The importance of Hegel to the analysis without doubt relates to von Wright's intention of freeing the problem of explanation in the historical and social sciences from the false problematic of a non-explanatory method of the understanding, which first came into existence as a reaction to positivism and its uncritical use of the concept of causality. The debate about the role of causality in history derives much of its confusion and obscurity from a failure to separate the question of the appropriateness of a certain (causal) *terminology* from the question of the applicability of certain (causal) categories or concepts to historical events.[18]

For Hegel, who was one of the first to criticize the unjustified use of a causal language in historical analysis, this methodical insight is obvious. The application of causal relationships to the interrelationships of life, natural and spiritual, remains inadmissible. What is thereby described as a cause always has a different content than the effect derived from it, because "that, which has an effect on a living being, is independently determined, changed and transformed by it, *because the living being does not allow the cause to reach its effect*, which means it is suspended as a cause".[19] To give as the cause of the works of Homer the Ionian climate or as the cause of the decline of the republican constitution in Rome Caesar's ambition means literally to have explained nothing. In both cases the effect has an entirely different content and does not have any relationship whatsoever to the cause: "In history generally the spiritual masses and individuals are in play with each other and under mutual influence; it is still even more true of the nature of the spiritual life than of the character of life generally, not to *assimilate another as original* or to allow a cause to continue in itself, but rather to break it up and change it".[20] Historical life cannot be represented through lineal causal relationships, but it requires a teleological language in order to be understandable in its movements.

The negative answer hardly means that Hegel thought a causal terminology was inadequate for history. This (obviously wrong) conclusion was first drawn by the hermeneutic method bringing with it the opposition of explanation (in the natural sciences) versus understanding (in the cultural sciences). According to Hegel's usage, the meaning of both words is determined partly by the common usage, partly by Kant's distinction between intellect (*Verstand*) and reason (*Vernunft*). In a certain way Kant can even be considered as the originator of the contemporary familiar usage of these words. The expression "to explain" is used by Kant with approximate consistency. It says as much as "to bring under laws", which can have "an object given in some possible experience".[21] "To explain" is synonymous with to substantiate, to comprehend, to deduce (derive), to explicate, – to grasp the conditions of the possibility of an object according to the (a priori) conditions of our own possible experience – such as the basic concepts of cause and effect.[22] According to theoretical philosophy (natural science), explanation is an achievement of the intellect, which Kant, impartial to language, called understanding. Experience says approximately "understandable perception", whereby the expression understanding means: "to know" an object given in perception "through the intellect by virtue of the categories".[23] This word usage changes in the practical philosophy and correspondingly in Kant's teleology (*Critique of Judgement*). The concepts of freedom and purpose are not constitutive for any objects of possible experience. What freedom means, can well be understood in a practical respect when someone is speaking of an ought, but it cannot be theoretically explained (in the sense of to substantiate and to explicate). With the moral principle we can create meaning (objective reality) for the concept of freedom: that means to comprehend the concept in its possibility according to the laws of practical reason. The beginnings of a terminological contradiction develop first with the introduction of teleology as a mediating principle between the intellect and reason. Causality and teleology are considered by Kant to be two kinds of explanation which have their peculiar principles, and which are not interchangeable. The problem is their unification in a common principle, denied by Kant: "These two principles are not capable of being applied in conjunction to one and the same thing in nature as co-ordinate truths available for the explanation or deduction of one thing by or from another. In other words they are not to be united in that way as dogmatic and constitutive principles affording insight into nature on behalf of the determinant judgement".[24] The principles of the teleological judgement (end – means) allow strictly speaking only a *discussion* (exposition) and not an explanation (deduction); Kant can also say that the intrinsic possibility of certain objects (in nature and in culture) is not explainable by means of a causality according to ends, but can only be made intelligible through the addition of teleological concepts.[25]

Hegel did not follow the Kantian terminology despite many points of similarity with Kant's teleology. The distinction between concepts of the intellect and reason appeared to Hegel awkward. All understanding that has not achieved rational concepts means nothing more than a representation of arbitrary contents of perception, – "as though someone, who has been told the way that he must go left at the end of the forest, perhaps answers: I *understand*, so *understanding* (*verstehen*) means nothing more than fastening in the imagination and memory".[26] Understanding in this pre-scientific sense remains bound to particular speech and action situations, while all comprehension of situations and actions with scientific terms such as cause and effect – despite the inadequacy of the means – goes beyond the mere representation and retention of arbitrary areas of perception. So Hegel – in principle not differently from Kant – accepts the validity of causal explanation in the realm of history: history "in general must be observed with reason, cause and effects must be made comprehensible for us".[27] The question of the adequacy of a causal terminology is not raised in the form of asking whether the use of cause and effect is suitable to historical events but rather of asking whether the historian correctly uses such concepts methodologically in correspondence with the norms of rational comprehension. Not without reason, Hegel criticizes the uncritical, unconscious usage of the concepts, which thereby eliminate the possibilities of historical explanation by dividing historical action into punctuated events, so that it will then necessarily – through external causal relationship – have to bind them together again. This concept – which characterizes the historical thinking of the Enlightenment as well as its opponent, the romantic glorification of history – contradicts the fundamental rule of scientific explanation (*causa aequat effectum*) which *reason* prescribes to nature and history.[28] That it was precisely the scientistic historical thinking which has not observed this rule was pointed out by Hegel in the *Science of Logic*.

When all events have their cause, then – according to the first argument – there cannot be any insignificant causes. Hegel speaks of the "common joke in history" permitting great consequences to originate from small causes.[29] Historical events – according to the second argument for a causal explanation – stand under the same universal laws according to which natural events occur; historical actions (revolutions, wars) are cases in a universal subsumption theory of explanation, under which the cases are subsumed.[30] On this basis (only extended by the positivists later) history will remain constantly incomprehensible. Mechanical thinking, which reduces historical events and interrelations to alleged necessity and this alleged necessity to causal relationships, has at its disposal no criterion for conceptual distinction. It finally degenerates into ingenious mental gymnastics that along with the contemplation of the understanding lose sight of all possibilities of rational comprehension. About this "arabesque picture of history", which allows a shaking stalk to go forth as a great figure, Hegel observed: "Present in the origination of the great out of the small is in general the reversal the mind (*Geist*) intends with the superficial (*Äußerlichen*); but exactly therefore it is not the *cause of it*, rather the reversal suspends (*aufheben*) even the relationship of causality".[31]

Natural and still more so historical life build the external appearance (*Äußerliche*) of nature in itself, *mind*, speaking with Hegel, *is active mediation with itself*. Structure and direction of this activity are no longer explainable in the language of linear causality. The result of critical examination reveals that the terms of cause and effect are insufficient as means of explanation for historical action, which requires the addition of another terminology more suitable to the conditions of human life.

The terms 'end' and 'means' are needed, the language of teleology, and not only that, but a principle is needed to make this terminology manageable and to prevent its uncritical use. Is Hegel, who convincingly refuted the unjustified transference to the realm of history of (natural scientific) categories of reflection, as he called them, in possession of such a principle?

III

Before that question can be answered the systematic position of the problem of teleology in the dialectical philosophy of history should be sketched first. According to von Wright, the methodical peculiarity of historical explanation lies in the combination of causal and teleological procedure. The explanation of historical events (wars, revolutions) often consists of simply pointing out *previous* events (for example an assassination, breach of a treaty and so forth), considered as their cause or partial cause. According to this schema *explanans* (the antecedent condition of a historical event) and *explanandum* (the event to be explained) are logically independent from one another. As opposed to causal explanation, they are not connected by a general law, but by a number of singular statements forming the premises of a practical inference. The conclusion is often not the explanandum itself, but some other intermediate action (for example an ultimatum or maneuvers by the armed forces), which leads to the premises of other practical inferences with further intermediate conclusions (mobilization), until finally after going through numerous practical inferences and their corresponding steps of action we reach the explanandum itself.[32] Historical explanations of this kind, which are without doubt characteristic for the methodical procedure of a historian at a particular level of research, von Wright prefers to call quasi-causal rather than teleological, although in his own opinion teleology essentially comes into the practical inferences connecting the *explanans* with the *explanandum*.[33]

This concept of historical explanation still incompletely describes what the historian does. Essentially it comprises all the working methods, which in the traditional language of historiography come under the methodical task of establishing the facts. That the premises of practical inferences embrace more than just the quasi-causal connection of events was explicitly conceded by von Wright, when, for example, he spoke of the teleological or motivational background of historical explanations. The goals and aims, as von Wright stated occasionally, in the background of an explanation of the type studied, are sometimes the rather subtle products of cultural, political, religious, etc. traditions.[34]

Traditions of this kind and their unique teleological productivity are the favorite themes of the dialectical philosophy of history. Hegel summarized both by the terms "objective" and "absolute" mind. His thesis is that historical action and historical events can be explained in their origin and progression only through a consideration of traditions, the background of the historically evolved "spirit of the times".

We can pass over the ambiguous meaning of these terms in so far as we concern ourselves solely with the methodical aspects Hegel also combined with the expression "mind". Above all else it is the viewpoint of teleology which determines the methodical handling of the problems of historical explanation in the dialectical philosophy of history. Hegel apparently did not regard the possibility of combining causal and teleological procedures. Already the immediate (*unmittelbare*) history – in scientific study of history the first and oldest step to historical explanation was achieved by the founders of historiography, Herodotus and Thucydides – is built entirely on the language of teleology, the terms of end and means. Teleology is justified by the fact that Hegel – in consensus with Kant's 'Idea for a Universal History from a Cosmopolitan Point of View' (1784) – sees the methodical task of written history and of the philosophy of history in the *explanation of action*. Human actions – generally – are motivated by purposes. Therefore the language of causality is inadequate. A causal explanation can only describe purposeful action from the outside. Historical actions can only be *explained* (by Hegel an expression almost synonymous with 'comprehend') when they are translated into the language of teleology and teleological background Hegel calls "mind".

We can clarify the Hegelian model for historical explanation with an example. How does Hegel explain the French Revolution? Of concern here is not just one among many tasks, but the *primary methodical task of the philosophy of history*. Whoever reads the final chapter of Hegel's lectures referring to this without bias will be surprised (as is usually the case with Hegel's presentation of events) by the poverty of causal analysis. References to previous events, descriptions of situations and actions, reconstructions of practical inferences are rarely made. Instead Hegel gives a representation, unsurpassable in its kind, of the teleological background before which the historical actions and events transpired. It was the goals and ends of the mind, those rather subtle products of the cultural, political, religious tradition, which brought forth the French Revolution: "The French Revolution had its beginning and origin in thought."[35] Hegel connects the revolution with the thought of reason in the most ancient (Anaxagoras) and most modern (Rousseau, Kant) philosophy. Anaxagoras begins with reason, he "was the first to say, that the *νοῦς* rules the world". But only with the start of the revolution, did "mankind reach the point where it recognized that thought should rule the spiritual reality".[36] Hegel draws a connection between the first (Greek) and second (modern) Enlightenment. The beginning and origin of revolution – according to Hegel's thesis in his attempted explanation, by no means original – lies in the Enlightenment, which itself came into existence with the rise of rational thinking in modern *natural science* and the modern Law of nature. These developments led to criticism of traditional determination of ends in human thinking and acting (the rule of authority) and therefore to the change in the teleological background from which finally the revolutionary events in France resulted.

The task of historical explanation is divided into two partial tasks according to Hegel. It consists of, (1) the systematic reconstruction of the 'spiritual' background, in this case: the substantiated aims and ends forming the premises in practical inferences, and (2) the description of peculiarities of the situation entering into the historical action and modifying – often importantly – the setting of goals. From the teleological background alone it is impossible to derive the event of the French Revolution. According to Hegel's representation the German and French Enlightenment were united through the same aims and ends; the substantiating principle of practical argumentation is the idea of *freedom*. To explain why it became revolutionary in France, and that means to explain the real causes of the French Revolution, it is necessary to have a description of the historical situation – singular sentences which have mutually exclusive premises for action (aims and ends) by individuals and groups and with that – again – have traditions as their content. The following text can serve as an example of such a description: "The terribly hard oppression resting on the population, the predicament of the government having to scrounge together the means for the pompous and prodigal count first gave rise to the dissatisfaction. The new spirit (the Enlightenment, M.R.) became active; the pressures forced inquiry ... the entire system of the state appeared as an injustice. The government, however, did not want to relieve the predicament by modification of the relationships and so the changes were necessarily violent, because the alteration was not undertaken by the government. It was not undertaken by the government, because the court, the clergy, the aristocracy, even the parliament did not want to give up possession of their privileges neither for the sake of necessity nor for the sake of implicit and explicit right, furthermore because the government as the concrete center of state power could not accept the abstract individual will as principle and reconstruct the state with it, and finally, because it was a Catholic state, the concept of freedom and rationality of laws were not regarded as having final and absolute obligatory power, because of the separation between them and the holy or religious consciousness."[37]

Hegel's answer to the question why the idea of freedom in France necessarily proceeded revolutionarily is as differentiated as it is in fact unsatisfying. The dissatisfaction with the Hegelian explanation does not lie in the fact that the dialectic philosophy of history – idealistically standing on its head – has as its principle the teleological productivity of the mind. Causes of historical changes, according to Hegel, are not only abstract aims and ends, but also material means, technical instruments and corresponding to these shifts in the economic relationships of a society.[38] Dissatisfying is the introduction and use of the teleological principles of explanation, that is, the generally unclear relation of the respective event or respective action to the teleology of mind, to which the philosophy of history lays claim. Hegel, for example, can say of the technical means such as gunpowder, which at the close of the middle ages leveled the feudal class differences, that mankind needed them and immediately they were there.[39] That this procedure does not always have to lead to pseudo-explanations is shown by the previous quotation. At the same time it reveals in detail what escapes Hegel's theory of historical explanation. The dialectical model sufficiently grasps *those* aspects making up the teleological background (in the widest sense of the word) of a historical situation. It is unable to grasp their particularity, however, the causal aspects of situations, actions and events. In contradiction to the methodical insight of the *Science of Logic* Hegel does not make any space available for the language of causality in the philosophy of history. What is lacking in the explanatory model are elements of causality joining the individual actions, circumstances and events with the *preceding* individual actions, circumstances and events. The apparent causal form of Hegel's answers (that happened, because ...) does not change anything in regard to that fact. The expression 'because' belongs to sentences describing purposes and intentions of participant groups; it is a component of that language, which, following Hegel's model, is supposed to explicate the teleological background of the actions.

The specific problematic of the dialectical philosophy of history consists in the fact that it – contrary to its intention – does not make plausible the mediation between the background and the historical event. As our example shows the Hegelian model of explanation does not explicitly include the practical inferences of those responsible through their argumentation and action for bringing about the historical context. This fact appears so conspicuous that it deserves a thoroughgoing discussion.

The teleological language has to orientate itself in the philosophy of history according to the *particularities* of the object of historical investigation. Hegel speaks of the "historical reality of an end generally",[40] which should be respected. The doctrine of principles in the philosophy of history is built around the teleological *terms* 'end' and 'means', 'purposeful activity' and its 'realization' in some material form. The philosophy of history makes further use – in agreement with Kant's doctrine of the principles of historical knowledge – of teleological *judgements*. But whereas Kant restricts the method of the philosophy of history to teleological judgements about singular actions and events, that is, systematically structured relations of events and actions, Hegel takes the step from judgement to teleological *conclusion*. Theme for the philosophy of history is not alone the individual practices of a historical person nor the partial practices of a group (the 'national spirit' of a time), but the successive whole practices (*Gesamtpraxis*), the historical totality of the human race. Hegel places the totality under the model of the practical syllogism. The principle of the whole historical practice (in Hegel's speculative language the purposeful activity of the world mind) is introduced as the *middle term* of an inference, in which members of the teleological relation are linked, here: the end (*Endzweck*) of world history is "reposing for the present in the deepest crevice of the mind", and the means to its realization, the individual and partial historical practices as well as their "material".[41]

What does this strange transfer of the idea of practical syllogism to the realm of history mean? Can Hegel, who does not explicitly use practical syllogisms in historical explanation, justify the use of teleological principles generally in the philosophy of history? And finally, how does his understanding of teleology relate on the one hand to the Aristotelian tradition and its doctrine of method, and on the other hand to the way of critical method in scientific reasoning opened up by Galileo and further entranced by Kant?

When someone approaches the chapter on teleology in Hegel's *Science of Logic* with these questions it appears at first that Hegel is orientated not to Aristotle's but to Kant's doctrine of principles of teleological judgement and in consensus with him gives clear, unambiguous information. The systematic continuation of Kant's teleology is recognizable by the *position* between theoretical and practical reason attributed to it. When Kant ascribes to that principle "a *reflective power of judgement*, he makes it a connecting joint between the *universality of reason* and the *particularity of perception*".[42] Hegel is referring to the problem raised by Kant of the particular that lies beyond explanation according to general laws and is therefore turned over to the faculty of judgement to be expounded through discussion (Kant also uses the term *reflection* for it). A similar estimation of the theoretical value of Kant's teleology is found in the parallel segment in the *Encyclopaedia of the Philosophical Sciences*. In his estimation of Kant, Hegel says that the concept of end "is correctly called a *rational concept* and held up in opposition to the understanding as the abstract and universal generally and the causal relationship in particular. The relation of the abstract and universal to the particular is understood as *subsumption* in so far as the abstract/universal does not directly include the particular: but thereby it remains an abstraction".[43] Causal explanation in the realm of the natural sciences – that is the insight first won by Kant – follows a *theoretical subsumption model*. The *determinant* judgement (Kant's expression for the subsumption procedure), however, includes only partial areas of experience and beyond that delivers only an incomplete picture of scientific conceptualization and methods.

For those areas left unaffected by the model Kant introduced a second procedure. When no universal law is given, just the particulars, for which a (empirical-contingent) law still has to be *found*, then the method of judgement is "simply reflective".[44] The procedure of reflective judgement – a unique way of explaining (in the sense corresponding to that given previously of making understandable the natural and spiritual life) – contains the alternative method to the subsumption theory of explanation. Kant developed it in view of a theory of humanistic and cultural sciences (aesthetics, biology, anthropology, history), which follow the model of *teleological understanding* in conformity with end and means in their conceptualization. Although Hegel is aware of his agreement in this important point of Kant's *Critique of Judgement*, it is exactly here that their systematic differences begin. Whereas Kant's method of procedure and terms of teleological judgement do not possess any constitutive function of determining the objects of those sciences (we are concerned here only with an idea of reason as a regulator guiding the conceptualization),[45] Hegel attributes to them an objective validity. Hegel's teleology does not limit itself to comprehensive subjective considerations (reflections) and tentative conceptual discussion (exposition) of the diverse contents of culture empirically given. It claims the objectivity of the determinant judgement and with it the theoretical value of *explanation* (deduction),[46] which Kant denied teleology on systematic grounds.

Hegel's criticism of Kant's doctrine of teleological principles is peculiarly ambivalent. On the one hand Kant is credited as having done a service to critical philosophy by purifying the teleological manner of speaking in metaphysics and, with the distinction between external and internal finality, of "opening up the concept of *life* to the *idea*".[47] With the correction of the Galilean method of conceptualizing universal scientific laws Kant prevented at the same time the degeneration of teleology in modern philosophy to a language capable only of expressing the uncontrollable gratification of human desires as its end: with this Kant returned philosophy to the Aristotelian tradition.[48] On the other hand Hegel accuses the Kantian teleology of inadequately unfolding the conceptual distinction between the two kinds of finality.[49] Without having any deep understanding for the achievements in method made by modern natural science and their substantiation in Kant's *Critique of Pure Reason*, Hegel measures the doctrine of principles of teleological judgement with Aristotle's comparatively naive natural teleology. Paradigm for the concept of an end is the living organism according to which the teleological relation between end and means is represented as the *self relation* or *internal determination* (internal finality) of an object, the natural thing itself, whereas Kant, despite reconstruction of the same facts, translated the concept which has become objective back into the language of causality again and makes its principle of judgement peculiar to our intellect.[50]

With this interpretation Hegel replaces the initiative taken by Kant to mediate between Aristotelian and Galilean conceptualization. What Hegel criticizes is the restriction of teleological judgements to subjectively valid principles (maxims) gained from reflecting on a subject area, which in conformity with Kant's insight can well be made methodically understandable and comprehensible, but which cannot be explained in the strict sense. In opposition to Kant, Hegel ascribes to teleology the value of an objectively valid principle determinant in relation to its objects and with that – paradoxically – the synthetic function of the Kantian categories of the understanding. The end is at the same time both systematic (*Idee*) and synthetic unity (*Kategorie*), the teleological relation is not reflective judgement, but the implicit and explicit existing truth that judges *objectively* and determines absolutely the external objectivity.[51] While Hegel in this section of the *Science of Logic* – in sharp contrast to Kant – introduces the conception of the end as complete determinate of the objectivity of the particular, he gives to the end-means relationship the logical form of a syllogism. The teleological relationship, following Hegel's central objection against Kant, is "thereby more than a judgement; it is the conclusion of the free and independent concept, joined together with itself through objectivity".[52]

The systematic points of criticism formulated here against Kant, according to which the terms of end and means are elements of teleological judgement, which are elements of inferences and thereby elementary functions of scientific proof, emphasize once again the reason Hegel's *Science of Logic* must grant to teleology the rank of scientific explanation. The understanding of teleological relationships as objective relationships corresponds with the conceptual dialectical context in which the terms of end and means are presented. The chapter on teleology is part of the logic of objectivity, the compendium of Hegel's philosophy of science. More precisely it is a logic of applied (technical) sciences. The kinds of objectivity (mechanism, chemism) prior to teleology were the subject areas of the natural sciences (physics, chemistry), constructed in conformity with the conditions of their subjects, to which teleological knowledge is then related. The logic of teleology has to do with neither particular subject areas nor objects in general, but with their relation to human interests and activities.[53]

The systematic position of teleological knowledge in the context of applied sciences indicates that we are apparently not dealing with a general theory of action, but a *special kind of action.* The basic terms of teleological knowledge and the associated forms of practical inferences were not developed in reference to a theory of humanistic sciences (*philosophy of the mind*). Rather they were developed in the realm of natural science. In short: the kind of teleology here is connected with mechanistic and chemistic objectivity and provides an explanation for *technical* actions having arbitrary ends. In the forms of practical inferences analyzed here we are not considering the substantiation of ends, but the means to use for ends already set. Technical action has as its subject the behavior of mechanical and chemical objects, the so-called external finality of nature, which can be useful for mankind through the intelligent choice of means.

It is therefore only consistent when the practical inference takes on the form of *technical reasoning* in Hegel's analysis. What is being established is the purposefulness of the means the agent inserts between the end he set and the given circumstances – a rational (*zweckrationales*) action that Hegel describes non-metaphorically as *power* and in a metaphorical manner of speaking as *cleverness of reason.*[54] The external finality only has the form of teleology; rational action that cannot establish its own ends remains formal, it obeys the logic of the means. Hegel therefore speaks of means as the "formal middle term of a formal inference" in place of which other means could also have been chosen.[55] This statement – apparently not sufficiently considered by von Wright in his reference to Hegel's teleology and its similarity with the idea of practical inferences is valid only for *technical reasoning*, as it can only be contended for the *means of technical action* that they are *external* (*äußerlich*) to the extremes of the intended and executed end. What the practical inference in the chapter on teleology in the *Science of Logic* reconstructs is the structure of technical action (the activity directed outwards), and the rationality of the means suitable to the attainment of established ends. The establishment of the ends themselves is the task of practical reason, the principles of which are introduced later in the section dealing with the *idea*.[56] The teleological principle of practical reason is called the *good*, the old Platonic-Aristotelian expression, which enters as premise in the inference of action (which is here impossible to be mistakenly considered technical and occasionally is called inference of good) that prevents the regression of the end-means relation through the concept of a final-end and allows the agent to transcend the level of merely natural history.

To examine Hegel's claim it would be necessary to pursue the continuing systematic argumentation and renewed discussion with Kant, which cannot be fully pursued here. At this point a few closing comments about the teleology in the analytical philosophy of science will be sufficient. Presupposing that (a) the idea of a practical syllogism only allows technical reasoning, and (b) the humanistic sciences are inherently practical sciences, then the answer to the question raised above is that the kind of teleology introduced by von Wright and explained with references to Hegel and Aristotle appears only conditionally suited to close its methodological gaps. Before they can be closed the entire range of theoretical and historical scientific problems brought together by von Wright under the perspective of the Aristotelian/Hegelian tradition would have to be resurveyed. What remains open here is stated in one phrase, the *relationship between methodology and question of substantiation involved in practical philosophy*, a problem obscured since the middle of the 19th century by the separation between the understanding and the explanatory sciences. The situation will probably only then be illuminated when the philosophy of science through extension of the perspectives opened up by von Wright returns again to Kant in its historical and systematic argumentation. What the criticism of the subsumption theory of explanation in end effects lays open is essentially the status of a problematic Kant formulated in the *Critique of Judgement* and left behind to his successors as methodical task. Dialectics and hermeneutics are not the only possible alternative to the analytical philosophy of science. The critical way which Kant pioneered in the teleology discussion still remains open.

University of Erlangen

NOTES

[1] Cf. the controversy between H. Albert and J. Habermas in: *Der Positivismusstreit in der deutschen Soziologie*, Neuwied/Berlin, 1969, pp. 139–305 (the reproach of *positivism reproach* p. 237, 208 footnote 27, 216, 281 among others). – The author wishes to thank Mr. Robert Pettit and Mr. John Insley, B.A., for their help in translating the essay into English.

[2] Cf. the synopsis by P. Gardiner (ed.), *Theories of History*, Part two: Recent Views concerning Historical Knowledge and Explanation, New York, 1959, pp. 344–515. Starting point is the well known essay from C. G. Hempel, 'The Function of General Laws in History' (1942).

[3] Cf. the sequel to the essay (1942) from C. G. Hempel/Oppenheim, 'The Logic of Explanation' (1948), is reprinted by H. Feigl/M. Brodbeck (eds.) in *Readings in the Philosophy of Science*, New York, 1953. The designation 'Covering Law Model' stems from W. Dray. The covering law model was first formulated by Karl R. Popper, *Logik der Forschung*, Vienna, 1935 and *The Open Society and Its Enemies*, London, 1945, Vol. II, p. 248f.

[4] G. H. von Wright, *Explanation and Understanding*, London, 1971, p. 11. Hereafter this is cited as *EU*.

[5] P. Gardiner, *The Nature of Historical Explanation*, Oxford, 1952, p. 91f.

[6] *Ibid.*, p. 60.

[7] Cf. *ibid.*, p. 103f. That Gardiner – in opposition to Hempel's intention – interpreted the subsumption theory of explanation causally is apparently connected with the fact that his criticism was predominantly influenced by Popper's formulation, which permits such an interpretation.

[8] W. Dray, *Laws and Explanation in History*, Oxford, 1957, p. 104.

[9] *Ibid.*, p. 119.

[10] G. E. M. Anscombe, *Intention*, Oxford, 1957, and Ch. Taylor, *The Explanation of Behavior*, London, 1964. – A good synopsis of the discussion since then is contained in A. White (ed.), *Philosophy of Action*, Oxford, 1968. Also compare the investigations of R. Bernstein, *Praxis and Action*, 1972, and A. Danto, *Analytical Philosophy of Action*, Cambridge, 1974.

[11] *EU*, p. 23.

[12] Cf. the historical synopsis in *EU*, 'Two Traditions' (I), especially paragraphs 1–4 and 10.

[13] *Ibid.*, p. 96.

[14] *Ibid.*, p. 97.

[15] *Ibid.*, p. 27.

[16] *EU*, p. 180, footnote 75. The reference to affinities between Aristotle and Hegel came from Juha Manninen.

[17] G. W. F. Hegel, *Wissenschaft der Logik*, Vol. 2 Section 2, Chapter 3 (in the English text and in the translation incorrectly given) from the Leipzig edition 1951, part 2, p. 394. Hereafter this is cited as *Logik*.

[18] Cf. *EU*, p. 200, footnote 4.

[19] *Logik*, Vol. 1, Book 2, Section 3, Chapter 4, p. 193.

[20] *Ibid.*

[21] Immanuel Kant, *Groundwork of the Metaphysic of Morals*, translated by H. J. Paton, New York, 1964, p. 127. I am referring here in part to a comment J. Blühdorn made in a discussion about my paper 'Positivismuskritik und Historismus. Über den Ursprung

des Gegensatzes von Erklären und Verstehen im 19. Jahrhundert' (1971). Compare J. Blühdorn and J. Ritter (eds.), *Positivismus im* 19. *Jahrhundert*. Beiträge zu seiner geschichtlichen und systematischen Bedeutung. Frankfurt/Main 1971, p. 102f.

22 The conclusive evidence concerning Kant's terminology is summarized in the *Critique of Judgement*, translated by J. C. Meredith, Oxford, 1969, Part. 2, § 78. Also compare the Introduction V (Explanation=comprehension) as well as Part 2, § 61 and 64.

23 *Ibid.*, § 78.

24 *Ibid.*, § 78.

25 *Ibid.*, § 78.

26 *Logik*, Vol. 2, Section 3, p. 407. Compare with it the corresponding usage of understanding in *Einleitung in die Geschichte der Philosophie*, published by J. Hoffmeister, Hamburg 1959, p. 30f.

27 *Die Vernunft in der Geschichte*, published by J. Hoffmeister, Hamburg 1955, p. 33.

28 Cf. *Logik*, Vol. 2, Book 2, Section 3, Chapter 3, p. 194.

29 *Ibid.*, The criticism refers to the historical causal explanation in the mechanical-materialistic philosophy of history during the 18th century. Cf. D'Holbach, *Système de la nature*, London, 1773, I. III, p. 163; Voltaire, *Essai sur les moeurs*, Paris, 1756, Chapter LIV.

30 Cf. Helvétius, *De l'esprit*, Paris, 1758, III 1.

31 *Logik*, p. 194.

32 Cf. *EU*, IV, 2–3, p. 142ff.

33 *Ibid.*, p. 142.

34 *Ibid.*, pp. 140, 144, 146.

35 G. W. F. Hegel, *Vorlesungen über die Philosophie der Weltgeschichte*, published by G. Lasson, Vol. 4, Leipzig, 1923, p. 920.

36 *Ibid.*, p. 926.

37 *Ibid.*, p. 926. With the reference to the religious tradition Hegel explained, as is well known, why the idea of freedom became revolutionary in France alone. Protestant Germany already had with the Reformation the revolution within itself. Also compare p. 922ff.

38 Evidence for this is Hegel's explanation of the rise of the cities and their consequence for the feudal social system. Compare the corresponding section in *Vorlesungen zur Philosophie der Weltgeschichte*, Vol. 4, p. 842ff, 855, 858.

39 *Ibid.*, p. 855.

40 *Die Vernunft in der Geschichte*, p. 86.

41 Cf. the draft of a teleological doctrine of principles, ibid., pp. 28–148, especially p. 93f.

42 *Ibid.*, Vol. 2, Section 2, Chapter 3, p. 389.

43 G. W. F. Hegel, *Enzyklopaedie der philosophischen Wissenschaften im Grundrisse* (1817), AIII: Die Lehre vom Begriff, § 153, *Werke*, Vol. 6, published by H. Glockner, Stuttgart, 1959, p. 123. In the second edition Hegel weakened his emphatic consensus with Kant's teleology (§ 204, *Werke*, Vol. 8, p. 413). Also compare with this the criticism in 'Glauben und Wissen' (1801), *Werke* Vol. 1, p. 315f.

44 *Critique of Judgement*, Introduction, IV.

45 *Ibid.*

46 *Logik*, Vol. 2, Section 2, Chapter 3, p. 390.

47 *Ibid.*

48 Cf. the chapter on Aristotle in *Vorlesungen über die Geschichte der Philosophie*. *Werke*, Vol. 18, p. 341ff., 346f., 349.

[49] *Enzyklopaedie der philosophischen Wissenschaften*, § 58, *Werke*, Vol. 8, p. 156f.
[50] Cf. *Werke*, Vol. 18, p. 342 with *Werke*, Vol. 18, p. 158.
[51] *Logik*, Book 2, Section 2, Chapter 3, p. 390.
[52] *Ibid.*
[53] Cf. the explanation of the concept 'Objektivität' in *Logik*. Book 2, Section 2, p. 358f.
[54] *Ibid.*, p. 397f., *Enzyklopaedie der philosophischen Wissenschaften*, § 208–209, *Werke*, Vol. 8, p. 419f.
[55] Cf. the footnote of this equation above in Section I according to von Wright, p. 10.
[56] Cf. *Logik*, Book 2, Section 3, Chapter 2, pp. 478–483.

MARIA MAKAI

AGAINST REDUCTIONISM AND PURISM: TERTIUM DATUR

Beginning *in medias res*: My point of departure concerning the age-long dispute between causalists and finalists is that I find the same fault with von Wright[1] as he with Peter Winch, namely the absence of the specific teleological aspect.

I also take the first step by referring to what von Wright presented for the illustration of the structure of action, the opening of the window, but then I take a different course, as I do not confine causality to its so-called experimental notion. What is the purpose of the opening of the window, an action intended to bring about something? In the given situation, it serves to keep me awake in the stifling heat and to refresh me so that I may go on with my work. This aim of mine is determined by the totality of circumstances: it is winter time, I live in a capitalist country at a relatively advanced technical level in the last third of the 20th century, I do intellectual work, but I am not one of those with an air-conditioned apartment; there is no energy crisis, so I need not economize on fuel. In these conditions, heat makes me (the subject of the action) feel drowsy. Thus, the specific synthesis of systems: virgin nature, nature transformed by man, class-relations and social-technical division of labour within a certain, historically determined socio-economic structure, the general political situation (its actual economic consequences) and my own nature, relatively independent (*e.g.*, it cannot bear heat very well), make me perform a certain action at a certain moment. When made to do so, I act as an element of infinitely open and relatively closed systems and, at the same time, as a relatively closed system myself. The means of the action is the opening of the window, which is the cause at the same time; its further means (and also the process of attaining my end) is that I consciously control the act of my letting the original and the artificial environments (outdoor cold and indoor heat) interact until the room (with myself in it) has cooled (refreshed) to a suitable degree, and then, as a final act, I close the window. The result of the action is the aim attained, my state of being refreshed. By anticipating this end from the

Manninen and Tuomela (eds.), Essays on Explanation and Understanding, 27–58.

beginning, I set off a chain (concatenation) of causal relationships which, as a closed process directed towards its effect as its goal, broke another chain of causal relationships otherwise open and (in the present case) disadvantageous to me. Through the favourable (*i.e.*, to me) interaction of the outdoor and indoor environments, which I controlled and brought about through the medium of a product of the present technical level, I took control over my own nature (since I liberated the natural forces driving away my sleepiness up to a point allowing me to work with a fresh mind, yet without my catching a cold), thus I acted as a formally free human being. However spontaneous the action of cooling a room may have become to everyone who enjoys that level of material wealth which enables him to possess a window opening to the fresh air and a window handle in working order, the effect of the specificity of purposeful action must always prevail in the sense that the anticipated effect acts as "a law determining the will" (Marx), and that control should be exercised all through the process in which existence changes from restlessness and activity into its quiescent state. If it is the wind that tears the window open, the process of interaction between indoor and outdoor environments takes place just as well, yet in this case it does so in its endless, open and 'blind' manner, running into the infinite instead of drawing to an anticipated end.

This example – in its necessarily simplified form, which cannot present the wealth of determinations involved – only points to the fact that a sound approach to the relationship between causality and action presupposes the structures to be viewed from a historical-social angle.

In analysing the relationship between causality and action, von Wright deems it necessary to distinguish between what we do and what we bring about. When we open the window, we do something which is the cause of effects. The cause is the result of action, and the effect (as a phenomenon which has been brought about, and which has to be kept distinct from action) is the consequence of action (pp. 66–69). But if doing something, as being the cause, is already the result of action, then the points of start and finish of the action coincide, and the action as such is annihilated. In this sense, what is really the anticipated effect (aim) of action is degraded into a simple consequence which does take place, yet does not attract the whole process in its direction. Action, however, must not be restricted to the first phase (the window-opening), which is – in its relationship to the aim – regarded merely as a means. The concept of purposeful action implies a mental process which anticipates the aim as a result, and also the result brought about. During this process we act not in the capacity of agents operating merely the antecedent (the 'cause') of a process but in the capacity of subjects anticipating and realizing a closed and directed system of causal relationships, as human beings, because we are able to objectivize ourselves through the modification of the object and ourselves, conquering "the one-sidedness of subject and object" in unifying them (Hegel). Not only action will lose its meaning when isolated from this context; the subject will itself become the "singularity of the moment" (Hegel): if separate causes have to be operated each time to make them act as consequences outside the action, then there would be need for several subjects for cooling the room: the last one for closing the window, more precisely, to turn the handle back to its original position. One pole is characterized by empty action void of content and 'subjects' operating the independent, scattered and dissociated causes, the other by an overpowering object, which has shaken off the influence of an active subject.

Is it possible to find such an explanation for action that presents something else than the comprehensive moment (*das übergreifende Moment*) of bringing about, which comprehensively arranges and directs the links in the chain of actions (events) both in abstraction and reality? The answer is definitely negative unless the phases of action disintegrate and the scope of causal explanation is limited to the antecedents put into operation and void of the attractive effect of the aim, and, furthermore, the teleological explanation (see later) renders itself independent of the ontic determinations of the act of realization. I am of the opinion that the subjectivization, emasculation and the deprivation of causal relationship of its 'dynamic', 'productive' (N. Hartmann, Bunge) ontic determinations constitute a direct ontological precondition to the fact that this conception leaves no room for a finalistic transformation of causality. Such a finalistic transformation would require the following theoretical preconditions: the ontic status of causality, as a genetic and lawlike connection, freedom of consiousness in the conceptual temporal-special dimension, consequently, the possibility of the subjective logical anticipation of the effect to be brought about, and, finally the transition of the subjective logical into the objective logical (the ontic in a narrow sense) (see later). But von Wright claims that the cause-effect relationship interpreted in terms of action is a logical one (p. 67), according to which the free subject is restricted to a consciousness that fails to work its way as a determinative factor of higher order into the system of causal concatenations.

In the 'punctualist' interpretation of action, the cause (that is, the result) is part of the action, whereas the action is not the cause of the result. The result, however, if interpreted as an end attained or an effect anticipated, is at the same time part of the action and the effect of action which finally penetrates, transforms the object, and thus materializes itself; it (*i.e.*, the result) is the moment of transition of subjectivity finally transcending itself and matter the form of which man has changed, and it is at the same time the end-point of action and the form itself that has come to existence. These relationships will only become evident if the relationship between causality and action is described in the tradition of the most eminent representatives of philosophy, as something more than the mere operation of a ready-made product of technology. (It should not be forgotten that such a product has itself been finalistically transformed, which fact is also demonstrated by the product acting as a cause when operated in a manner proper for its function. In this case, however, it may seem that action is confined to the operation of the phase which serves as the first cause.) It is not as if I were to restrict the right of choosing the example at will, however, the basic model of human practice is the working activity, in a wider sense the activity of social production which model (genetically) contains at the same time the source of all other activities, and all those basic characteristics of being and consciousness (*Seins- und Bewusstseinsbestimmungen*) which reveal the basic features of human activity in general.

The interdependence of action and causality is also elucidated in von Wright's conception by a unique interpretation of experiment.[2] This is the point where causality and finality disappear at the same time. The succession of states in a system does not involve causal relationship; causality must not be reduced to states following one another.[3]

Furthermore, experiment, as the practical organ of theoretical cognition, aims at obtaining (new) information (correcting or dismissing knowledge grown outdated), and exploring hitherto unknown laws of reality. Throughout the experimental process, the agent has to interpret the process in its relation to his hypothesis, as an instrument of obtaining new informations, and as its mode of verification, and so his teleological attitude towards the effect as the anticipated aim of process is doubly justified.

However, the aim of obtaining information pervades and directs the whole course of the process, (otherwise) taking place in artificial, undisturbed conditions; the alleged passivity of observation theoretically deprives the experiment of its teleological aspect.

Even this first approach enables me to draw up my chief counter-argument, namely that the deprivation of action of its ontic (causal) components necessarily results in the annihilation of the specifically teleological aspect. Of course, von Wright's intention is nothing else than to break the rule of positivist mechanistic monism and elaborate the *sui generis* nature of purposeful action. In defiance of the naturalist reduction of the teleology, he offers the direct opposite of reduction, *i.e.*, the purist interpretation of teleology, liberated from causality as a lawlike connection.

Von Wright gives evidence of his straighforwardness as a scientist in indicating in more than one instance that there are heavy counter-arguments likely to be brought up against his position, and this is really so, since it provides no possibility of settling the old dispute between finalists and causalists. On the other hand, the undecided debate is no proof of what von Wright refers to as his ultimate 'argument', *i.e.*, that the ultimate position in this dispute is 'existential', or else we would irrationalize the foundations of science, giving way to the tyranny of subjectivist views. The undecidedness of the dispute rather proves that bourgeois philosophers have so far been unable to get over their attempts at solving problems in the mechanistic-monist, Humean or purist-finalist ways, which experiments, in their specific one-sidedness, have time and again evoked criticism from, and the recurring emergence of, the other pole, and, consequently, have given rise to efforts at uniting tendencies which are in fact contradictory only on the surface. (Naturally, we cannot even make mention of the social, political and ideological motives concealed behind these shifts in position.) Von Wright has left out the analysis of the gigantic enterprises for, and successes in conquering one-sidedness – from (one of the tendencies of) Aristotle through Vico, and Hegel to Marx, Darwin and N. Hartmann.

Von Wright's position – elaborated with the utmost clarity, unambiguousness and, above all, with a view to achieving comprehensiveness – exemplifies the crisis of today's, so far mainly positivistic, philosophical analysis, a stage in a search for new roads, where it has become evident that the analytical trend, though with the preservation of certain essential positivist premises, has turned for its weapons to *Geistesgeschichte*.

When asserting at a methodological level the interrelations between understanding and explanation, von Wright adopts a position that rejects the mechanistic and monistic features of positivism, especially the notion of law as reduced to mechanical causality as regards the interpretation of human activity. At the same time he rejects the open subjectivism and intuitivism of *Geistesgeschichte*, and links the understanding of aims with the task of reconstructing individual chains of causality. However, his position still remains on the basis of neopositivist ontology: states entirely independent of each other ('the sun is shining', 'the door is open', 'it is raining', *etc.*) constitute thus the sole ontological elements "building-bricks" (p. 45) of his world – from where the ontological entities (pp. 42–45) of things, qualities and relations are excluded. This is something more than a "*quasi-ontologische Entscheidung für subjektive Vernunft*"[4], and it forms the basis of both analytical philosophy which also disguises its antimaterialism by rejecting the mechanistic *monistic* forms of positivism, and – even if indirectly, *mutatis mutandis* – the basis of *Geistesgeschichte*, of the metaphysics of the direct knowledge of aims. It is exactly this 'world' that is left out of this ontology, things, qualities and relations with their interdependence determined by a great variety of laws; this world, *das Seiende als Seiende*, which is the indispensable ontological condition of, among other things, the generalization of the problems teleology. Teleology thus stiffens into a state of affairs, and assumes the form of thesis, being something given simply as the content of direct knowledge. The disappearance of this 'world' makes self-contradictory all those experiments which von Wright uses for trying to find the explanation of an object in its objective logic, though his most fruitful discoveries are directed against subjectivist ontology. Broadly speaking, his position is centered upon a kind of methodology, which is not the concentrated expression of the essence of its object, thus being the only adequate method of its attainment (*Aneignung*), but a mode of explanation which has necessarily become dissociated from its object, and, consequently, is objectless (*i.e.*, it fails to reach beneath the superficial phenomena of the object). Yet von Wright sometimes insists upon the primacy of the ontic status of the object,[5] whereas according to the mainstream, which manifests itself through self-contradictions, the characteristics of the object depends on the way it is approached.[6]

Instead of taking up his position on the basic issue concerning the ontological primacy of causality or finality and correcting the roundabout ways in the history of cognition, von Wright projects upon existence the epistemological appearance ensuing from these roundabout ways. It is well known that man realized causality by the mediation of his own practical activities, but if one should be led to think that "no proof can decide ... which is the more basic concept, action or causation"[7] then this would really mean that an archaic gnoseological relict penetrates this modern ontological conception, determining, or rather disarranging, the order of categories. Although it is true that each *new* relationship must be subjected to the test of practice and experiment before asserting their nomic relation, it would be a great error to deny that these relationships, as determinations of being (*Seinsbestimmungen*), preceding conscious ideation, are primary to the mode of conscious ideation, and also determine it. But von Wright's method of making conscious the causal nature of new relationships leaves no room for ontological inquiry, and freedom, considered primarily on an epistemological level, makes the issue of ontological primacy undecidable. The reference to the epistemological roundabouts implies a rather absurd defence of human freedom, as von Wright presents causality in relation to man as a supreme power of restraint, something responsible for human disabilities and incapacitations (p. 81), whereas we can safely say that it would be totally impossible to speak about human ability and capacity but for causality that is recognised and, to some extent, directed by ourselves.

It is obvious that only a materialist ontology on a firm historical foundation is capable of illustrating the process in which the hordeman completely subjected at first to the laws of nature gradually harnessed it through action, which operated chains of natural causal systems even though the links were first only instinctively 'planned'. Von Wright himself is of the opinion that his position is rather vulnerable to arguments maintaining that human action cannot be understood unless causality is taken into account as well. To my mind, such arguments have quite often been set against the one-sidedly finalistic (and one-sidedly mechanistic-causational) idea of action, and it ought to have been von Wright's task to take them into account and refute them, thus laying the foundations for his own position. As has been indicated, these considerations fall outside the scope of the author's interest.

This is also an inevitable consequence of his ontological attitude: once *das Seiende als Seiende*, in its historical-dialectical interpretation, is absent, the theoretician is unable to describe either the genesis of higher-order determinations of being and consciousness from those of lower order (the presence of those of lower order in those of higher order), or the dependence of the latter upon the former. This, however, inevitably produces a purist version of teleology, the central point of which is an analysis of practical discourse and, within it, of such sentences in which some elements of individual aims get their linguistic expression.

Cristicism here must content itself with making mention of the presence of causality in the finalistic structure. This is all the more necessary, since recent philosophical analysis, which strives to rid itself of certain positivist premises and traces itself back to Aristotle, has disregarded (provided we give full credit to von Wright's analysis of this issue) N. Hartmann's contributions to the Aristotelian description of finalistic relationship.[8]

That finalistic determination is even more complex than the way N. Hartmann described it in explaining and developing the Aristotelian analysis is made obvious by the Hegelian idea, which – implicitly – sets out to illustrate a double motion being realized also in the first act, as, with respect to the setting of the aim, the subordination of the aim to existing matter and, inseparably of the former, the subordination of matter to the aim[9] take place at the same time.

Furthermore, the selection of means in the second act takes place simultaneously against and in the direction of the time-stream, which necessary correction ensues from our intention of putting a certain matter into shape, and this form-giving can only be realized, even at a noetic level, in a dialectic interaction between the anticipated form and what it is the anticipated form of. If we were to say that the selection of means were a conscious process starting only from the anticipated effect, then we would exaggerate the independence of the would-be form of the matter to be shaped (which would inevitably bring its own punishment during the working process). What is more, in this case we would need yet another plan, which, reversing the stream of time, would anticipate the process of realization, and which would determine in the order of realization the relationship between the means themselves and between the means and the aim. In this sense, the finalistic structure, as interpreted by Hartmann, should contain four acts instead of the present three. There is however, no need for this, as the act of choosing the means takes a single leap from *what is* to *what is to be brought about* subordinating at the same time the given matter to aim and aim to the given matter.

The finalistic structure itself is a product of history, as is also proved by the two noetic acts coming to being and gaining independence. The half-animal action, lacking in real teleological content, when proving successful in its effect after thousands of millions of repetitions, slowly disintegrated and changed form. Action, leading to its result, changing the form of matter and materializing itself in form, interiorized and gained a noetic mode of existence as a would-be form to be realized on the one hand and as an anticipated, only possibly existing image (idea) of form-realizing action on the other. These noetic actions, on gaining independence, 'unfolded' the elements still in germ in instinctive action: this process of unfolding resulted in something qualitatively new. There came into being a new, a noetic existence, from then on preceding in time material action as the conscious anticipation of its process and result. Thus material action acquired an internal mental meaning: it became the realization of the aim.

The finalistic structure historically indicates the actual degree of the conquest of nature, since aims are always bound by the achieved stage in the development of means. Possibilities that can be anticipated as aims are in fact newly unfolded and specified possibilities so far hidden in matter perpetually transformed by man. It is only in this specification of aims that freedom, manifested in domination over nature, can be further developed and realized, to which the continuity between the past and future modes of form-giving is indispensable.

Thus it is outlined here that the subject and starting point of final actions is something other than an isolated, aim-setting consciousness (not to mention the major premise of a given practical syllogism); it is mankind (in its antagonistic structure) as the subject of domination over nature. Each stage in conquering nature, the existence of virgin nature *and transformed nature*, as the "inorganic body of man", in one word, all the results of his activity are at the same time preconditions to the ever renewing process of his actions (Marx). Such results and preconditions are production-relations (social forms of the act of conquering nature, which also determine the forms of social-natural subjects, objects and actions) which also pre-form the act of setting aims. Consequently, the finalistic structure is part of a whole-determination, and, considering its concrete form of existence, it is preconditioned at every phase both historically and socially. For instance, the advance of capitalism disrupted the former unity of the finalistic structure (the original unity of planning and execution, realized in a collective society or – *mutatis mutandis* – in guilds of artisans). The first two acts of the finalistic structure had become separated in the course of development from the working process, went over into the possession of science which developed later in the social form of capital, and science materialized these noetic acts in the form of machinery, which gradually replaced the heterogeneous arsenal of individual tools, and so it also subjected the mode of execution (the third phase) to mechanical motion.

Thus in the finalistic structure there appears social division of labour as the basic condition for class stratification and within it the separation of material and intellectual kinds of labour in their concrete social forms.

However reasonable an abstraction the finalistic structure is in its interpretation by Hegel and Hartmann, – and this finalistic structure also appears in a similar form in the Marxist analysis of the working process, when viewed in its simple and abstract elements[10] – the totality of the Marxist *oeuvre*, and particularly the Marxist treatment of the process of social production, leaves no doubt about the fact that though the finalistic structure reflects the objective generality, this generality can only exist within specific social formations, and the characteristics of these social formations appear in the historically determined varieties of this structure.

If von Wright wishes to find his explanation of social activity in a teleological structure not to be reduced to mechanic causal relationships, then this cannot be done, to my mind at least, without performing the above mentioned analysis of the finalistic structure. An analysis of the specific social-historical form of existence of the final structure will certainly shed light on the specific whole-determination of the second (transformed) nature and the given social order, which manifests itself for man acting in a given period of history as the "real prius" (Marx) of his activity. Man, who is made to act by this prius, and who makes this prius act, can only have a causal effect on primary and already transformed nature through his means of production if he realizes and takes advantage of the laws relevant to (the given sphere of) nature, means of production and technological processes.

If we put into brackets what forced man, at a certain stage in the development of his 'real prius', to operate (or to make others operate) his means of production, if we put into brackets the mechanism of which the essence is to direct elements and processes causing each other in real a temporal succession towards the anticipated effect (aim), if we thus disregard what man himself, is, as the 'actor' of a certain, historically determined stage of domination over nature and as its 'author' as well, (though only by right of being the former)[11] and also disregard what he transforms teleologically and how, by what means and laws he does so, then we lose teleology itself.

Thus teleology is confined to the explanation of action *in actu*, which means that the subject matter of causal explanation – nomic relations between phenomena – become constrasted with that of teleological explanation (as phenomena of non-causal, non-nomic character).

The main supporting pillar of teleological explanation is practical syllogism. In this schema

> *A* intends to bring about *p*,
> *A* considers that be cannot bring about *p*, unless he does *a*,
> Therefore *A* sets himself to do *a*,

the end-state of the syllogism is the action functioning as a means for the attainment of the aim.[12]

The practical inference, presents the characteristics of the previously analysed finalistic structure as a kind of ready-made results. It emasculates and makes one-sided these characteristics, as it presents in a ready-made aim-means-aim relationship what the finalistic structure illustrates in its manifold complexity. Considering that analysis centered on practical discourse virtually fails to reach beyond the analysis of the premises of this syllogism, furthermore, it treats this inference as an 'upside-down' model[13] of the 'teleological explanation' of action, we can say that this sort of model and this analysis serve to evade methodologically the conceptualization of the real ontological and epistemological characteristics of consciousness and being. (Here we are speaking only of this concrete sort of practical inference and of its specific interpretation as the 'teleological explanation of action'. Practical syllogism in another interpretation as one aspect, moment of the finalistic structure naturally requires a detailed analysis.)

In his time, N. Hartmann paid great attention to the so-called 'determinative neutrality' of finalistically determined processes: according to him, it is not possible to read off whether or not the phases and stages of a process are real means, *i.e.*, *ad finem* determined. This determinative neutrality may serve to stabilize the Humean notion of causality equally as well as to finalize neutral processes, or even to dissociate theoretically the finalistic processes from their ontic determinations.

The agent may really be mistaken in saying whether *a* is relevant to the bringing about of *p*. However, the philosophically important question is not what he thinks but what he has to *do* after all, provided that he does not wish to give up his aim. He is forced to find the processes causally relevant to the bringing about of *p*. When explaining practical inferences, we must not disregard the ontic relationships (and causal ones, at the core) contained in them.

After all, the dispute between intentionalists and causalists is connected with the determining force of practical syllogism.[14]

First of all, this logically binding validity continues or ceases to exist according as the phases of the action prove to be directed at the aim or not. The fact that practical necessity is of a feebler character than theoretical necessity indicates that existence to be transformed determines action and consciousness: if *in actu* action turns out to be non-adequate, the whole course of action is consequently suspended, and the syllogism ceases to possess a logically binding validity. *It proves to be ontically binding* to change the concrete contents of the logically binding validity (*i.e.*, the premises of the practical inference). Thus this logical character is not a primary given, but the subjective logical expression of teleologically transformed causality which, in all probability, will prove successful. We may not leave this specific ontic mediation out of consideration, as it is a characteristic feature of the practical syllogism that it is such a transitory formation between cognition and practice as contains at the same time the ontic interrelations relative to what is to be brought about and the conscious modes of establishing such interrelations.

Thus the only reason why necessity contained in the practical syllogism is weaker than theoretical necessity is that it has to take into account the necessity of being to be transformed, and so its determination must be a *double* one: the logical being dependent upon the ontic. Further, ontic determinations become relevant not solely in the concrete determination of a plan for action, as, at the point where a given practical inference begins to come being, the determinations of existence independent of the mind already exercise their effects: namely, my choice in wishing to pursue an aim – one that is perhaps in contradiction with the others – to be realized here, and in this particular instance, is determined not only by my past existence but the present situation and future, as I foresee it, as well (obviously not in a mechanic way but through the mediation of my subjectivity). (This was also pointed out by Aristotle, though on a more abstract plane.)

Whether or not logically binding validity survives in the practical syllogism viewed in the process of its coming to being depends on a variety of conditions: the type of action, the given circumstances, the skill of the agent, his foresight, the foreseeability of the actual possibilities, the possibility of creating a closed system of causality directed at the aim, the available means, the possibility of correcting individual links in the chain of events during action, *etc.* These conditions will then determine the logically binding character of the premises: there are cases in which the premises need continual correction in accordance with practice, where necessity will only be manifested as the result of a series of 'trials and errors', only permitting a retrospective generalization, whereas the other extreme is when the agent knows in advance that the logically binding character of the premises will have a coercive effect, as the action is another recapitulation of a chain of events ending in a result expected with good reason, since it has proved successful thousands of millions of times, and the agent has himself performed it in a great number of instances.

We can say that with the appearance of large capitalist factories and mass production in general, and also during the process of science becoming a force of production the necessity included in the practical inference grows up to the theoretical-logical necessity of which it (the practical syllogism) represents the relatedness to the practice of production.

The necessity contained in the practical syllogisms of production tends to assume an unambiguously binding character as the theoretical and practical rule over nature increases (while it is the duty of natural scientific and technical experimentation to produce steadily improving methods for the practice of production, and to bring them to the level of unambiguously binding technological instructions.) The contradictory character of the necessity in the theoretical and in the practical inferences could only exist in relation to the working activity as long as work was realized mainly by artisans. During this time, the working process had to be suspended now and then in order to improve it and its result. The logically binding validity has thus to be broken during the process in order to try alternative solutions. There is, however, a tendency, though by no means of universal extension, in mass production in our age that the binding validity of some technological instruction cannot be suspended during the process of production, and that a suspension of this kind can only come from outside the field of production. Thus the historically matured characteristics of the suspension of logically binding validity reflect the given stage of scientific-technological progress (and its social-economic form), and it gives rise, indirectly-though, to the fact that the 'narrow-minded totality' of the direct (immediate) producer disappears, and the worker, being a mere tool of realization, will become opposed to that social intellect, (as embodied in machinery) which nevertheless serves private interests. However, the necessity still survives in its weakened form in areas not yet connected to mass production, and in the field of social activity (both in collective and individual actions).

Is it then possible to draw the conclusion from von Wright's conception, which disregards these comprehensive relations and centers on the syllogism of a man running after a train, that the necessity included in the practical inference can only be interpreted *ex post actu?* I am of the opinion that I can only regard something *post actu* as necessary if its necessity has already become visible either *ante actum in mente* or *in actu* (when being corrected, *etc.*), (except for the case when the direction of the action proves necessary later as a non-pursued effect of action, in which case there is no practical syllogism in question at all). If we only consider *post actu* necessity, then we have placed necessity in the mind of the agent who passes judgment, or, more generally, in our subjective knowledge of the performed action. In this case, however, necessity is not on the side of the active agent but on the side of the one who judges. As has been mentioned, the finalistic aspect of action, as a specifically interwoven web of determinations of being and consciousness, is independent of the agent's self-reflection, and of the judgement the judger has passed on the action. Adequate judgement can only be passed on action, as being the object of self-reflection and the reflection of others, if it really possesses the finalistic structure.

Von Wright seems to give me a feeling of uncertainty: he sometimes treats the practical inference as an anticipated plan for action, then again (and this is the dominant feature) as a model of the explanation of action already performed (or in the process of being performed). Considering the circumstances, this is by no means a problem of secondary importance, since if I regard it as the model of explanation viewing it from the outside, having *e.g.* no exact information of the circumstances, then I can only regard it as necessary if it has already been performed. If, on the other hand, I lay down a plan for the action as an active agent, then I must form its links and their interdependence in view of necessity. If the agent is content with learning about the necessity of the action *ex post actu*, then he does not need the first two premises at all, and we have then taken up the position of accidental, instinctive causation (see later).

The indistinctness of, and the rather dubious interaction between, the ontological aspect and methodological questions (with the primacy of the latter) may also be explained by saying that the practical inference illustrates as a conclusion drawn from the premises (*i.e.*, conclusion drawn solely by consciousness) what is in fact the transition of mental anticipation into practice, the decisive step from the sphere of consciousness into a chain of actions performed on objects, from a mentally anticipated causation into real causation. If we were only to regard this syllogism as a form of consciousness, and not also as a form of actually stepping out into reality, which step is obviously more than, and different from, a conclusion of a syllogism, then there can only exist a logical relationship between the premises and the conclusion. But if we interpret it also as a phase in a process taking place in reality, as the transition of the 'cognitive will' into a chain of real events, as the act of self-determination and the determination of objects and processes, then we shall find the relationship between the premises and the conclusion to be logical and causal at the same time. It is logical in that the cognitive elements contained in the premises, *i.e.*, a series of causal relationships directed at the aim, in its anticipated form, determines the structure, character and direction of action. The anticipated subjective logical appears in the concrete process, the mode of formation the objective logic of action. However, it appears in such a way as enables it to retain a certain measure of relative independence (this makes it possible for us to change the way we do things in the course of work if we notice that the method adapted does not (or not entirely) lead to the end desired, which also signifies the subsequent correction of the second premise). The logical is thus transformed into the objective logic of causation directed at the aim. Further, the relationship between the premises and the conclusions, as an action, is not only a logical but also a causal one: purposive cognition, as a derived, noetic cause, operates action: the aim (in its concrete form determined by the second premise) directly operates self-determination for action and the finalistic determination of objects. At the same time, this direct beginning, (in respect of individual action) considered as an element of a relatively closed system, *i.e.*, a link in the chain of ontic relationships, is only a partially conscious moment within more comprehensive determinants. These determinants exist prior to the aim, they penetrate it, 'exist in it', and transcend it (as was mentioned, with inevitable simplification, in the example at the beginning of our analysis). Thus, the logical and causal elements in the final structure cannot gain independence, their essential feature being their origination in one another, penetration and development into one another: if isolated, they both lose their meaning.

The finalistic structure expressed (and hidden) in the practical inference is not a concealed form of a nomological deductive model (according to which the event would certainly come about on the basis of the precedents and the law): von Wright is fully justified in rejecting this mechanistic monist, positivist view. At the same time, I think he is wrong when he takes the intentionalist position, which presents the logical connection only, excluding law from the finalistic structure.[15]

The practical syllogism contains a law in a specific form. I again refer to von Wright's example, which illustrates mechanic causal explanation with the bursting of the radiation in a car: on the given occasion it is my aim to prevent this very radiator, here and now, from bursting from the frost, *etc.* I have to know the precedents and conditions which make the radiator burst. However, my starting point is not the law in its generality, but this particular instance, which serves as a basis for the concretion of the law. During this process of concretion, the law will become definitively general, *i.e.*, specific. On the other pole, I generalize the singular event, disregarding what is of no consequence – thus negligible – in my radiator in the given circumstances. What I have to find is the point of intersection of the particular and the general, the territory where their scopes of action overlap. As is well known in classical philosophy, the scope of teleological activity is the specific (*das Besondere*). This co-occurrence, however, is dynamic and points to the future: I wish to bring about something hitherto unrealized (or to prevent its realization), and so the case, the law, will gain its concrete form as related to 'what is to be realized'. I bring about or prevent the realization of those conditions in which the law is valid (or not valid). Law, in its real ontic existence, naturally precedes the finalistic structures that make it concrete. It operates in them, by them or against them. And if it fails to be enforced, though agent intended to have it enforced, the agent himself is also to be held responsible, as he failed to recognize cognitively or practically, or failed to make adequately concrete, the system of conditions, the closed, effect-directed process of causal sequences, which make the effect of the law possible and necessary. The specificity of human teleology is something other than independence of laws, whereas teleology cannot be reduced to the spontaneous effects of laws in the total absence of human interference or direction. The causalist position contains a partial truth by leaving room for the nomic relationship within the finalistic structure, whereas the intentionalists are right in emphasising the *sui generis* character of the finalistic structure, and, in particular, the importance of conscious determination. However, both conceptions theoretically disintegrate the finalistic structure itself, and neither is able to illustrate the 'determinational plus' characteristic of this structure: the causalist because his position is based on his ignorance of this *plus*; the finalist because he wants to grasp only the *plus*, negatively disregarding what this *plus* should in fact be related to. If the motivational mechanism is of a teleological and not of a causal character, and if von Wright recognizes the role of certain factors acting as causes of actions, but only regards them as purely formal ones, then he can only comment on the impoverished concepts of the common sense through this purist interpretation of finality, instead of elaborating unfolding, the *sui generis* character (which he himself made a point of) of action. In this case, teleological explanation is restricted to establishing a relationship between an observed event, as the means, and the aim (deemed) to be pursued. In certain cases – the man running after the train, the man seeking shelter from the rain, etc. – this is nothing but a tautological repetition of something that the common sense registers in a split second. There is no particular need even for a special activity of the eyes, these 'theoretical organs' of perception, to realize, *e.g.*, what purpose running, as a means, serves. It is not the task of philosophy to register at the level of ordinary consciousness facts we already know. The *in actu* directedness of each action towards its aim is only an element which reveals its full significance even to common sense in the relative totality of its relations. Even common sense uses a manifold and somewhat profounder teleological explanation, reaching beyond the purist interpretation of the finalistic structure. This common sense may raise the question: why did the man have to run after the train, what secondary interest does his running serve? This kind of thinking will instinctively inquire about a possible interest that had been transformed into the aim of travelling, by that particular train. And these interests, producing their effects *a tergo*, cease to be simply external ones and become immanently contained in the finalistic structure. (In certain cases it is just these 'external factors' that determine that one has to catch a certain train, for *e.g.* he has no car, and the reason for having to run at all may be that he wishes to apply for a post and arrive sooner than any other applicant, in order to ensure a living for himself and his family. Or he has undertaken blacklegging, and is pressed for time.) A whole world seems to be unfolding itself in front of our eyes if we break through the 'teleological explanation', which is restricted to the common sense interpretation of facts, and it must be so restricted provided the validity of purist teleological explanation within the frames of practical discourse is to be maintained – and this is the world that is shut to us, this is the way, to quote Shchedrin, "we close America" once we are content within this limited sphere.

Is it then true that the specificity of teleological explanation is that it points towards the future?[16]

Even common sense goes farther than to ask about the aim of a given action. The main object of its inquiry is "purposefulness externally manifesting itself", something that has been and is to be, realized: "How did you do it?" This is especially easy to see in the case of the working activity: since the materialized aim itself becomes a source and the means of new aims, it becomes a materialized form stimulating further action, a solidified form of human activity, a permanent moment in the process of active man surpassing every boundary of existing forms. Thus what common sense seeks most is an explanation for the realization of this materialized aim, the recapitulation of often, and successfully, applied laws effective in the finalistic process. Von Wright's view of future-directedness, which is the specific feature of teleological explanation, which, in its turn, breaks itself independent of the laws inherent in its object, comes to a halt at the door of abstract possibilities lying ahead in the future, and stays confined in the subjective intention of the agent just about to step out into the realm of objectivity. The adequacy of the means can only be verified at a point where action has reached its quiescent, materialized state: the agent who is just beginning to act and who is watching this action at the moment of its realization may think an activity purposeful, which otherwise may prove purposeless and *vice versa*, – both get stuck in the field of subjective feelings. Can we ignore the laws inherent in the finalistic process when analysing teleological explanation? It is obvious that the more I approach the surface in explaining a phenomenon or a process, the more it is likely for me to disregard inherent laws and, at the same time, to fail to explore the essence of the process. In reality, to refer again to the case of the man wishing to catch the train, the meeting of two objects running at given speeds and intersecting each other's courses at a given point becomes a necessity. The running man acts according to this physical law, and this is why he does not sit down or walk instead. We can see that even the *in actu* explanation of the action – provided it aims at a comprehensive application – cannot render itself independent of the laws acting within the finalistic structure, *i.e.*, of the fact that action directed at an aim is only a phase appearing at a given moment in the finalistic structure relying on a law, which may not perceived by the observer but which nevertheless is actively made use of in practice by the agent. Behind a teleological explanation, which dissociates itself from the laws of nature and interprets action *in actu*, in a phase where it has not yet become materialized, characters of Kafka seem to come in sight, – the topic cannot be dealt with here for lack of space – whose actions (Gracchus the hunter, the envoy of the Emperor of China, *etc.*) have moved out of real space and time, and represent the bad infinite. There is, however, a substantial difference in that Kafka's tragedy is replaced here by the self-confidence of common sense of itself, a conciliation with a self-conflicting, irrationally atomized world.

So far we have used a simplified method for examining the relationship between the aim and the means with which to attain it: we have taken for granted that this relationship mostly appears to the agent of cognition in an evident, direct way. However, there are cases where the meaning fails to appear so distinctly in the action, what is more, what can be immediately deduced from the action turns out to be an appearance blurring the original meaning.[17]

Where action, as the object of cognition, does not reveal its meaning directly, we are first faced with the task of finding out 'what it means to the agent'. It is obvious that only then we can evaluate the action, as being a means at a certain state of adequacy of achieving the aim, and only then we can place it into the whole of the finalistic structure. This kind of explicative understanding necessarily precedes the explanation of the totality of the action in case the object of cognition (may it be individual or collective action) being the object of the 'theoretical organs', is given as a whole of which the meaning does not reveal itself directly. This object, appearing as a perceptible whole, can just as well be the object of common sense as that of a correspondent being present at a socially relevant event on which he is entitled to comment by profession (*e.g.*, sociographer, ethnographer, on a theoretical plane), or it may be the object of cognition for a foreigner getting acquainted with the forms of activity prevalent in a country unknown to him. The innumerable examples, which it would be a waste of time to mention here, all amount to this: the only case where the starting point of cognition is the understanding of the aims of the agents is when these aims are first of all given as objects of the 'theoretical' organs, and when they do not directly reveal their subjective meanings to the agents.

Von Wright treats "the intentionalistic understanding of behavioral data", as being a process usually preceding "the teleological explanation of action", as a problem of explanation in social and historical sciences (p. 132). Part of the contents that von Wright has taken out of the finalistic structure now have to function as the object of an incomparably more complicated theoretical cognition. On the other hand, "the intentionalistic interpretation of behavioral data" can only be one of the final stages in scientific social research. Only after exploring objective material factors, which determine, partly exist in, and reach beyond the aims, factors, whose existence is independent of consciousness considering their relative totality and tendencies of development, (*i.e.*, all the processes, relations, and so forth which determine the genesis of aims, pre- and post-exist as their basis, and which, during their transformation, bring about the transformation of the aims through ideological, political, *etc.* mediations), so only after theoretically following up on the events and as a result of this, can we interpret the aims of those taking part in the events of history. That is what makes this kind of interpretation stand in the first place in the explanation of action functioning as the object of 'theoretical organs', and what puts it in the *last place* in the study of the totality of social movements, which can only be grasped through abstraction: since this kind of research (if it is scientific) starts from "historically determined social life conditions" (Engels).

Of course, we do not consider here the mode of the theoretical presentation of the object but the logic of its attainment (*Aneignung*). The presentation may, however, take its start from the interpretation of the antagonistic world of subjective aims if it has already explored the objective logic of the object (the – relative – totality of social processes), as this latter can only make it possible for us to reveal the contradictions between the subjective meanings of aims and their objective meanings which they obtain within the totality of social processes. As long as man is an unconscious author of history, we shall be unable to grasp directly the subjective meanings of aims, since these meanings are but subordinated elements and aspects of objective meanings determined by, and only *post festum* manifested in, a non-conscious mass of social processes. But the moment we go further than the subjective meaning of aims, we have to give up 'teleological explanation', since what we are to explore does not appear as an intention, or appears as an intention different from the subjective meaning of the action which inevitably makes us examine something of which the false consciousness is this consciousness.

Thus if von Wright holds the view that it is not sufficient to identify ourselves with the subjective aims of the agents – as is done, however, in the classical method of hermeneutics –, and that, consequently, the intentionalistic understanding of the interpreter should be deepened by the addition of causal explanation, then his position proves useful.[18]

If, for example, we see a crowd of people shouting and waving flags, and at first we don't know whether it is a religious procession or a folk festival, *etc.* that we witness, then the explicative interpretation of facts – von Wright writes – develops, through their perpetual reinterpretation always producing new qualities, into the explanation of the causes of events: "Something which used to be thought of as a reformatory movement in religion may with a deepened insight into its causes come to appear as 'essentially' a class struggle for land reform... From the study of the causes of religious dissent we may be led to an inquiry into the origin of social inequalities as a result, say, of changes in the methods of production in a society." (Pp. 134–5.)

What this all amounts to is that not only is hermeneutics at fault when presenting as a final instance the subjective states of consciousness, the aims and values of the authors of history, but also all those neopositivist, logical atomist, analytic ontologies which cannot go farther than recording 'states of affairs', *was der Fall ist* in a positivist-descriptive manner. This very much so, as the given event, evaluated on the basis of hermeneutics and logical positivist 'monadology', can be interpreted only as a reformatory movement in religion. And if von Wright happens to find the essence of the reformatory movement "within and without it", *i.e.*, in our terminology, in the given mode of material production, productive-ownership relations which this reformatory movement expresses in the form of false consciousness, then he should draw the conclusion from the position he rightly adopted that this idea is in contradiction with the ontological basis on which he (among others) stands: the exclusion of the 'entities' of things, qualities, relations from ontology. This is the most comprehensive reason for the disappearance of a whole world (and the concomitant annihilation of the specific teleological aspect) from this conception, which only recognizes superficial elements torn out of the totality of being raised into the domain of consciousness, reduced to certain judgment of facts and smoothed out so as to be considered merely the products of subjective *Verstand.* These elements will then become unintelligible (*i.e.*, intelligible at will) as their genesis, their supporting and transformational basis and their results have been theoretically excluded.

If we associate ourselves with some theoretical opinions about the methodology of social cognition which maintain that, unlike in the case of mechanistic-causalistic relationship, where laws mediate between events (the spark makes the powder-barrel explode), in history it is motivations that mediate between events,[19] we inevitably reduce historical interpretation to mere chronology, *i.e.*, the recording of decisions made by individuals or groups in power. This issue will also be dealt with later, as I should now rather like to concentrate on the theoretical background of methodology. The idea that man in pursuit of his ends is a conscious maker of history has been known ever since Vico. But Vico and, later, Hegel also surmised ingeniously (if loaded with mystification) what has only been known since Marxism came to existence: that the reason why the results of activities do not correspond to the original intentions is that in *naturwüchsig* (Marx) societies not yet based on conscious planning, both aims and results are determined by social laws unknown to man which operate, even though concealed in action.

Even the representatives of *Geistesgeschichte* and the philosophy of life felt compelled to admit that the result of conscious, live (*lebendig*) action may become an alienated and dead form, which accounts for the appearance of a number of antinomies in their works.[20]

The passionate opposition to the *Jenseitiges* related to conscious aims is confronted with the same *Jenseitiges* in the aims already realized. "Causality, law, constancy" forced to exile in nature invades society. "Living, historical, free and developing life" hardens into dead, stagnating, undeveloping forms in the impenetrable objectivity of society, which resists the subject (*Gegenstand, gegen-ständig*). If we bracket this resistant, material, objective and alienated power, we shall come to the 'living historical life' of *Geistesgeschichte*, as being the adequate object of understanding. If we consider this 'resistant' power as being a coercive mechanism impenetrable to the mind, then we shall come to the (classical) positivist view of society, to the historically-socially unspecified object of causal explanation. The former conception expresses the supposed power of man in a given historical period, the latter illustrates a real lack of power (if the inherent contradictions are for the moment disregarded). Further on, let us oppose the open intuitionism and subjectivism of *Geistesgeschichte*, let us recognize the exceptional role of individual aims even more by individual causal explanation (and let us deepen understanding at a methodological level by this kind of causal explanation), and we shall arrive at the theory of 'purposive-rational action' (*zweckrationales Handeln*) formulated for the first time by Max Weber. Von Wright's position is also related to this theory, however, his conception is deficient in what is the strongest point of Weber's theory and what produced with in his *oeuvre*, too, a series of antinomies: the problem of unintentional processes, tendencies as opposed to 'purposive-rational actions'.

By reconstructing the linear relationship of individual causes and aims taken out of the multitude of social processes, we remain on the surface of events: we can only register what everybody knows, *e.g.*, that the Sarajevo incident was a direct cause (pretext) of the outbreak of the First World War.

It is usually common sense that considers the events of daily life by establishing linear relations between causes and aims ("Today I have done this and this, for such and such reasons"). However, the activity of the subject of common sense, considered in its own ontic system of interrelations, realizes more than indicated by the intentions in its concious aims. Also Marxist social sciences and realistic novel-writing strive to explore this 'hidden world', which, developing behind the backs of individuals (or groups), decisively determine the individual causes and aims entering their minds. "Only the whole is true" (Hegel) – and we can only find out the meanings of facts, as significant (or non-significant) elements of certain social processes, if we explore this 'whole' with the greatest possible accuracy.[22]

If the theory fails to take its start from the whole, *i.e.*, the analysis of the historically determined social structure, then it can only describe what appears on the surface of society and in the decisions of the ruling classes. This, too, necessitates methodological individualism and the principle of charisma whereas the theory is bound to leave unmentioned those who, forced by a class-determined monopoly of decisions, are tools for realizing aims alien to their interests. This is where the latent, implicit apology of the mechanism of decision-making in class societies originates from: 'relativistic rationalism', which views the activities of the ruling political representatives of all ages, including the era of monopoly capitalism, in the light of aims and cognitive behaviour, regards the capitalist system, this specific form of social existence, as being directed by general human intellect and teleology. In this sense, it is not a particular form of social existence that appears in these aims and cognitive behaviours but – implicitly – the ever reasonable form of society. However, the position maintaining that is sphere the subjective intentions where the main source and relations of activities are to be searched for can hardly be reconciled with the economic structure of this society, with the coercive law of competition imposed from outside even, upon monopolies. There is no doubt about the fact that this law is enforced through certain subjective intentions, that aims are also the 'transmission belts' of the realization of these instinctive, *naturwüchsig* (Marx) tendencies. Yet it is not enough – or too much, rather – to consider aims and cognition in social aspect without specifying them socially-historically (and according to class), *i.e.* without exploring the way the forms of aims, etc. are socially-economically determined.[23]

A specific law of the still *naturwüchsig* forms of social existence rather than a general law of nature operates in this society. Both the naturalist reduction and the individual-psychological immaterialization of social laws imply some kind of apology: the former by assimilating in theory a specific form of social existence to the general operation of natural laws, the latter by liberating the specifically human (teleological) from its *naturwüchsig* characteristics representing human existence in capitalist society as being generally human. In this way it qualifies as aims what in reality nobody has set as aims, what is the objective, material and alienated determinant of the wills and unwills (and the false, partial consciousness related to them) of man.

As has been indicated, the absence of thinking in terms of historically determined social systems serves as a comprehensive ontological and methodological basis for the metaphysics of isolated subjective meanings. The logical atomist view of nature and society is only a form of philosophical actualism, dissolving-immaterializing social substance (as the given economic structure) in the interrelations between indeterminate subjective aims. Where von Wright attempts to analyse society on the analogy of a cybernetic model, his system-based thinking could have broken through the barriers of his preconception: that human actions are linked through 'subjective logic'.

Strikes, a form of communication, interpreted as a negative feedback

(p. 158), presuppose certain social conditions of production and ownership as primary determinative factors of social existence. They do not operate on the basis of mechanistic natural causality but on the whole-determination of capitalist society, in which the two necessarily opposing poles are the state of being deprived of possessing means of production and the state of controlling them, together with the expropriation of socially relevant coercive decisions, and the class-determination of the actual rules of the social game and the institutionalized forms of communication. From the time of the Luddites to the strikes of our days, the representatives of the 'secondary systems' have always been determined by their existence to act in communicative forms 'often contrary to the existing rules of the game'; thus this form is primarily determined by a socially determined 'mute compulsion'. Writing about the forms of compulsion, von Wright only recognises commands, requests, threats, physical violence, *etc.* (p. 146), but he leaves unmentioned what is most important: the primary compulsion stemming from the relations of production and ownership, codified by the state and the law, which are carried into effect also by organs of power enforcement. These determine in a comprehensive manner what sorts of action are to be necessarily realized and cannot be put into separate practical inferences. Such an enormous material-spiritual power seems logically irrelevant, more precisely, following the logic of singular practical syllogisms, nevertheless this power determines the premises of colonizers, arms manufacturers and those who are manipulated into getting slaughtered in the wars with enthusiasm and 'of their own accord'. This power also manifests itself in the fact that, in relatively stable periods of capitalism, the logically binding validity of these (also) compulsorily predetermined practical syllogisms is not broken. Any scientific study of society in monopoly capitalism has to be a critical one, which cannot take its start from the level at which von Wright inquires into the logical necessity of action. At this level, the power determining the premises disappears, and all that remains is to find out whether what consciousness approves is binding on action.[24]

The real question, which rises from a deeper layer of determination, is whether a division of society, as described by the cybernetic model, according to powerfulness and powerlessness can be considered necessary at a given stage of material reproduction. The view focussing on logical necessity presents the last empty moment, stripped of its relations; it gives a picture of a stagnant *status quo*, indifferent to the powers that have so determined the premises of the masses. Logical necessity is thus indifferent to that matter of which it is the regulating principle. The reduction of the necessary into the logical demonstrates as a final instance the point where individual consciousness determines the action, ignoring that this consciousness itself is also determined (and not only by an individual cause entering consciousness), and that this determination takes place within a historically determined social system with its active members not knowing about determinations which come to existence behind their backs and which form their premises and also manifest themselves in the unconscious aspect of their actions.

It would probably be unnecessary to point out here that the present criticism is directed not against the presentation of the specificities of logical necessity but 'only' against a view interpreting teleology on a social plane and disregard the ontic mediation in the logical necessity contained in the practical syllogism. It is no longer possible to complete a formal study that disregards ontic determinations of teleology by adding to it a content-analysis which does consider ontic determinations once the theoretician has stood up for the primacy and immediacy of the logic of consciousness.

Von Wright attempts to explain individual action, socially relevant actions and history itself where he can disregard the laws of social being. To avoid any misunderstanding: it *is* possible to describe at the level of separate singular practical syllogisms, *e.g.*, the process in which three adventurers took a nation of 36 million by a surprise coup, the way the obsession of Louis Bonaparte come true (as was done, *mutatis mutandis*, by Victor Hugo). But it is also possible to present this historical process as a necessary outcome of class struggles basically determined by economic conditions, exploring the "law of motion" and the "strife of historical unrest" (Marx-Engels), arriving at the laws regulating the preconditions to the process and the resultant of individual volitions and non-volitions.

The dichotomy of this explanation, set against the nomological one, thus being in fact an idiographic one, describes the abstract non-identity of nature and society as the opposition of determinism of predictability and intentional determinism (pp. 166–167). These two aspects are, however, incompatible. Predictability is an *epistemological* criterion (and can only be the consequence of a nomic determination independent of the subject who predicts), whereas intentional determinism suggests a *misunderstood ontic* character. As I have pointed out: the object of social sciences is not an element which exists without its ontic basis (I mean the last mediated moment in the process of determination of individual action) but it is the omnipresent whole, existing also in the analysis of concrete events, which, everything considered, is the given structure itself with its laws of development, *i.e.*, the social-economic formation.

The practical inference, as a model proposed by von Wright for an analysis of historical-social events, is unworkable not only because it disregards determinative layers of social being that form the aims *a tergo*, but also because it unjustifiably smooths out hostile, and alienated character of class societies. This kind of *reductio ad hominem* (pointed out also by Adorno) is deceptive: it makes one accept the unacceptable and heals the stigma imposed upon the individual from outside to appear as some consciously accepted teleology. The formal individual freedom liberated from natural and social laws becomes the absolute, final instance, and mankind, only to be realized in the future, disappears behind a 'teleology' reduced to being the object of a positivist description of given states of affairs.[25]

Translated from the Hungarian by
István Ambrózy

Philosophical Institute
of the Hungarian Academy of Sciences,
Budapest

NOTES

1 G. H. von Wright, *Explanation and Understanding*, London 1971.

2 "The discovery of causal relationships presents two aspects: an active and a passive one. The active component is the putting in motion of systems through their initial states. The passive component consists in observing what happens inside the systems – as far as possible without disturbing them". (p. 82).

3 Cf. Mario Bunge, *Causality: The Place of the Causal Principle in Modern Science*, Cambridge, Mass. 1959, Ch. 3.3.1–3.3.3.

4 See Th. W. Adorno, *Aufsätze zur Gesellschaftstheorie und Methodologie*, Frankfurt am Main 1970, p. 172.

5 For instance, if the action is determined by intentions and cognitive elements, then it can be explained teleologically (p. 165).

6 In this sense, "behavior gets its intentional character from being *seen* by the agent

himself or by an outside observer in a wider perspective, from being set in a context of aims and cognitions" (p. 115).

[7] See p. 74. In opposition to the mainstream of his idea, he once asserts that causality operates also in places forever inaccessible to man (p. 73), maintaining at the same time that "it is only through the idea of doing things that we come to grasp the ideas of cause and effect" (pp. 81–82).

[8] In his analysis, N.Hartmann writes that Aristotle distinguished two kinds of motion within the finalistic structure: the noetic one, taking place in consciousness, and the real one, taking place in the external world. The first kind explores the means of the realization of the *eidos* (taking its starting point from the *eidos*, from the ultimate means, moving towards the primary, the directly tangible means). This motion can only be performed by consciousness, as it moves contrary to the stream of time; the second, the motion of realization, starts from the temporally primary means, realizing one means by the other, moving with the time-stream, whereas the temporally ultimate means realizes the aim itself. "What is decisive and of permanent value in this Aristotelian analysis is that he understood that the finalistic structure involves a stratified, *e.g.*, at least twofold interdependence..."

The finalistic structure, in Hartmann's contribution, contains three acts: "(1) the setting of the aim by leaping over the time-stream, *i.e.*, the anticipation of something in the future; (2) the choice of means in consciousness begun from the aim set (retroactive determination); (3) realization by the series of means chosen: a real process, moving forward outside consciousness". Characterizing the finalistic structure as a transformation of causality, Hartmann writes: "If each means did not cause the next one to come existence, then it would be impossible to realize any aim at all... What is meant by the suitableness, 'purposeful' of the means is that in the case of a certain application of the means – *i.e.*, the utilization of its specific causal effect – is the 'cause' of the aim set... finalistic relationship presupposes causal relationship... the recurrence of causal relationship may be considered to be the typical transformation of causal relationship" (Nicolai Hartmann, *Teleologisches Denken*, Berlin 1951, Ch. 7).

[9] Cf. G. W. F. Hegel, *Enzycklopädie der philosophischen Wissenschaften im Grundrisse*, Leipzig 1949, § 208.

[10] See K. Marx, *The Capital*, Sect. 3, Ch. 5.

[11] It is important to emphasise that man can only be the author of history if he is its 'actor' as well, as philosophical actualism, and the philosophy of praxis, consider man as the author only.

[12] Von Wright's method is generally characterized by the abbreviation of the descriptions of objective and subjective processes, which brings its own punishment by depriving these processes (certain phases of theirs) of their complex meanings. He holds for instance that the first premise in the practical syllogism already contains the fact that the agent knows how to bring about the object of his intention. "Thus the intention includes a cognitive element" (p. 103). Naturally, there are a great number of cases like this; *e.g.*, if my intention is to open the window, I need no separate act of thinking to relate its means (the turning of the handle) to my intention. But here lies a trap set by the theoretical generalization of habitualized and instinctive action, namely that it blurs or even totally conceals the necessity of exploring and teleologically transforming causal relationships in the second premise.

[13] "The starting point of a teleological explanation (of action) is that someone sets himself to do some-thing or, more commonly, that someone does something. We ask 'Why?' The answer often is simply: 'In order to bring about *p*'. It is then taken for

granted that the agent considers the behavior which we are trying to explain causally relevant to the bringing about of *p*, and that the bringing about of *p* is what he is aiming at or intending with his behavior. Maybe the agent is mistaken in thinking the action causally related to the end in view. His being mistaken, however, does not invalidate the suggested explanation. What the agent *thinks* is the only relevant question here" pp. 96–97). Teleological explanation – as against causal explanation normally pointing towards the past (this happened *because* that had occurred – points to the future. "This happened, *in order* that should occur" (p. 83).

14 "... those who think that the intention can be a humean cause of behavior (are) causalists, and those who regard the connection between intention and behavior as being of a conceptual or logical nature (are) intentionalists" (p. 95). "If one regards practical inferences, when properly formulated, as logically binding, one takes an intentionalist position. If again one accepts the causalist view, one would say of practical inferences that the truth of their premises insures the truth of their conclusions, but that this is a 'causal' and not a 'logical' entailment" (p. 97). Von Wright, who adopts an intentionalist position, does not deny that desires and wishes may have a causal effect on behaviour, and in interpreting behaviour, he does not exclude the understanding of the roles habits and inclinations, *etc.*, play in its formation. In opposition to the causalist view, he maintains that, on this interpretation, in the case of a causal relationship between intention (von Wright deliberately treats this word as synonymous to the setting of an aim, to the will and the cognitive elements in this sphere, *i.e.*, the factors that formulate the first two premises) and behaviour, there exists a law, a nomic relationship, in which case teleological explanation (or the practical inference) is merely a disguised form of a nomological deductive explanation.

15 Here, the limitations of the practical inference, as the alleged explanation model pointing towards the future, are especially obvious. – "The validity of the 'genuine' teleological explanation does *not* depend on the validity of the assumed nomic relation involved in it." For instance, if someone is anxious to catch the train, we suppose that he thinks it necessary and, perhaps, sufficient to run in order to reach the station before the train has left. "His belief, however, may be mistaken – perhaps he would have missed the train no matter how fast he ran. But my explanation of his running may nevertheless be correct" (p. 84). – See later at the *in actu* explanation of action.

16 *Op. cit.*, p. 83.

17 Krupskaya writes in her memoirs on Lenin: "We recalled a simile that Tolstoy uses somewhere: going in the street, one sees in the distance another man squatting on his heels, performing some stupid, meaningless movements with the hands; he must be mad, one thinks, but on coming nearer to the man, one sees that the man is really sharpening his knife on the pavement. This is the case with theoretical disputes as well. To those listening from outside, they seem to be mere twaddle – whereas those who go deep into the matter will know that the very essence of things is being dealt with".

18 However, there are two reservations to be made here: one I have already indicated, namely that teleological understanding cannot be a basis for social explanation. The other is that causal explanation, which helps von Wright over hermeneutics, can only be a phase in the study of social totality determination, which must also have two other important objects of exploration: economic conditions and the given state of class struggles (basically determined by the former). Though the examination of these two levels of whole-determination (*Ganzheitsdetermination*) also requires the study of causal relationships, it is not restricted to it. Yet the process of explanation, as presented by von Wright, leads, even if from different motives, to the Marxist thesis that, *e. g.*, reli-

gious events are in the last analysis results, not causes of the economic structure standing at a given stage of development.

[19] What connects action, "is not a set of general laws, but a set of singular statements which constitute the premises of practical inferences". Writing about the assassination in Sarajevo preceding the First World War: "We have a sequence of independent events: the assassination, the ultimatum – the outbreak of the war. The events are linked, we said, through practical syllogisms" (p. 142, etc.).

[20] Dilthey emphasises on the one hand that historical categories originate in the aims, wishes and values of the individual, whereas on other hand the individual is only a point of intersection of cultural systems existing independent of him. Ideas always conceal violence, facticity, and a basis never to be idealized. Everything originates in the aims of the individual: "nothing is solid or strange here", yet life, action and energy confront us as results in the form of rigid substance. Wilhelm Dilthey, *Gesammelte Werke*, Leipzig 1958, VII, Bd. 251, 287, 148.)

[21] Cf. in connection with the latter remark Th. W. Adorno, *op. cit.*, p. 140.

[22] In this respect, the Sarajevo assassination, for instance, represents the Serbo-Croatian national aspect of the war, which Lenin considered to be of secondary importance, underlining that the national element does not change the general imperialist character of the war.

[23] It is not enough because we then are confined to the philosophical anthropology of man as an abstract, superhistorical being. It is too much because, as we have seen, it is compelled to describe the capitalist development of society as its ever-reasonable form. It necessarily ignores that the basis and results of individual 'rational' subjective aims is objective irrationality manifested in overproduction, wars, famine, manipulated consumption for the profit of private production, culture producing idiocy for the masses and sophisticated elite culture for the few, *etc.* 'The natural position of the animal' is reflected in this 'relativistic rationalism' as the reasonable position of man.

[24] Using a cybernetic model of society, von Wright describes the process in which the representatives of the secondary system, as distinct from the primary power group, wish to change the given mechanism of the primary system through strike, protest, sabotage, and other channels of communication. It is accidental whether or not those in the primary system will respond to the challenge by the secondary system and it is also accidental whether or not the information of the former will influence the cognitive behaviour of the latter. "But once the premises, *i.e.*, the volitions of agents in the one and the cognitions of the agents in the other system, have been formed the actions which follow become, in the light of the new premises, logically necessary" (p. 159).

[25] The view appearing in the analytical trend at a methodological level can also be found in more than one aspect in a so-called "Marxisant" trend. In this latter, however, it has assumed the form of ontology. The (explicit or implicit) principal theses of today's 'philosophy of praxis' ('social-ontology') are the abstract non-identity of nature and society, and an actualistic view of society, a negative abstraction of the social-natural (matter transformed by man, above all the means of production) and of specifically social determinations of existence, of those production – (ownership) – relations, into which man enters in his act of transforming nature. Thus the essence of society becomes reduced to the aims of anthropologic individuals torn out of their real social being. All this obviously necessitates the indeterministic and decausalized notion of teleology. The well-known representative of this position was the Austro-Marxist Max Adler: he introduced neo-Kantianism into Marxism right after the turn of the century, building his theories through decades in opposition to Plekhanov, Kautsky

and, above all, Lenin, *i.e.*, all those who saw in Marx's and Engels's materialism more than "a simple protest against contemporary speculative idealism". This "Marxisant" trend has since subsisted upon the *Geistesgeschichte*, the philosophy of life and esistentialism (Dilthey, Simmel, Bergson and Heidegger in particular), and these joint sources explain why there prevail such similar views within this trend, philosophical analysis and existentialism of today. To analyse this question would need a study far more extensive than this. However, it is certain that a study of all these points of contact and intersection might, even if on a philosophical plane, shed light on some essential social-ideological relations of our age.

RÜDIGER BUBNER

IS TRANSCENDENTAL HERMENEUTICS POSSIBLE?

0. Introduction

I'm afraid that my title, 'Transcendental Hermeneutics', may impress an English-speaking audience as one of the typical verbal abominations which are supposedly used by teutonic philosophers to intimidate rather than clarify. I will make every effort *not* to confirm this widespread prejudice. To begin with I should make clear the scope and limits of this paper.

The concept of hermeneutics has, for some time, played an important role in the recurrent methodological controversies in the so-called cultural sciences or *Geisteswissenschaften*. In these debates hermeneutics, as the theory of interpretive *understanding*, is usually contrasted with methods of scientific *explanation*. A reformulation of this distinction in terms of analytic philosophy is one of the great merits of Professor von Wright's recent book, *Explanation and Understanding*. At the present time several influential German authors are introducing a new element into the debate. Philosophers such as Gadamer, Apel, and Habermas are seeking to link hermeneutical theory with the transcendental claim passed down through the legacy of Kantianism. Hermeneutics itself would then overcome the traditional distinction between explanatory and interpretative modes of procedure and extend into a comprehensive, fundamental theory of knowledge. In what follows I will examine this program of a transcendental hermeneutics.

The body of my paper consists of four sections: the first and second dealing with the concept 'transcendental', the third with 'hermeneutics' and the fourth with the possibility of their conjunction and some neglected logical problems therein.

In order to attain some initial clarity about the concept of the transcendental I find it useful to devote the first section to an elucidation of the concept which avoids any exegesis of Kantian texts. Instead, I draw upon writings of Wittgenstein, Quine, and Strawson, philosophers who surely cannot be suspected of dogmatic Kantianism. These authors, while cer-

Manninen and Tuomela (eds.), Essays on Explanation and Understanding, 59–77.

tainly not to be taken as 'transcendentalists', provide us with three – relatively unrelated – examples of the modern usage of the term. These usages, as we shall see, have a common characteristic which allows us then, in section two, to throw some light on Kant's use of the term 'transcendental' without entangling ourselves in the intricacies of his entire system.

Section three consists of some brief remarks on the history of hermeneutics as a background for the question at issue. On this basis we will then be able to proceed, in the fourth section, to a critical examination of the program of a hermeneutics which raises the transcendental claim. In the summary and conclusion (section five) I will, for the most part, point out difficulties and pose questions rather than provide answers. But, fortunately, this is permissible in philosophy.

I

I shall begin with an example from Wittgenstein. In the *Tractatus* we read the following sentences: "Logic is not a theory but a reflection of the world. Logic is transcendental". (6.13)

What does the use of the word 'transcendental' mean here?[1] Apparently logic, the elaboration of which from within the inner structure of meaningful language is central to the whole *Tractatus*, is not to be treated like an autonomous formal theory with the claim to analytical truth. But logic does not indifferently confront the world of facts, thereby abandoning the discovery of truth to empirical observations. Logic is the reflection of the world: in it, those conditions are revealed which constitute the world of facts. Wittgenstein expresses this in the following manner: "The logical propositions describe the scaffolding of the world, or rather they present it. They 'treat' of nothing. They presuppose that names have meaning, and that elementary propositions have sense. And this is their connection with the World" (6.124). The connection between logic and the world lies in this presupposition. Logic does not, as it were, overtake this presupposition; it builds upon it.

In the metaphor of the reflection (*Widerspiegelung*) an inescapable presupposition is expressed, namely, that formal logic with its pure tautologies or analytic truths does not hopelessly depart from all the factual truth of the empirical, but rather it reflects precisely *those* structures according to which the world is ordered. The ontology which underlies this interpretation in accordance with the principle "The world is everything that is the case", need not concern us here, since it represents not the reason for, but the consequence of the transcendental presupposition. The presupposition in question must be made, if logic is to have any meaning at all for the regulation of meaningful statements, in which the existence of the world as a world of facts, is stated so adequately that an empirical examination in terms of 'true or false' becomes possible. This presupposition is called transcendental, since it assumes a structural identity in the relationship between logic and reality, upon which the concept of meaning depends. A logic which did not make this presupposition would renounce all its competence with regard to the meaning of statements. It would simply be an arbitrary calculus, the language-game of specialists.

But neither can the presupposition be introduced subsequently at a higher level, for in order to introduce it, a language would be required in turn, which already possessed the competence in question with regard to the concept of meaning. This results in a circle. The alternative to the circle would be the infinite regress of ever new metalanguages which Plato was the first to recognize in the dilemma of the Third Man, and which Wittgenstein explicitly puts forward against Russell's hierarchy of types. Russell, however, felt so sure of agreement with Wittgenstein in the basic interpretation of logical atomism, that he did not notice the irony of suggesting in his introduction to the *Tractatus* precisely that hierarchy of metalanguages, against which Wittgenstein had formulated his transcendental presupposition of an original link between logic and world.

Wittgenstein falls back upon the traditional concept of the transcendental to ascribe to logic the function of making empirical knowledge possible. This cannot be derived from any superordinate principle, but must be secured in advance for analysis which intends to clarify the meaningfulness of statements as the capacity to reproduce reality. The analysis must presuppose something without which it could not operate. With the term 'meaning' it presupposes a certain relationship between logically ordered statements and the construction of reality. The relationship is such, that only on account of it, is it possible for a statement to reproduce a given piece of reality, and this means to make a meaningful proposition. The meaningfulness of statements or their pretention to empirically testable truth or falsehood does not then descend from heaven, nor does it stem from an ungrounded 'dogma of empiricism', but rather it derives from presuppositions which can be apprehended. The presupposition of a meaning-relation between statements and reality must be revealed by analysis, for it is this relation alone which also makes the analytic procedure itself meaningful. Consequently, analysis reveals something, which it must presuppose if it intends to perform the task of clarifying meaningful statements. It must accept the relationship between language and reality as one which exists prior to the analysis, and for this reason cannot be produced in an arbitrary fashion. Nevertheless, with the revelation of the relationship which makes statements meaningful, the analysis of language penetrates through to the conditions, upon which it itself rests as a logical analysis.

We are faced then with an intricate complex here, insofar as a relationship must be presupposed, in order that statements referring to reality may be possible as such and this relationship simultaneously represents a presupposition by means of which logical analysis first becomes possible. The clarification of the logical presupposition for meaningful statements is self-instructive in its capacity of clarification concerning the limitations and possibilities. In referring to a logical presupposition for language the analysis is referring to itself. In my view it is this complex formal structure which induced Wittgenstein to use the concept of the transcendental. I shall designate the essential structural element by tentatively calling it *self-referential.*

My second example is borrowed from Quine's essay 'Ontological Relativity'. Here he brings his old topic of an ontological commitment inherent in all forms of language under a strict principle of relativity. He takes up Carnap's notion of a linguistic framework for ontological questions. In 'Empiricism, Semantics and Ontology', Carnap suggested that ontological questions should only be considered with regard to a given linguistic framework. From this internal aspect another aspect has to be distinguished which is called external because it goes beyond the given framework by asking whether something mentioned in a language 'really' exists. The choice between different linguistic frameworks is supposed to be relatively free and more or less a matter of tolerance, so that a careful distinction of internal and external aspects would put an end to the permanent quarrel about ontology.

Quine,[2] however, is of the opinion that ontology is not a matter of tolerance and of distinguishing aspects. To him this seems but a new version of Russell's hierarchy of types eluding the ontological commitment. In 'Ontological Relativity' he claims that relative to every form of language one is bound to assume certain entities designated in the language. Ontological questions only make sense with an underlying linguistic system, a 'background language'. External aspects or worrying about Being in an absolute manner become therefore superfluous. As Quine states, the inescapable relationship to a background language and the emptiness of every ontological question detached from the latter, confirm in "a suddenly rather clear and tolerant sense that ontology belongs to transcendental metaphysics".[3]

The concept of the transcendental is evoked at a point where the relativity of the ontological question concerning a linguistic system is recognized at the same time as being incircumventable. That a given language enters into obligatory commitments concerning the being of what it signifies, is valid in relation to that language and consequently does not represent a universal requirement from an absolute standpoint. Nevertheless, the ontological commitment, relative as it is, must be held to be necessary for all linguistic systems. Kant coined the term 'transcendental' for such a logical relation between the relative and the necessary. Even if the striking analogy with 'transcendental metaphysics' is a marginal note, it seems that Quine not unconsciously has taken over the term, for he shares with Kant the interest in the legitimation of the scientific knowledge, which we already have – or can have – at our disposal.

In Quine, certainly, this interest springs from the un-Kantian soil of pragmatism, but does not as a result fall prey to the arbitrariness of ever changing practicability, or to a margin of tolerance which is in principle left open. Besides the given speech context and the appertaining ontological commitment, ontological relativity in the strict sense does not permit a third position, from which the advantages of utility might be weighed and margins of tolerance could be determined. The recourse to pragmatic standards only liberates the ontological question from a false dogmatic absolutism, in order to tighten in this manner all the more the connection between de facto existent linguistic systems, right up to a high level of formalisation, and the essential assumptions, made simultaneously, about the reality signified in this language.

This uninterrupted ontological relativity presses for application even to the linguistic forms in which it is stated. Quine's pragmatistic conception apparently seeks to evade the problem of a methodological circle which worried Wittgenstein, although it was devised precisely with regard to the formalised languages of philosophy and logic. In accordance with a relativity principle, understood transcendentally, it would have to make reflection upon its own bondage to the law of ontological commitment explicit. As far as I can see, however, it remains unclear just how consistently in the case of ontology, relativity is extended to the thesis itself, or how the language of the philosopher, which asserts a general ontological commitment, is distinguished from all the languages concerning which the commitment is stated. In a word, the structure of *self-referentiality* seems unavoidable but is not explicitly worked out.

My third example of a revival of transcendental terminology within the domain of analytic philosophy is provided by Peter Strawson, who apparently draws upon Kant's *Critique of Pure Reason* much more than either Wittgenstein or Quine. Strawson's book *Individuals*, which bears the subtitle "an essay in descriptive metaphysics", investigates amongst other things, the identification of certain empirical data within the spatio-temporal framework of reference, which must be established beforehand.[4] The clarification of the conditions for such identification, which underlies the semantic meaning of linguistic expressions that can be mediated through dialogue[5], plays a paradigmatic role for what Strawson calls a 'transcendental argument'. The logical analysis of this particular type of argument does not prove to be too clear. This has recently led to an extensive discussion of Strawson's comments.[6]

Strictly speaking, Strawson only says the following: 'With transcendental arguments, a problematic question is not solved in a way such that from a given premise about a given middle term an inference results which solves the initial problem, but rather the process of argumentation runs in the opposite direction: only because the solution exists, does the problem emerge at all'. Strawson elucidates this with an example borrowed from Kant, when he says: 'One does not infer from a given spatio-temporal framework of reference for given empirical objects that these objects fulfill the conditions of that system, but instead starting from the fact of the objects being given within a system of reference the function of the system as a system is first thematised'.

One could not say that this construction of a reversed direction of inference really brings to light the distinctive feature of transcendental reflection. As far as I can see, in the ensuing debate only Hintikka gave, with reference to Kant himself, the due correction. He made clear that not only the way back to given preconditions be called transcendental, but rather that the only argument worthy of the name is one which in doing this goes back to the conditions of its own operation. In other words revealing the conditions for the possibilities of using certain concepts must simultaneously show how the revelation is possible.[7] To put it paradoxically: A transcendental argument states what it states and says something about itself.

To sum up the result of our first consideration: *Self-referentiality is characteristic of a transcendental argument.* Despite the differences between the three witnesses summoned more or less by random the self-referential structure is prominent in all the cases discussed.

II

At this point recourse to Kant himself seems sufficiently prepared, so that we can extricate his conception of transcendentality from manifold entanglement with the rest of his system and examine it for its essential feature. I preferred not to start the examination of the concept of transcendentality by quoting classical definitions of Kant and thereby taking the risk of being biased by the Kantian systematic. Rather, it seemed promising to shed new light on familiar insights from a tradition which is not stamped by the Kantian influence.

I suggest that we concentrate solely upon the concept of the transcendental,[8] without thereby reproducing the entire construction of Kant's theory of knowledge. Kant expresses himself quite clearly concerning the particularity of transcendental argumentation. He writes in the *Prolegomena*[9], for instance: "For me the word 'transcendental' never means a relating of our knowledge to things but rather to the capacity for knowledge." In analogy to this, the *Critique of Pure Reason*[10] urges that "not all cognition a priori must be called transcendental, but instead only that by means of which we recognize that and how certain ideas (perceptions and concepts) are a priori applied and possible. Consequently, neither space nor any geometrical determination of the same a priori is a transcendental idea."

Transcendental knowledge and apriori knowledge cannot therefore be posited as identical. Only such knowledge is transcendental in which empirical knowledge is related to the capacity for knowledge, *i.e.*, in which the relationship between the conditions for the possibility of cognition on the one hand, and empirical knowledge on the other, is itself the object of knowledge. What is apprehended in this transcendental manner naturally precedes the actual process of cognition. But one cannot as a result claim that everything which precedes cognition counts as transcendental conditions. A whole series of physical, psychic, societal and historical preconditions can be adduced without which cognition is impossible. The mere antecedence of such conditions in no way qualifies them as transcendental conditions. One may very well count sufficient nourishment and a well stocked university library as preconditions of knowledge without however calling them transcendental.

On the other hand a form of knowledge is conceivable which can exist independently of all experience. Kant had geometry in mind. But a pure theory of linguistic competence such as Noam Chomsky's generative grammar on the basis of innate ideas could also be called an apriori theory.[11] The apriori nature of this type of knowledge, taken simply by itself, does not however make it into a transcendental type. For transcendental knowledge primarily takes as its object the preconditions relevant for cognition. From these two considerations it follows that one should not talk of transcendentality if only an unspecified precondition for knowledge is at stake, nor if only some knowledge is meant, independent of the empirical and consequently prior to all experience. According to Kant, only that knowledge is transcendental, in which knowledge is thematised concerning its specific possibilities. If this is true, then that knowledge which is called transcendental takes as its object, together with the general conditions of knowledge, the conditions of its own genesis and functioning. *Self-referentiality characterises the transcendental argument.*

As far as the formal construction is concerned, the three examples taken from the school of linguistic analysis conform in varying degrees to the Kantian line of thought just developed, although they all reproduce the structure of self-referentiality. All of them, however, deal with the linguistic interpretation of reality instead of the intellectual cognition which Kant had in mind.[12] Wittgenstein goes furthest of all here, since in his view the logical analysis of meaningful sentences does not seem possible without reference to the analytical procedure itself. In Strawson's work, the debate on the reformulation of transcendental arguments has advanced, at any rate, to the clarification of the moment of self-referentiality, whilst recall of transcendental philosophy on the basis of Quine's relativity-thesis makes self-reference in fact inevitable, but Quine does not expressly reflect upon it.

One might justifiably object that this is a highly unorthodox interpretation of Kant; that in the transcendental deduction Kant is concerned with the legitimation of the objectivity of our knowledge of the world. This is correct, but it is not what we are discussing here. In the present connection only the form in which such a proof is carried out is at issue. The strategy which claims to be transcendental, is at issue. With regard to the transcendental character of the deduction I believe it can be demonstrated that the structure of self-referentiality is decisive, a structure which we have emphasized. It is a different question whether a line of argument which must be called transcendental in terms of self-referentiality merits the strict title of a deduction. This concerns a detailed interpretation of Kant[13] and need not concern us here, since we only wished to know what the concept 'transcendental' can mean. Nor does the appeal to the programme of transcendental philosophy within the framework of present day hermeneutics ever refer to the idea of a deduction in the strict sense, but rather to the moment of self-referentiality.

III

Here I shall take up my discussion of hermeneutics. A short historical survey may serve as a sort of introduction to the second part of my paper. Usually hermeneutics has been seen in the old dualistic scheme of natural and cultural sciences. Hermeneutics is the theory of understanding (*Verstehen*) and has been assigned the methodology of the *Geisteswissenschaften* (cultural or social sciences) which are supposed to represent a contrast to the explanatory natural sciences. It is well-known that the opposition is a legacy of the nineteenth century and stems from historians and philosophers such as Droysen and Dilthey, who, emanating from the tradition of German idealism, protested against the attempts to deal with historical and social phenomena adopting the natural-scientific model with its law-like explanations (A. Comte, J. S. Mill, T. Buckle). Later the neo-Kantian school set up the distinction between nomothetic and idiographic[14] methods (Windelband) the former meant to establish laws of nature while the latter focusses on the particularity of unique historical events. It was then H. Rickert who with the means of Kantian transcendental philosophy sought to derive logically the "limitations of natural-scientific concept formation"[15], to make room for a cultural-scientific understanding of meaning. Rickert's ideas have entered sociology via Max Weber. The most recent example of this influence is Peter Winch.[16]

It is useful to recall briefly the history of the problem since the old dualism still shows unabated force in the contemporary discussion of the methodology of social science. This is true of the philosophy of science which has developed from Karl R. Popper[17] (Thomas Kuhn, Imre Lakatos, Paul K. Feyerabend) insofar as historical understanding has become an essential topic of the philosophy of science. Here I see a convergence with the hermeneuticians of Continental philosophy.[18] In the present-day methodological controversy, exemplified by the 'positivism-debate' of the sixties, the neo-Marxist dialecticians (Theodor W. Adorno, Jürgen Habermas) frequently stand on the hermeneutic side. In their struggle against the 'positivistic' deformation of social science they forget the old dispute over the 'idealism' of the theory of understanding.[19] Cutting across the various schools and camps within the theory of science and methodology, the old distinction between explanation and understanding continues to prove that it is still active.[20]

I do not wish to discuss the problem of the methodological dualism here in detail.[21] Instead, I shall concentrate upon the strongest thesis with which hermeneutics has engaged in the discussion – the one which raises the transcendental claim. Hermeneutics is here no longer concerned with establishing or legitimating one particular method as opposed to another, namely establishing the method of understanding which has its specific place in the humanities. Transcendentally conceived hermeneutics surmounts the dualism of methods and clarifies the meaning of the possibility of all methodical cognition.[22] As a final consequence, therefore, it must also provide insights into the constitutive presuppositions of the explanatory sciences and their methodological procedure. It is not possible for me to deal with this aspect adequately. Very detailed studies would be required on the antecedent horizon of understanding in explanatory science, in particular, natural science. At present such questions are being discussed from many perspectives in connection with Thomas Kuhn's concept of the 'paradigm'.[23] I am omitting this and other aspects important for a substantiation which could make the hermeneutic program convincing. For the rest of my paper then I shall restrict myself to the transcendental claim as a logical problem of hermeneutics.

IV

For hermeneutics, understanding means a fundamental apprehension of truth which takes place in intersubjective processes of communication and in the mediation through history. The social and historical basis of understanding provides a more comprehensive foundation for all cognitive acts in such a way that theory as it were is embedded in the practice of life. Understanding must be conceptualised as a process which transcends the subject-object division in so far as one recognises oneself in the other. The social or historical object which seems alien at first, becomes familiar in this process of understanding. So the process itself can be regarded as an attempt to recover what potentially belongs to oneself, but appears under an objective mask. The hermeneutic act of recovery, as far as this aspect is concerned, could be interpreted in terms analogous to what platonic epistemology called anamnesis. Since no intellectual activity can proceed beyond these elementary acts of understanding, hermeneutics takes itself to be fundamental. According to this view scientific knowledge as a methodically regulated and controlled special form of cognition can be derived from elementary acts of understanding.

The grounding of all cognition in original understanding has induced hermeneutics to raise the transcendental claim. The demonstration of this theory's transcendental character lies in the structure of self-referentiality and finds itself in complete agreement with our analysis of the transcendental argument. Here 'self-referential' means that the hermeneutic uncovering of the conditions for the possibility of knowledge, in original acts of understanding, must also ground its own insight into such understanding. What forms the object of investigation in hermeutics is also the precondition for this investigation. Let us first consider the role played by self-referentiality in the framework of hermeneutics.

For hermeneutic reflection it is not sufficient to approach the matter from outside, and point out certain historical conditions which remain obscure and guide our cognition from behind, in the way that the paradigm à la Kuhn indicates the direction of normal science without being a topic itself. Hermeneutically, the revelation of certain presuppositions as being relevant for cognition is only plausible if this function is applicable to hermeneutic reflection itself. Hermeneutic reflection finds itself involved in the very conditions of understanding it is out to demonstrate. Thus the self-referentiality shows how inescapable and far reaching the influence of certain conditions is which stamp a social or historical situation. The structure of self-referentiality is used to demonstrate that there is no unconditioned understanding from an allegedly independent point of view.

But the transcendental demonstration encounters a difficulty at this point. Hermeneutical reflection has to refer the same limiting conditions it uncovers to its own functioning. In fact it seems only possible to relate the general character of knowledge's being restricted to the very hermeneutical reflection, while this reflection must not accept the particular restricting factors and historical conditions to the same extent. The model of Kantian transcendental philosophy did not find itself exposed to this aporia, for transcendental logic within the *Critique of Pure Reason* in any case analyses only general formal structures of knowledge, which are valid once and for all a priori. The program of a transcendental hermeneutics, however, requires the historical concretion of the particular traditions which determine a given historical state of affairs. Reflection upon the antecedently determining role of these factors certainly suspends the unconsciously binding influence of concrete traditions, and subjects itself merely to the principle of dependence upon tradition, since the revelation of historical influences can only succeed by virtue of a direct contact with them. In short, hermeneutic reflection does indeed take place within a context of tradition, but no longer blindly belongs to it, because it ipso facto extricates itself from the specific historical situations. The self-referentiality of the transcendental argument can only be realized for hermeneutics *in general terms*, and *not at the concrete level of historical reality*. But this contradicts the demand for concretion raised by hermeneutics. Here we are confronted with a logical difficulty to apply the structure of self-referentiality to the hermeneutical act of understanding.

Now, secondly the self-referentiality of the transcendental argument further reveals two sides, which are essential to hermeneutics. I should

like to call them '*task of legitimation*' and '*genesis of the problem*'. The task of legitimation refers to the old Kantian theme of the validity of that knowledge of the world which we factually have at our disposal. In canonical terminology this is the question *quid juris* as opposed to *quid facti.* In contrast to the deduction of the *Critique of Pure Reason*, however, hermeneutics does not fulfill the task of legitimation in accordance with the model of a stringent proof, but instead by means of a demonstration of the lack of alternatives.

The possibility of alternatives cannot be rejected a priori, there is no metaphysical position available which might stand over all empirical knowledge, in such manner that it was entirely independent of the preconditions for the latter, and yet was capable of valid statements about empirical knowledge. On purely logical grounds possible alternatives to the form of knowledge we actually have cannot be ruled out. It can, however, certainly be demonstrated from case to case that in order to be considered as competing candidates at all, potential alternatives to the given form of knowledge must reveal a minimum of common elements with that form, for which they are proposed as alternatives. To a certain degree alternatives must be capable of being translated into one another. For such a translation and for such comparability to become possible, essential preconditions must be shared in all cases. Hermeneutics undertakes to reveal in alleged alternatives such common preconditions, which ultimately may not be forfeited if the general claim to knowledge is to remain meaningful.

But hermeneutics takes a decisive step further, for the lack of alternatives appears especially convincing if even reflection upon knowledge and its possibilities proves to be dependent upon the preconditions which determine the real processes of knowledge. If there exist certain common preconditions for knowledge in elementary acts of understanding, then they are also the preconditions under which a clarification of these preconditions itself stands. The self-referentiality of the transcendental argument corroborates, therefore, the lack of alternatives, whilst the demonstration of the lack of alternatives supports the intersubjective binding force of the preconditions for knowledge. In this way one would find a hermeneutic solution for the task of legitimation of our knowledge.

For the purpose of illustration I should add that in Kuhn's scheme of scientific revolutions this requirement should be placed at the point of paradigm-change. That there is a gap in the scheme has been proved sufficiently by the debate on Kuhn's theory. The alternative paradigms which change in a revolution must at least be comparable in their function of making concrete knowledge possible. The comparison itself can only be carried out upon those foundations, which the comparison shares with the paradigms compared. A quite similar necessity confronts the task of legitimation within the framework of hermeneutics, a task which is solved with the aid of the self-referential structure of the transcendental argument.

The elucidation of the *genesis of the problem* is more difficult. Here, a clarification is required of the fact that it at all occurs to us to pose such questions about knowledge, its preconditions and its legitimation. The problem here is how knowledge becomes a problem. This question may sound simple, but the epistemological discussion nevertheless indicates that the genesis of the problem represents an unsolved puzzle. This is seen only rarely, since people either move along the well-trodden paths of the academic discussion or they meet the enquiry with certain pragmatic trivialities, such as the notoriously unhelpful doctrine of trial and error and consequently forego logical analysis. Hermeneutics has induced us not to take for granted that there are such things as philosophical problems.

Kant was very well aware of the new enterprise he was undertaking by a radical search for transcendental foundations. On the other hand he was unable to contribute anything really plausible to the elucidation of the genesis of transcendental philosophy. In this connection, he draws attention to the inexhaustible investigatory power of reason itself, which calls everything into question and proceeds to the last principles. We simply possess this reason and cannot rid ourselves of it. Reason as such compells us to transcendental investigation. This explanation does not seem satisfying. Here we see that critical transcendental philosophy is based upon a *dogma* of reason, where it runs up against its own limitation. Husserl's revival of transcendental reflection on the basis of the phenomenological analysis of the life-world, undoubtedly struck upon the unclarified problem-genesis. In my view, however, Husserl did not take this problem of the genesis of philosophical reflection a stage further, but instead covered it over with an empty terminological abstraction. The term 'transcendental-phenomenological epoché' replaces the exploration of the ground of reflections upon the precondition for knowledge. Finally, Heidegger, in a characteristic transformation of the concept of phenomenology[24] made the 'revelation of Being itself by itself' responsible for this and cut through the problem like a Gordian Knot. Does modern hermeneutics promise to bring more clarity to problem-genesis?

A clear-cut answer is difficult, although I do not conclude, therefore, that the question is nonsensical simply because it is hard to find an answer. Hermeneutics interprets the raising of a problem, the process of problematisation quite historically. This means that hermeneutics grounds it on experience. The experience in which something becomes problematical is the experience of alienation. If this is to be called a historical experience, then what unexpectedly appears alien must at one time have been familiar. A situation of uninterrupted understanding has been alienated through a historical process and this experience initiates the genesis of the problem. I shall not speculate here about the causes for this process of alienation. But I am in no way satisfied with the unhistorical remarks of Wittgenstein who states somewhat mythically that philosophical problems arise where "I don't know my way about", since language "goes on holiday".[25] Language is not an autonomous entity which governs over our heads without our being able to control it. History is certainly a process, however, in which we stand and to which we are bound in so far as we have our experiences in it.

Having comprehended the problem-genesis, in the hermeneutical manner, from within a historical process of alienation, we are still confronted with the question of a supposedly transcendental solution. One is tempted to assume that this difficulty will also be countered with the argument of self-referentiality. The historical experience of what is alien instigates the quest for better understanding and is at the same time the origin of hermeneutic theory itself. Understanding the object which appears strange and which becomes a problem, and the understanding of understanding coincide in the question concerning problem-genesis. Within this framework, the individual act of understanding and the general theory of understanding seem to collapse into each other. For hermeneutics cannot explain the genesis of a problem in any other manner than by recourse to the historical experience of concrete problem formations. No general grounds of explanation are available if one abandons a comprehensive philosophy of history, which might possibly be capable of deducing the geneses of problems once and for all. Since, however, this claim exceeds the limits of transcendental hermeneutics there remains merely the paradox of the coincidence of the concrete understanding of problems, and the understanding of understanding in terms of the elucidation of problem-genesis.

V

I have now reached the end of my discussion of a possible association of hermeneutical theory with the program of transcendental philosophy and shall summarise the results of this last section of my paper.

A transcendentally constructed hermeneutics corresponds to the structure of self-referentiality which we have shown to be a general and formal characteristic of transcendentality. It corresponds to this structure in so far as it recognizes in general the binding ties of tradition even for itself. In individual cases of concrete understanding, hermeneutics reveals the particular binding ties, whilst in doing so hermeneutics frees itself from any dependence concerning precisely these given conditions.

Transcendental hermeneutics fulfils the task of legitimation by conversely emphasizing the dependence of its own reflection upon the concrete prevailing preconditions of knowledge. Thus by means of a demonstration it presents in an *ad hoc* manner the lack of alternatives with regard to the knowledge which is to be legitimated.

Transcendental hermeneutics is only able to clarify the genesis of the problem in so far as the understanding of one problem and the understanding of understanding coincide paradoxically in their origin.

Thus self-referentiality is recognized as being only generally valid. The task of legitimation, on the other hand, is solved in a concrete fashion and *ad hoc*, whilst the problem-genesis is elucidated by means of a paradoxical identification.

I have examined the conception of a transcendental hermeneutics only in terms of these three aspects, but they seem fundamental for the intended goal and for the procedures. It goes without saying that study of other aspects is not thereby excluded. Our results show, at least, that the argumentation for a transcendental hermeneutics does not proceed consistently, but instead varies from aspect to aspect. This need not be an objection in principle, since argumentations which do not proceed along one line and end in monolithic systems can certainly enrich our insight and prove productive for theories. We need not consider the implications of this conclusion here, since my intention was simply to point out certain difficulties, whose solution must surely bear upon the entire conception. For up till now it seems to me that not even these difficulties have sufficiently held the attention of either the protagonists or the critics.

But allow me to conclude with a personal avowal. No matter what shortcomings the conception of a transcendental hermeneutics may have, it opens our eyes for unsolved problems, moreover it indicates problems even in fields where our normal understanding would not suspect anything problematical at all. Hermeneutics, therefore, has the virtue of not regarding resignation as the ultimate goal of theory. Such resignation, together with a prohibitive norm, I detect in the following statement by Wittgenstein: "We call 'understanding' not the action – no matter what action – which shows us the understanding, but rather a state for which this action is a symptom. And that is a proposition about the grammar of the denotation of such a state."[26]

Johann Wolfgang Goethe University,
Frankfurt am Main

NOTES

[1] Ever since E. Stenius (*Wittgenstein's Tractatus*, Oxford, 1960, Ch. XI) who compared Wittgenstein and Kant, there has been talk about the 'transcendental' element in Wittgenstein. However, it has never been made clear what was meant by this. Though the observation is not new, I don't think that the following remarks are useless, because they try to explain transcendentality within the framework of Wittgenstein's own philosophy. For the continuity of this topic in the late Wittgenstein, cf. my article 'Die Einheit in Wittgensteins Wandlungen', *Philosophische Rundschau* **15** (1968) 160–185.

[2] 'Carnap's Views on Ontology', in W. V. Quine, *Ways of Paradox*, New York, 1966.

[3] *Ontological Relativity and Other Essays*, New York and London, 1969, p. 63.

[4] London 1964, I, 1 and 3, especially p. 29f, 38ff.

[5] Cf. Strawson, 'On Referring', *Mind* **59** (1950) 320–344.

[6] B. Stroud, 'Transcendental Arguments', *The Journal of Philosophy* **65** (1968) 241–256; M. S. Gram, 'Transcendental Arguments', *Nous* **5** (1971) 15–26; J. Hintikka, 'Transcendental Arguments: Genuine and Spurious', *Nous* **6** (1972) 274–281; Gram, 'Hintikka and Spurous Transcendentalism', *Nous* **8** (1974) 'Categories and Transcendental Arguments', *Man and World* **6** (1973), Must Transcendental Arguments be Spurious?, *Kant-Studien* **65** (1974).

[7] Hintikka, 277f. Cf. also Hintikka, *Logic, Language-Games and Information*, Kantian Themes in the Philosophy of Logic, Oxford 1973, 114ff. With this demonstration of the self-referentiality of transcendental arguments with regard to what they reveal, it seem to me that the controversy between Körner and Schaper is overcome. (Cf. St. Körner, 'The Impossibility of Transcendental Deductions', *The Monist* **51** (1967) 317–331 and

E. Schaper, 'Arguing Transcendentally', *Kant-Studien* **63** (1972) 101–116.) This controversy comes to a halt at the thesis that the necessity and uniqueness of a presupposed system of concepts must be proved from within the latter itself. Stephen Körner wanted to explain the impossibility of transcendental deductions by showing the impossibility of legitimating the validity of a hypothetical cognitive scheme without recourse to other reasons. Despite the subtle opposition by Eva Schaper in her plea for the original Kantian way of thinking I am in sympathy with Körner's thesis that a deduction senso stricto is impossible. Nevertheless, this controversy does not take into account the inner connection between a system of concepts and the very attempt at legitimation which we have termed 'self-referential'. Given this basis one can indeed make sense of Kant's transcendental argumentation withdrawing a strong deductive claim. See below.

[8] N. Hinske deals with the history of the concept in *Kants Weg in die Transzendentalphilosophie*, Stuttgart, 1970. Cf. I. Angelelli, 'On the Origins of Kant's Transcendental', *Kant-Studien* **63** (1972), and also Hinske's reply in *Kant-Studien* **64** (1973) 56–62.

[9] §13 (A71).

[10] *Einleitung zur Transzendentalen Logik* II (A 56).

[11] Cf. Th. M. Olshewsky, 'Deep Structure: Essential, Transcendental or Pragmatic?', *The Monist* **57** (1973) 430–442.

[12] In a little known passage Kant himself considered the possibility of a "transcendental grammar, which contains the basis of human language". An investigation of how the linguistic forms "lie in our intellect" would be a kind of preparation followed by formal logic and then "transcendental philosophy, the theory of general concepts a priori" (*Vorlesungen über die Metaphysik*, ed. 1821, Repr. Darmstadt 1964, p. 78).

[13] I have dealt with this question in somewhat greater detail in the final section of my essay 'Zur Struktur des transzendentalen Arguments', Proceedings IV. Internationaler Kant-Kongress, *Kant-Studien*, Sonderheft I (1974) 15–27. I make extensive use of this material here. (Addendum: An enlarged English version is about to appear in *Review of Metaphysics*, March 1975).

[14] It is perhaps of interest that this term was meant to indicate the particular (greek: idion) and should not be misread as 'ideographic'.

[15] This is the title of an important book by H. Rickert, *Die Grenzen der naturwissenschaftlichen Begriffsbildung*, Tübingen, 1902, first edition.

[16] It is not surprising that the hermeneuticians have taken special notice of his book *The Idea of a Social Science and Its Relation to Philosophy*, London, 1958. Cf. the critique by A. MacIntyre in *Proc. Arist. Soc.*, Supl. Vol. **41** (1967).

[17] Cf. Popper, *The Poverty of Historicism*, London, 1960, especially his 'hermeneutic' turn in *Objective Knowledge*, London, 1972.

[18] Useful as first information, G. Radnitzky, *Contemporary Schools of Metascience* II, Gothenburg, 1968.

[19] Cf. the translation of *Der Positivismusstreit in der deutschen Soziologie* (Adorno, Albert, Dahrendorf, Habermas), London, 1974.

[20] Besides the book by G. H. von Wright already mentioned we should also point out the ambitious project of Stephen Toulmin, *Human Understanding*, Vol. 1, Princeton, 1972. Here Hegelian influences – via Collingwood – can be detected quite distinctly. They stretch as far as an incorporation of the topos 'the cunning of reason' (final chapter).

[21] Cf. the third and fifth essays in my book *Dialektik und Wissenschaft*, Frankfurt am Main, 1973.

[22] Hans-Georg Gadamer, *Wahrheit und Methode*, Tübingen, 1960, Engl. transl.

Boston, 1975, Jürgen Habermas, *Knowledge and Human Interest*, London, 1972; see also the postscript to the second German edition (Frankfurt am Main, 1973), English version in *Philosophy of the Social Sciences* **3** (1973); 'Toward a Theory of Communicative Competence', in H. P. Dreitzel (ed.), *Recent Sociology*, No. 2, London, 1970; Karl-Otto Apel, *Transformation der Philosophie*, Frankfurt am Main, 1973, 2 vols. Cf. also Apel, *Analytic Philosophy of Language and the Geisteswissenschaften*, Foundations of Language, Suppl. Series, Vol. 4, Dordrecht, 1967, 'The Apriori of Communication and the Foundations of the Humanities', *Man and World* **5** (1973).

[23] We shall simply refer to one recent publication which stands for many: Tendenzen der Wissenschaftstheorie (with contributions by L. Krüger, E. Ströker, G. Radnitzky, H. Pilot), *Neue Hefte für Philosophie* **6** (1974), Göttingen.

[24] *Being and Time*, §7.

[25] Cf. *Philosophical Investigations*, §38, 123.

[26] *Philosophische Grammatik*, Frankfurt am Main, 1973, §41, p. 84.

JAAKKO HINTIKKA

THE INTENTIONS OF INTENTIONALITY

1. Intentionality and Conceptuality

The contrast flaunted in the title of this colloquium, "Explanation and Understanding",[1] is predicated on an important assumption. This assumption is best known as Brentano's thesis. It says, roughly, that there is an irreducible conceptual difference between two kinds of phenomena which I shall refer to as intentional and nonintentional phenomena. The nonintentional or physical phenomena are subject to explanation, the intentional ones to understanding.

Corresponding to these two classes of phenomena, we have intentional and nonintentional concepts. If Brentano is right, neither of these two classes of concepts reduces to the other. According to him, every intentional phenomenon "is characterized by what the scholastics of the Middle Ages called the intentional... inexistence (*Inexistenz*) of an object (*Gegenstand*), and what we could call... the reference to a content, a direction upon an object..., or an immanent objectivity."[2] This directedness is what according to his lights distinguishes the intentional from the nonintentional.

Brentano identifies the contrast between the intentional and the nonintentional phenomena with the distinction between the mental and the physical. This identification is highly dubious, however. There are lots of mental events, for instance twinges of pain, which do not seem to involve any directedness to an object or to a content.[3] They point to no way beyond themselves. Brentano tried to argue that even such mental events are accompanied by a presentation, a *Vorstellung*, and are therefore intentional.[4] But this term is dangerously obscure in that it does not distinguish images or mental pictures from conceptualizations. And it is in my judgement only the second, conceptual interpretation that really counts here. According to the views of Husserl and other leading phenomenologists, the notions of intentionality and conceptuality are intertwined. The world of intentions is the world of concepts, and vice versa.

Manninen and Tuomela (eds.), Essays on Explanation and Understanding, 79–110.

Brentano seems to think that the fact that we for instance experience pain as localized is enough to show that it is accompanied by a *Vorstellung* and is therefore intentional. But even if this were admitted, does it make the experience of pain itself intentional? Using a well-known Wittgensteinian ploy, one can try to argue that insofar as we can speak intersubjectively of our pain-experiences, they must be accompanied by public criteria by means of which they could presumably be conceptualized.[5] But does it follow from this that the raw sensations of pain and pleasure are themselves intentional in the intended sense? I don't see that it follows. It is much wiser to say with Husserl that, although intentionality is a universal medium of all conscious experience, there are within it experiences that are not themselves intentional.[6] In other words, it seems that the best course to follow here is to 'bracket' Brentano's assumption and to count as intentional, not all and sundry mental events, but only those characteristic of conscious, conceptualizable human experience. They seem to be precisely what Husserl referred to as 'acts'. Husserl in any case unmistakably identified the intentional with the conceptual. He went so far as to hold that all 'meanings' relevant to intentionality can be expressed in language. The vehicle of intentionality *apud* Husserl are the noemata, and every noematic 'meaning' is according to him conceptualizable. "Whatever is 'meant as such'", he writes,[7] "every meaning in the noematic sense (and indeed as noematic nucleus) of any act whatsoever *can be expressed conceptually (durch 'Bedeutungen')*."

This viewpoint is not without consequences for the dichotomy figuring in the subject matter in this colloquium. Presumably an account of human behavior given exclusively in terms of pleasurable and painful experiences would count as an explanation. (Similar accounts could be given, say, of the behavior of lower animals.) However, as soon as a conceptual element is imported into the account, we surely have to practice understanding and not only explanation. Or so it seems.

2. INTENTIONALITY AND INTENSIONAL LOGIC

In recent years, several studies have been carried out concerning the logical structure of various particular concepts which are intentional by any reasonable token, such as belief and knowledge.[8] It may be time to survey some of the vistas opened by these studies so as to see what they imply for the distinction between intentional and nonintentional. In this paper, a suggestion will be made concerning the nature of this important distinction. I shall not examine systematically the question of the reducibility of one half of the dichotomy to the other, but rather the ways of spelling out the dichotomy and of relating it to certain important philosophical problems.

3. Intentionality as Directedness

Conceived of in the way as we have done, there are few questions more important than this problem of characterizing the nature of intentionality. For the question is then: What is characteristic of conscious, conceptualizable human mental life and mental experience? This question is intimately related to the salient philosophical questions: What is man? and: What is thinking?

It is of course quite possible that none of these questions can be given a brief nontrivial answer. But even if that were the case, it would be important to examine and to criticize the answers that actually have been given to them and especially to our question of the nature of intentionality. We have already seen the answer Brentano gave. Answers of the same general type still enjoy a wide currency. To come extent, they are encouraged by the etymology of the term 'intentional'. According to this type of answer, intentionality equals *directedness*. Intentional phenomena, we found Brentano affirming, are characterized by their directedness to a content or to an object, and they contain in a sense this object as existing or, rather, inexisting in it. On views of this sort, an act or other phenomenon is intentional if it has an object to which it is directed ('aimed at', one is tempted to say) and if this object is somehow present in the act itself. Of course, this is not so far removed from what 'intending' originally meant, although its career as a philosophical term (which Brentano reminds us of in so many words) is perhaps not so easily predictable as its etymology suggests.[9] In the field where this colloquium is supposed to move, views of this sort are represented by frequent assimilations of intentional, *verstehende* accounts of action to the so-called teleological explanations.

4. Intentionality as Intensionality

This putative solution to the problem of intentionality is mistaken, I shall argue. To it I shall contrast a different answer to the same question. Formulated in blunt terms, this solution says that a concept is intentional if and only if it involves the simultaneous consideration of several possible states of affairs or courses of events (in brief, involves several 'possible worlds', to use this metaphysically loaded term). In other words, possible-worlds semantics is the logic of intentionality, and intentional is what calls for possible-worlds semantics.

Several minor explanations are in order here. The word 'simultaneous' refers of course to logical parity rather than to contemporaneity in the literal sense of the word. By 'involves' I refer to the semantical explication of a concept, not to the overt features of its use. The 'possible worlds' contemplated here are not grand histories of the world but usually only what a theoretical statistician à la Savage might call 'small worlds',[10] that is to say, alternative courses of events which are rather short in duration and which concern only a minuscule part of the universe, for instance alternative courses that a single experiment might take. What is crucial is only that several such alternative courses must needs be considered within the same 'logical specious present'. We shall also find that the word 'possible' in my phrase 'possible world' has to be taken with a grain of salt, too,[11] and that the possible-worlds semantics in question has to be of the right kind and even so will exhibit different degrees of intentionality.

With these provisos, however, my thesis seems to be intelligible enough to be discussed at some length. Since we shall need a handy label for this thesis, I propose to dub it the thesis of *intentionality as intensionality*. (As you can see, I am also not above exploiting etymology for my own purposes.) In spirit, the thesis is perhaps not so far removed from Hazlitt's dictum: "Man is the only animal that laughs and weeps; for he is the only animal that is struck with the difference between what things are and what they ought to be." According to it, intentionality is not a matter of relations obtaining within the world. Its gist lies in comparisons between several possible worlds. It is an interworldly business, not an intraworldly one.

5. Artistic creation not directed

Perhaps the most persuasive counterexample to the identification of the intentional with the purposive is offered by acts of artistic creation. In the eyes of the Greeks, they would not have offered any challenge to the identification, for most of the ancients seem to have been committed to the idea of an artist as an imitator, even if the paradigms he imitates are perhaps not perceptible.[12] We moderns tend to emphasise the relative or absolute novelty of an artist's conception. Even if the process of concretely realizing this conception is purposive, the genesis of the conception itself cannot be. But this dichotomy of conception vs. realization is already a dangerous admission to the identification of the intentional with the purposive, for it is in effect an attempt to maximize the scope of the purposive in the aesthetic area. The crucial feature of artistic creation is that what is most truly novel in it does not come about through a goal-directed process. To paraphrase Picasso's inimitable formulation, a creative artist does not seek: he find (finds without seeking, that is).[13] But this very non-purposiveness of artistic acts is often found paradoxical and perplexing, for unfortunately we prefer more familiar teleological models of human action. The recalcitrant element in artistic creation which cannot be accommodated in this teleological model is frequently subjected to various mystifications, ranging from theories of the unconscious to interpretations of an artist as a 'medium' of a genius 'possessing' him. (It is a sobering thought to note how repulsive such speculations would have appeared to Plato.[14]) These mystifications must not cloud the paramount fact that artistic creation, which is by any token one of the freest and most human activities one can hope to indulge in, just is not purposive (conceptually speaking). No prototype of a truly novel artistic conception 'inexists' in the act that gives rise to, nor is this act 'directed' to it. Its outcome can be a surprise even to its originator. Yet artistic creation surely ought to be counted as intentional in the sense which was intended by Husserl and with which we are here concerned. It is a form of free, conscious activity which typically even involves a clear intention on the artist's part, though not an intention to produce any particular, already defined *objet d'art*.

And such artistic creation is indeed intentional in the sense of my thesis. The very descriptions that emphasize the spontaneity of artistic creation involve concepts that are intentional in my sense. Perhaps the most characteristic descriptions involve the notion of surprise whose analysis involves especially clearly a contrastive comparison between several 'possible worlds' – those that someone expected and the one that did in fact materialize so as to surprise him. Nor are intentional concepts of this kind unrelated to our aesthetic evaluations, for such evaluations inevitably involve tacit or even explicit comparisons between the details of a work of art and what its creator might have executed instead. All aesthetic evaluations involve comparisons between the actual and the possible, and all artistic creation involves choices between alternatives of which only one can be actualized.[15]

6. Is perception intentional?

The contrast between the different interpretations of intentionality is especially acute in the area of perception. Over and above the intrinsic interest of this group of problems, it is lent a historical importance by Husserl's frequent use of acts of perception as paradigms of (intentional) acts in general.

At first blush, it might seem that perception is not intentional at all in any reasonable sense, and certainly not in the sense of intentionality as directionality. What happens in perception seems to be determined completely by physical stimuli and physiological processes, not by our purposive strivings and searchings. Perception seems to be a purely passive matter which scarcely merits the term 'act' even in its extended Husserlian sense. Even Immanuel Kant, who otherwise emphasized the role of human activity and human planning in his philosophy, spoke of sensation as *giving* us intuitions.[16] And rightly so, it might appear. One does not choose what one sees, nor is seeing a process with an end or aim. As Aristotle observed, 'I am seeing' implies 'I have seen': there is no end or aim which is realized through seeing.[17] Should the authority of Aristotle be insufficient, we may recall Quine listing perception as an unequivocally non-intensional concept.[18]

7. Noesis as the source of the intentionality of perception

Yet there is a deep truth in the insistence by phenomenologists like Husserl that perception is an intentional act. Only they carried out this idea wrongly, partly because they could not completely disentangle themselves from the wrong idea of intentionality as directedness. Husserl did not correctly diagnose the intentionality of perception as intensionality. On at least one plausible interpretation of what Husserl says, based largerly on the formulations he uses in the *Ideen*, Husserl in effect retained the nonintentional character of 'pure' perception, and introduced the intentional element only secondarily, in the form of an act of *noesis* superimposed on the perceptual rawmaterial.

Of course this is not precisely Husserlian terminology. However, what his jargon amounts to is in effect just what I said. Husserl locates the 'given' in what he calls sense data (*Empfindungsdaten*) or hyletic data or simply *hyle*. This terminology already illustrates a difficulty in Husserl's position. For most of the philosophers who have employed the sense-datum terminology, sense-data are *what* is experienced. For someone like Moore, they are what judgements of immediate experience are *about*. Husserl makes it clear that his sense-data are not *what* is experienced. They are merely components of perceptual acts. These are the experiences we have whenever our senses are stimulated. However, sense-data alone are "not by themselves experiences *of an object*" (Føllesdal's exposition of this line of interpretation,[19] my italics) and hence are not by themselves intentional by to Husserl's criteria.[20] Hyletic data nevertheless "normally occur as components of more comprehensive experiences, acts, which in addition to the hyle contain experiences of an intentional kind, the noeses. The noesis 'informs' the hyle, so that this multitude of visual, tactile and other data is unified into a set of appearances of one object"[21] and only thereby made intentional in Husserl's sense.

Thus quite literally only an additional noesis or thought-element (*Auffassungsmoment*) makes the hyletic data intentional. Thus in a sense, which is problematic but not Pickwickian, raw sensation (unedited perception) is *not* intentional according to this reading of Husserl.

What this amounts to in more mundane terms can be roughly explained by saying that our sense-impressions only become intentional when they are organized by means of one's expectations, memories, etc. These (and the like) are what the noesis relies on that makes perception intentional. The objects of our senses are reidentified from moment to moment largely by means of the continuity of the beliefs we attribute to them.

8. Conscious illusions as counterexamples to Husserl

As a corollary, we can conclude that when one's beliefs concerning what one perceives are correct, one perceives correctly. In other words, illusions (incorrect perceptions) are false beliefs induced by senses. They are discrepancies between what sensation usually leads us to believe and what further experience (*e.g.*, measurement) will show. For instance, in the case of a pair of Müller-Lyer lines "I may be aware that I am having an illusion and expect that the two lines will come out the same length when measured by a measuring rod. In this case I am, at least so far, correctly perceiving an object."[22] But these corollaries to the views of the phenomenologists already offer clear indications of the insufficiency of their line of thought. The views of illusions under scrutiny are not addressed to one of the most important 'phenomenological' problems we face here. They have the effect of neglecting the crucially important phenomenological distinction between not having a perceptual illusion at all and having it knowingly (and being able to compensate for it in thinking). Yet this is surely a vitally important distinction here. To be able to correct a sensory illusion *in thought* just is not the same thing as to be able to correct it *in perception.* There is a very real difference between seeing or not seeing a curved surface *as* the front side of a three-dimensional object, even if one knows that it is not;[23] seeing or not seeing a chair through a peephole in an Ames experiment when one knows in either case perfectly well there is none there;[24] perceiving or not perceiving a causal connection between the movements of two lightspots in a Michotte experiment even when one is aware that there cannot be;[25] and so on. Such distinctions are persuasive proofs that there is a kind of truth and falsity and therefore a kind of intentionality in a perfectly good sense even in spontaneous, unedited impressions largely independently of what current beliefs (memories, expectations, etc.) we associate with them. They correspond or fail to correspond to facts independently of what we know or believe these facts to be.

Nor can Husserl be excused by saying that the expectations and memories we associate with an object may be unconscious and that to speak of a noesis which structures the hyle into an aimed act is partly just another way of speaking of the effects of the unconscious editing and organizing process to which the sensory input is subjected by our central nervous system. Such an interpretation would destroy an even more important cornerstone of the system of phenomenology, viz. the accessibility of all noemata to phenomenological reflection.[26] In order to avoid this disaster, Husserlian phenomenologists must consider the noesis which structures ('informs') the hyletic data as falling within the purview of one's consciousness. Husserl goes in fact further and says in so many words that even the hyletic data can always be 'grasped':

> However, we can always grasp them directly as they are themselves, without being interested in the fact that with them something different, more specifically, something objective and spatial, appears to us. (*Husserliana*, Vol. IX, p. 163, lines 17–19.)

This shows that according to Husserl we can in some sense attend also to the hyletic data. What is even more important, it shows that the hyle can according to Husserl be attended to in abstraction from the three-dimensional objects which appear to us. This is the view I want to criticize here.

9. The True Phenomenology of Perception

To Husserl's (apparent) interpretation of the intentionality of perception I want to contrast another one. According to the former, sense-data enter our consciousness as an unstructured mass which is given a form (and hence made object-directed) by the mind's activity, largely through comparisons with past and future sense-experiences.[27] According to the latter, the most primitive sense-experience which can show up in one's consciousness is (in normal perception) already experience *of* certain objects, their properties, their interrelations, etc. In other words, one's unedited sense-impressions are already structured categorially. This is of course the outcome of complicated physiological processes which are conditioned by past experience and future expectations. However, the point is that these formative processes and their background information leave a trace in the conscious end product.[28] On this view, the most primitive layer of sensation we can reflectively behold is already directed, *i.e.*, organized so as to be *of* definite objects.

All this is compatible with saying that the noema of a perceptual object is not unlike a complex of expectations concerning this object.[29] However, on the view I criticized above (whether it is in the last analysis Husserl's or not) these expectations are superimposed on sensory raw-materials by a nonperceptual *noesis*. On the view advocated here, they are built right into the data of sensation themselves. If we want to describe truthfully our most spontaneous sense-experiences, we already have to use the language of those expectations. One does not perceive a hemispheric surface, and expect it to go together with the rest of a soccer ball because one recalls past experiences of it. One literally perceives a soccer ball, period.[30] The backside of a tree one sees is not brought in by apperception, but is already part and parcel of one's unedited perceptions. The examples given earlier of 'illusions' which persist even when one perfectly well knows that they are illusory shows how badly this kind of description is needed if one is to capture the true quality of one's sensory impressions.

I cannot argue here for this view of the phenomenology of sense-perception as fully as it deserves. It seems to me to be what such psychologists of perception as J. J. Gibson and David Katz have been arguing for.[31] Some reasons for this connection between the contemporary psychology of perception and my interpretation of the intentionality of perception are indicated in my paper 'Information, Causality, and the Logic of Perception'.[32] The relationship between the views put forward here and those defended by the psychologists of perception is especially close when a psychologist like Gibson emphasizes the informational character of perception.[33] But even apart from this special link, the object-directedness of perception has been strongly emphasized by several leading psychologists.[34]

All this is not incompatible with saying that we can occasionally have 'pure' (unstructured, not object-directed) sense-data and that they can for instance be studied in psychological and sense-physiological experiments. If anything, the need of elaborate experiments argues for the view presented here rather than what I have taken to be Husserl's view. For if we could always grasp the hyletic data directly, apart from their object-presenting function (cf. the last quote above in Section 8), such special setups were redundant. We could always attend to the sense-data directly. The truth seems to be quite different in that it takes special and unusual situations (*e.g.*, colored spotlight on the wall of an otherwise completely dark room) to break the object-directedness of our spontaneous perception.

10. Perception is Intentional because Informational

But why do I say that this view is based on a conception of intentionality different from Husserl's? Because it is a direct corollary to perception's being intentional in my sense of intentionality as intensionality. In this sense the intentionality of perception means that it involves a comparison between several possible states of affairs. And it is indeed obvious what these states of affairs are in the case at hand. Perception is intentional because it is informational,[35] and all talk of information involves several different possible states of affairs or courses of events in that it involves a distinction between states of affairs compatible with this information and those incompatible with it.[36] Since to specify what one perceives at a given moment of time is (I have argued) to specify the information one's senses then convey to one about the object of one's perception, this specification involves several unrealized states of affairs, *i.e.*, is intentional in my sense. And of course sensation cannot convey even putative information about its objects in this way unless the contents of sensation are to be specified in terms of the same realistic concepts as apply to its objects. Thus perception's being object-directed (in that its contents have to be described by speaking of its objects) is just the other side of the coin from its being intensional, *i.e.*, specifiable only by means of several different 'possible worlds'. Husserl thus used the right word in a wrong sense. Perception involves in-formation, not in the etymological sense of form-giving, but in the modern sense of telling the perceiver something about his environment – or at least appearing to do so.

11. The Intentionality of Perception Implies Its Conceptuality

The intentional and informational character of perception is not unconnected with the role of conceptualization in perception. On my view, the intentionality of perception presupposes that sense-impressions are dealt with by the perceiver as conveying information and that they do not operate merely as signals for triggering responses. This of course requires a certain level of conceptualization of the part of the perceiver. However, it does not require that this conceptualization is something over and above our primary sense-impressions. Rather, the conceptualization is built right into these unreconstructed sense-impressions themselves. This is reflected by the fact (already argued for above) that to describe my sense-impressions is to specify which states of affairs are compatible with them and which ones are not. For, as was also alluded to, these states of affairs have to be described in realistic terms – in the same terms as we ordinarily use of speaking of perceptible objects, not in terms of special imponderable entities like sense-data or sense-impressions. But the appropriateness of such a description depends on the concepts the perceiver has available to him. In this way, one's concepts enter into the very sense-impressions one has, not as the result of a further noesis performed on them, but as a part and parcel of their very texture. It is in this way that perception can be partly culture-dependent, it seems to me.[37] All these observations are consequences of the analysis of (spontaneous) perception in terms of the class of states of affairs it admits of, and this analysis is little more than spelling out the claim that perception is intentional in the sense of the thesis of this paper.

12. Exegetical problems

It is in order to add a few interpretational warnings to what was said above about Husserl. My account of his views is closely geared to what he says in the *Ideen*, but it must be added that Husserl himself indicates that the account given there is only a provisional one.[38] It seems that he was bothered by doubts not unrelated to the problems I discussed, and introduced various qualifications to the oversimplified account outlined above.

The main qualification we sometimes seem to find in Husserl is that we cannot separate the hyletic data from the noematic sense and consider it in isolation. In a passage to which Professor Føllesdal has drawn my attention, Husserl writes:

> We cannot place side by side two components in intuition, sense and filling. We can only obtain the difference by contrasting the empty and the filled sense, that is, through a synthesis of intuition and empty consciousness. Perhaps we might put it thus: the abstract identical in several different acts of consciousness which is called sense is an essence (sense-essence) which particularizes in its special way, indeed in two basic modes, in the mode of intuition ... and in the mode of non-intuition, the empty mode.[39]

It nevertheless appears that Husserl did not reach anything like full clarity in the problems confronting him. The quotation just given is not very easy to bring in step with passages such as the one quoted toward the end of section 8 above. Passages like the one just quoted also do not rule out the possibility that Husserl is only saying that we cannot meaningfully speak of hyletic data in their virgin state, unsullied of any *noesis*. If this is what he means, what is involved is merely a consequence of the fact that hyletic data are according to Husserl only conceptualized (made intentional) through a superimposed noesis. Since language presupposes conceptualization, we cannot speak of the hyletic data until they are subjected to the noesis. And even then we cannot really speak of them alone, only of the intentional experiences to which they belong as components.

If so, Husserl is not really modifying the dichotomy hyletic data vs. sense but only signalling the limitations of our language and conceptual thought in speaking of it. He is not saying that there are no unstructured hyletic data. He is only saying that we can speak of them only insofar as they have already been structured by the noesis (and even then only at the second remove, that is, only in so far as they are components of our experiences of those objects which are the only rightful subject matter of conceptualized discourse.) So understood, Husserl's second thoughts do not belie my criticism of him, but merely introduce a cautionary footnote as to how this criticism ought to be formulated.

Husserl is so elusive a thinker that I cannot put forward this interpretation with complete confidence, although it seems to me the likeliest account of his position. In particular, it seems to me that Husserl was too deeply committed to such contrasts as matter (hyle) vs. form, *Empfindung* vs. *Wahrnehmung*, etc., to be able to accommodate fully the idea that the most primary ingredients of our sensory consciousness are already 'informed', already perceptual and not mere complexes of sensations.

It may also be that two different ideas are run together in Husserl's concept of hyletic data to the detriment of this notion. In addition to the alleged formlessness of these data in Husserl, they are also closely related to what is 'given' to us in the sense of being what can 'fill' a noema, that is, roughly, what meets or fails to meet the expectations which are built into this noema. I cannot suppress the suspicion that Husserl may have thought that the role of the hyletic data in 'filling' noemata somehow implies their nonintentionality.

Hence the contrast between the hyle and the filling-component mentioned above in note 39 is nevertheless relevant. Husserl never actually says that filling is done by the hyletic data. But the two must somehow be bound inextricably together, even though Husserl never spells out their connection.

It is to be noted, though, that there is in any case one clear difference between the two in that the filling in Husserl belongs squarely with the object and therefore appears to be already conceptualized. However, just in so far as Husserl considers the filling to be done by what is most primitively given to us, the fillings become difficult to speak of and difficult to conceptualize, in brief, become like hyletic data.

If anything, my point is thus strengthened by making this hyle-filling duality explicit. For then we have to ask: Are the expectations (or expectation-like *Unsättigkeiten*) that can be 'filled' expectations concerning our sensations, or concerning the objects of these sensations? The whole tenor of Husserl's thought argues for the latter answer. But if so, the fulfilling data must already be articulated in the same terms as the expectations they meet. Thus Husserl faces a dilemma he never seems to resolve satisfactorily. Either he must disassociate 'fillings' from the hyle, or else admit that the hyletic data are already object-directed (intentional). Both horns of the dilemma appear unacceptable to him.

Even apart from these different problems Husserl's ubiquitous matter-form terminology exposes him to devastating criticism. There just is no matter-form contrast in normal perception, in so far the phenomena of our conscious experience are concerned. Whatever 'form' there is in one's perception is present already in the most spontaneous sensuous 'materials' that can surface in one's awareness. Speaking of 'matter' and 'form' in perception thus appears not only empty or problematic, but positively misleading.

Although important questions are being begged by these brief remarks, they may nevertheless suffice to indicate my overall interpretation of Husserl's position, provided only that a due allowance is made for its ambiguities.

13. Directedness and Conceptuality

Another comparison between my interpretation of intentionality and the views of the phenomenologists is possible. My thesis of intentionality as intensionality shows what the precise relation is that obtains between intentionality and conceptuality. Conversely, this application serves to bring out the nature of my thesis more fully. Again a comparison with the idea of intentionality as directionality is instructive. This idea in fact seems to have its strong point in this area, for it seems to bring out precisely the connection between intentionality and conceptuality which we are looking for. Briefly, concepts are the *meanings* (Frege's *Sinne*) of our linguistic acts and more generally the generalized meanings or noematic *Sinne* (in Husserlian terminology) of all our conscious acts. Acts as it were aim at objects. In linguistic acts, these objects are the *references* of our expressions. Now a noematic meaning is what establishes the *direction* of this aim. A noema is the vehicle of directedness. It "determines what the object of an act is, if the act has an object", as Føllesdal says, "just as the meaning of a linguistic expression determines which object the expression refers to". [40] As Føllesdal also puts it, according to Husserl "to be directed is simply to have a noema."

It is hard to find anything specific in such formulations with which one can disagree. Yet such remarks tend to convey a wrong idea of the relationship between intentionality and meanings which we are considering. What is wrong with such accounts of meanings as vehicles of directedness is that they are partly metaphorical. They do not really say how the determination of the object of an act through its noematic *Sinn* takes place. And in so far as something is said, or suggested, concerning the nature of this determination, the ideas associated with it are insufficiently abstract and insufficiently generalized to pass the muster as an answer to our question. A noema, conceived of as the 'vehicle of directedness', far too easily becomes like a concrete aid to a rifleman's aiming in that it becomes a single entity, however abstract. Frege would say of noemata in general the same he appears to tell us of his *Sinne*, *viz*. that they are what he calls 'complete' entities.[41] Husserlian *noemata* and Fregean *Sinne* are thus easily reified into *individuals* in logician's general sense of the term. (It may very well be that a detailed analysis of Husserl shows that, appearances notwithstanding, this is not a part of his full-fledged doctrine; but it is in any case how he is easily read.) A noema may be an *abstract* entity; but it is a on the view under discussion still definite entity in the etymological sense of the word, an one-tity.

14. Intensions as functions

This is simply a category mistake, it seems to me. The deep true idea in Husserl no less than in Frege is of course *not* that a *Sinn* (noematic *Sinn*) is an *argument* which functionally determines an object (*e.g.*, a reference), for the two different noemata would only rarely be directed to the same object. Rather, the true idea is that the *Sinn* is itself the *function* which accomplishes this. It includes, Frege said without following up the consequences of his statement, besides the reference also the way in which this reference is given.[42] And of course all such talk of 'ways of being given' must in the last analysis be understood functionally. Meanings of expressions and meanings of acts are simply the functions which determine their references or objects, respectively.[43] Although neither Husserl nor Frege appreciated the point, this is the only general thing one can say of the logical status of noemata or *Sinne*. No matter how interested one may (for perfectly valid reasons) be in the specific recipes for determining the values of such functions or in other aspects of their concrete realization, one must not forget their categorical status as functions.

But what are these meaning functions functions *of*? What are their *arguments*? Again one may in special cases try to look for specific candidates for this role. The question before us here is, however: What are all the different things which these functions depend on as their arguments or components of such arguments? On the general semantical level on which we are here moving, the only plausible answer is the apparently trivial one: everything. In particular cases, they depend on some special features of the possible world where the reference is located, but they cannot depend on anything more than this world as a whole. Concepts, meanings, are therefore functions from possible worlds to references (objects), for the whole idea of 'possible world' is that it comprises everything (at least everything that is relevant to the particular question we are asking).

This, however, is precisely the general answer given by possible-worlds semantics to the question of the status of the concepts.[44] Concepts, as meanings, are according to possible-worlds semantics functions from possible worlds to references (extensions). This is their logical type, and it is of course literally worlds apart from the logical status of individuals ('entities').

What we have just done is that we have as it were deduced this view by criticizing the flaws of the Husserlian and Fregean answers and by correcting their shortcomings. It exhibits the connection between conceptuality and intentionality which I promised to show to you. Concepts, being functions from possible worlds, are intrinsically intentional in that they involve a multiplicity of possible worlds (as their arguments), which in turn was precisely my definition of intentionality. Even if all uses of a concept do not have to involve all possible worlds (if such totality as much as makes sense), any nontrivial use of a concept implicitly involves more than one possible world.

Thus a closer analysis of the very idea of intentionality or conceptuality as directedness (directedness to an object) leads us back to my thesis of conceptuality as intentionality in the sense of involving several possible worlds at one and the same time. Moreover, once we see this, all emphasis is shifted away from directedness as a special relation of an act to an object *in this world*. The same function which in our world gives us one object will give us another object in another one. The relation of this function to its values is no more remarkable in one world than in another. This shows in further detail in what sense intentionality is an interwordly affair, not an intrawordly one.

On earlier occasions, I have in effect criticized illicit reifications of intensional objects (*e.g.*, sense-data in the Moore-Broad-Price sense) into alleged denizens of the actual world, while in reality their peculiar status lies entirely in crossworld comparisons.[45] Somewhat analogously, I am now criticizing what looks to me like an illicit reification of intentionality, not into intrawordly objects, but intrawordly relations.

15. The Primacy of Possible Worlds

Perhaps the most important systematical implications of the thesis of intentionality as intensionality are seen by noting what follows from it concerning possible-worlds semantics itself. The thesis can be taken to imply that the only 'raw materials' we are allowed to use in constructing the semantics of intentional concepts are possible worlds. Everything else must be constructed or, to use a perhaps more familiar jargon, constituted from the materials presented to us by a suitable set of possible worlds.

But what else could there be which it might be tempting to try to postulate as existing independently of the possible worlds? Several answers are possible, but only one such rival answer is important. There is a strong temptation, motivated both by philosophical arguments and by considerations of logico-mathematical elegance, to think of the particular entities, the individuals in the logical sense of the word, as somehow being primary with respect to possible worlds whose members they may be, and in this sense as independent of these worlds. Several logicians have in fact presented semantical treatments of intensional logic by starting out from a fixed set of individuals which can then make their appearance in different possible worlds.[46] Admittedly not all the results they obtain are necessarily any worse off if this starting point of theirs turns out not to be the ultimate starting-point of semantical and philosophical analysis. However, the philosophical weight of their results is apt to be diminished by reductions to more basic elements. And such a reduction is not only possible but (for many philosophical purposes) indispensable.

To the alleged primacy of individuals we can contrast a view which admits that each possible world comes to us already analyzed categorially into individuals, their properties, their relations, etc. However, according to this view the identity of such entities, especially the identity of individuals from one world to another is not fixed by any absolute logical principles but is at least partly constituted by our comparisons between the two different possible worlds whose denizens the two respective individuals are.[47]

In order to have vivid terminology at our disposal, let us imagine that the manifestations (roles, embodiments) of the same individual in different worlds are tied together by a line, the *world line* of this individual. (David Kaplan has jokingly called them TWA's, that is, trans world heir lines.) Then I can formulate my claim by saying that the world lines of individuals are not fixed by immutable laws of logic or God or some other equally transcendent power, but that they are as it were drawn by ourselves – of course not by each individual alone but by tacit collective decision embodied in the grammar and semantics of our language.

Certain caution is nevertheless necessary here. David Lewis has suggested that the world lines joining to each other 'counterparts' (his term) in different worlds are based on the similarity of the counterparts in question, this similarity being something like a weighted average of many different kinds of similarity considerations.[48] This is a misleading view, I want to argue (and will do so in greater detail in my Quine paper). It is misleading because by far the most important vehicle of trans world comparisons is given to us by various continuity principles. This diminishes greatly the psychological and sociological arbitrariness of our cross-identification principles, but it leaves these principles largely at the mercy of the laws of nature which serve to guarantee the continuity of our 'natural' individuals (*e.g.*, physical objects) in space and time. Moreover, the precise kind continuity in question can in principle still be chosen differently. (Is the identity of persons based on bodily continuity or on the continuity of memory?) Hence this reliance on continuity does not eliminate our choices in drawing the world lines.

16. THE BEHAVIOR OF WORLD LINES

The question of the primacy of world lines vs. the primacy of 'prefabricated' individuals is by any token an important philosophical problem concerning possible-worlds semantics. Now we can see that it is closely related to the general thesis of intentionality as intensionality and that the thesis implies a definite answer to this question.

It is also seen that the thesis is indirectly but strongly supported by whatever evidence there is for the primacy of world lines and for their being due to our (*i.e.*, the language community's) constitutive decisions. Here I cannot survey exhaustively recent work in possible-worlds semantics and its applications to philosophical and semantical problems so as to spell out all the evidence in this direction. It is nevertheless in order to indicate what kinds of evidence one can find.

From the assumption of prefabricated individuals, certain restraints follow on the behavior of world lines. Clearly, on this assumption world lines never split in two or merge into one. On the assumption of the primacy of possible worlds there is nothing that would prevent such splittings and mergings. Even that bugbear of the critics of possible worlds, quantification into intensional contexts, makes sense as soon as the world lines have been drawn, no matter whether or not they are allowed to diverge and converge.[49] Hence the phenomenon of world lines splitting or merging offers potential evidence for a decision between the competing assumptions.

This evidence is fairly unambiguous. It seems to me clear that, for certain epistemic notions at least, splitting must occasionally be allowed.[50] Merging is not even ruled out in many actual treatments of intensional logic.[51] Hence there are good reasons to distrust ready-made individuals and prefer to think of them as fabricated from the raw materials of possible worlds.

This result is all the more surprising as it involves (as you can easily see on a closer scrutiny) the occasional failure of the most solid-looking version of the logical principle which is known as the substitutivity of identity and which has been alleged by several philosophers to be a *conditio sine qua non* of any satisfactory interpretation of identity.[52]

17. Different kinds of world lines

But even more striking evidence is forthcoming. If world lines are in the last analysis man-made, it ought to be possible in principle to draw them in (at least) two different ways. This expectation is dramatically confirmed by the (by now well-established) result to the effect that in our actual conceptual and linguistic practice we are all the time operating, not with one system of world lines, but with two different ones.[53] One of them is established by the methods which were already alluded to and which rely essentially on continuity. It is said to establish *descriptive* cross-identification. The other mode of cross-world comparisons identifies individuals having the same perceptual relations (or other direct cognitive relations, as the case may be) in different worlds to the person in question – meaning the person whose acts are being considered. This I will call (taking a cue from Russell) cross-identification by *acquaintance*. The genuineness of the distinction, and the reality of its both halves, is attested to by their linguistic counterparts. The logic of descriptive cross-identification is to all practical purposes the logic of subordinate interrogative wh-questions with an epistemic main verb like 'knows', 'remembers', 'sees', etc. The logic of cross-identification by acquaintance is the logic of direct-object constructions with the same verbs (or, rather, with some of them). This linguistic distinction thus serves, when its semantics is laid bare, as strong evidence for the primacy of possible worlds and therefore also for the thesis of intentionality as intensionality.

18. Acts *de dicto* and acts *de re*

A third major way in which possible-worlds semantics helps to adjudicate the rivalry between different conceptions of intentionality is based on the insight that not all acts of knowing, believing, remembering, etc., have to be directed to a particular individual, or perhaps rather on the new insights into the semantical roots of this time-honored distinction. It is in fact honored in terminology and not only in time, for it was already in the middle of the Middle Ages dubbed the distinction between acts *de dicto* and *de re*.[54] Let's suppose that you can be truly be said to believe that the present Prime Minister of Denmark is a social democrat. Then you might either have a particular Danish politician in mind of whom you believe this. (In fact, the identification of this person as the P.M. of the happy nation of Denmark might be mine, and no part of *your* beliefs.) In such circumstances, your belief is *de re*, the *res* in question being the person whom the belief is *about*.

However, the identification may be a part of the specification of the content of your belief. Then you need not have any particular politician in mind, but would be willing to express your belief by saying, "Whoever the Prime Minister of Denmark is, he is a social democrat". (You might for instance believe that no other party is capable of forming a cabinet right now). Then your belief is *de dicto*.

The possible-worlds semantics of this distinction is clear enough. In the example at hand, a *de dicto* belief is about the different politicians who are in their respective possible worlds the P.M. of Denmark. These possible worlds are all the worlds compatible with everything you believe. Hence these persons are usually different, and are all joined by one and the same world line only when you have a belief as to who the present Prime Minister of Denmark is.

In contrast, in a *de re* belief an individual member of the actual world is chosen satisfying the condition of being the P.M. of Denmark. Then we hang on to this individual and follow him along his world line to your different belief worlds. The individual constituted by this world line is the one whom your belief is about.

Thus the distinction is clear enough, and it has even attracted a great deal of attention on the part of contemporary linguists. It nevertheless presents a formidable challenge to the whole phenomenological conception of intentionality. For *de dicto* acts just do not seem to be directed to a particular object in any natural sense of the world.

19. HUSSERL AND *de dicto* ACTS

I find it very hard to tease out from Husserl's text any very satisfactory answer to the question as to how he proposed to deal with such in a sense undirected but yet unmistakably intentional acts.[55] The nearest I have come is the suggestion that acts *de dicto* are not directed to individual objects but rather to *facts*, in our example to the fact of the Danish Prime Minister's being a social democrat.[56] But this suggestion is most unsatisfactory. It destroys the obvious parity of acts *de re* and *de dicto*. It is also very hard to reconcile with the fact that an act *de dicto* is turned into an act *de re* when the believer acquires an additional opinion concerning the identity of the reference of the *dictum* in question. (In the example above, this means your forming a belief as to who the P.M. of Denmark currently is.) If the respective objects of an act *de dicto* and an act *de re* are entirely different, such a change would involve a major 'explosion' of the noema *apud* Husserl. However, reflection betrays no sign of explosion at all in such cases.

Furthermore, this apparent Husserlian attempt to solve the *de dicto* problem merely shifts the problem instead of solving it. For the cross-world identity of facts is quite unclear as compared with the cross-identification of individuals. How can a *de dicto* act your belief in our example be directed to one and the same fact, if this 'fact' involves different person's being the Prime Minister? Surely the fact of *A*'s being the P.M. is different from the fact of *B*'s being the P.M. Hence it is not clear that Husserl's stratagem would save the directedness of *de dicto* acts even if it were viable.

It is also hard to reconcile the alleged role of facts as the objects of certain acts with what Husserl says of the structure of noemata. There are any number of things he says of the determinable *X* which simply are categorially different from what he would be expected to say if facts could be objects of acts.

Hence the undeniable reality of *de dicto* acts is a further argument against intentionality as directedness, for such acts are undirected but yet intentional. Again, my thesis of intentionality as intensionality works out very well, for act *de dicto* certainly involve a multiplicity of different worlds. Indeed, the very distinction between what is *de re* and what is *de dicto* collapses if we are merely dealing with the actual world.

20. Perceptual individuation intrinsically *de re*

If a historical conjecture is admissible here, I wonder whether Husserl's oversight here might possibly be due to his use of perception as a paradigm of acts. The distinction *de dicto* vs. *de re* cuts right across the distinction between cross-identification by acquaintance and descriptive cross-identification. It applies to both. But in the case of cross-identification by acquaintance, especially in the case of perceptual cross-identification, a new aspect of the conceptual situation emerges. In a sense everything that is perceived at all is *perceptually* cross-identified. This is essentially the point discussed above in connection with the intentionality of perception. Normal perception is *about* individuals; its content is spontaneously articulated into objects with more or less definite locations in perceptual space. Since it is this space that serves as the frame of reference in perceptual cross-identification, all the objects of normal perception are perceptually cross-identified. One does not always see *what* or *whom* it is that one is seeing, but one cannot help seeing *it* (direct object construction). There are no perceptual acts *de dicto*, only acts *de re*, if one goes by perceptual cross-identification. In this sense, all acts of perception are directed to particular (perceptually cross-identified) objects.

Hence the mistaken but plausible thesis of intentionality as directedness is made especially seductive if one uses perception as one's paradigm case, as Husserl in fact did to a considerable extent.

More exegesis is needed to see whether this diagnosis of Husserl's oversight is correct. But independently of it, we have obtained additional, strong evidence for the thesis of intentionality as intensionality.

21. THE FORM-MATTER CONTRAST VINDICATED

The contrast between two kinds of individuation methods, the perceptual and the descriptive, can be used to put Husserl's ideas into a new perspective in still other ways. I emphasized earlier that cross-identification is not fixed by any absolute but is in principle carried out by ourselves. In this respect there is an important difference between the two kinds of individuation methods. Not only is perceptual individuation *de re*. It is determined by factors over which we cannot exercise any conscious control. As can be deduced from what I said in criticizing Husserl's theory of perception, one's perceptual individuals are in a sense given together with one's primary, consciously *un*edited perceptions. One is free to use other methods of individuation besides this perceptual one, and one can try to dispense with it altogether, but one cannot change it at will. This mode of individuation is as it were built right into the way one's perceptual apparatus functions. No separate noesis is required for perceptual individuation.

In contrast to perceptual individuation, descriptive cross-identification typically relies on such clues as can (in principle and by hindsight) be recovered by memory and thinking from the sensory flux. It involves comparisons between different possible states of affairs or courses of events, which often may be thought of as different perceptual situations. Such comparisons are not given together with the possible situations themselves. Descriptive cross-identification is in fact very much like Husserlian constitution of objects which he describes in terms of a noesis acting on sensory materials or 'hyletic data'.

Husserl is thus right in the deep sense that the constitution of the objects which our experience is about is neither automatic nor given together with our sense-impressions. However, he fails to distinguish between the two different modes of cross-identification, and in effect applies to the perceptual one what only holds true of the descriptive mode. Thus in so far as Husserl took his insight to imply that sensation somehow has only the role of supplying the hyletic materials to be operated on by a noesis, he was badly mistaken. As I emphasized, unedited perception is in a perfectly good sense as fully articulated and as intentional as, say, thinking. It comes closer to truth to say that perceptual individuals and perceptual data, even though they are in themselves fully structured into solid three-dimensional objects and states of affairs, supply materials for the construction (constitution) of descriptively cross-identified individuals.

Husserl's mistake on this score is dramatized by the spontaneity of perceptual individuation, *i.e.*, its being part and parcel of the rock bottom materials of sensory awareness. But now it also turns out that Husserl's form-matter contrast is not so much wrong as misplaced. Only the matter in question is not sensory raw-material, but sets of possible worlds. In the last analysis, they are such stuff as our ontology is made of, if by 'ontology' we mean the class of descriptively individuated objects of our propositional attitudes.

22. Constitutional problems

These brief references to the actual development of possible-worlds semantics will have to suffice here. Since I suggested that this semantics is virtually tantamount to a general theory of intentionality, it is nevertheless in order to try to make good this claim and to show how possible-

worlds semantics helps us to understand better the problems that have figured in earlier discussions about the nature of intentionality. Even though we cannot carry out more than a cursory discussion of a few selected topics, the brief glimpses of the nature of possible-worlds semantics offered above perhaps suffice as a basis of comparisons between my theory and certain views that have come up in the phenomenological literature.

For one thing, we can now see what precisely is involved in the constitution of the objects of our acts.[57] Speaking of 'constitution' easily evokes idealistic associations. They would be misplaced here, however. No doubts are thrown by possible-worlds semantics on the reality of the actual world or of its inhabitants, nor is it incompatible with this semantics to adopt the same realistic attitude to other possible worlds. Constitution is not a domestic matter, as it were, but a matter of foreign policy, that is to say, a matter of cross-world comparisons. What is brought about in constitution is literally neither here nor there – neither in one world nor in another. Constitution does not create inhabitants of any possible world, only methods of comparing entities in different possible worlds for their identity. In this sense, the possible-worlds theory of intentionality is compatible with a strong form of realism.

Notice that the point I am making here is a generalization of what was said earlier in connection with the problem of perception (section 15). It was pointed out there that we must distinguish sharply between on the one hand the problem of splitting up one given sensory manifold into individuals and on the other hand the problem of an individual's identity between different sensory contents. It was argued that only the second problem matters phenomenologically. This is just a special case of my general thesis that the crucial problem of constitution (making one's act to be directed to one particular individual) is a matter of interworld comparison.

23. THE NOEMATIC OBJECTS OF ACTS

The false appearance of idealism in connection with the constitution of world lines is brought about by the fact that in traficking with intentional concepts we are (according to my main thesis) dealing with individuals as members of more than one state of affairs or course of events. What a logician calls individuals and what he needs for instance as values of his bound variables are therefore often not elements of this or that particular world but individuals considered as members of several possible worlds. But such a consideration of individuals as members of different possible worlds already presupposes cross-identification. In such cases, the individuals needed in one's logic and semantics are more like world lines than their nodes.

Since these world lines are in principle, to borrow a phrase from Hobbes, "drawn and described by ourselves", the logician's individuals relied on in using intentional concepts are *in the specific sense which appears from these remarks* constituted by ourselves. (To repeat myself, such a constitution is of course not up to an individual's decisions but is codified in the conceptual rules of the whole speech community.) What makes this interesting but in itself ontologically innocuous observation striking is the fact that in a sense the only viable candidates for the role of the *objects* of our acts of knowledge, belief, etc., are these cross-identified individuals. In this sense, the objects of our knowledge and belief are therefore *selbsttätig* (as Kant would have said), that is, man-made. However, we have already seen that this observation does not militate against the realistic tendency of possible-worlds semantics.

It is thus the need of cross-identification that necessitates the constitution (in the Husserlian sense of the world) of the objects of our propositional attitudes. In the same way as we according to Husserl can intend one definite object only by means of a noema, in the same way we can on my view have knowledge of one definite object only if we can re-identify it in several possible worlds. In fact, I submit that the two processes, the cross-identification of individuals and their Husserlian constitution, are at bottom identical.

It is to be noted that in thus speaking of the *objects* of certain acts the term is not to be understood in the sense of the phenomenological distinction between an act and its object (in the actual world), but rather in the sense of that component of the noematic *Sinn* which establishes the identity of this object, the '*gegenständlicher Sinn*' in Husserl's terminology. There thus obtains a close relationship between my notion of a world line and Husserl's notion of the '*gegenständlicher Sinn*'.

24. CONSTITUTION NOT ONTOLOGICAL

This serves to bring out the sense in which the very Brentano dichotomy, the whole contrast between the intentional and the nonintentional is not ontological in character. Ontology is a matter of 'the furniture of the world', that is, of what there is in the actual world or for that matter in this or that possible world. Such questions are not affected by the constitution of world lines. One reason why this plain truth is frequently overlooked is philosophers' tendency to reify world lines into alleged denizens of one particular world. Elsewhere, I have criticized a few of the many ensuing confusions.

25. The Transcendence of Objects

Although the connections are perhaps a shade less clear, it is also worth pointing out several other relationships between Husserlian ideas and possible worlds semantics. Husserl's idea of the transcendence of certain objects is closely related to the inexhaustible multiplicity of the possible worlds which are compatible with our beliefs and as a member of which such objects can occur. Our experience can narrow down this class but not boil it down to one world only, if we are dealing with a truly transcendent (in Husserl's sense) object. The same class of worlds can change in other respects in the light of further experience without changing the identity of the object in question. That is to say, its world line can often be continued to new worlds. But not always. When it cannot be so continued, the result is what Husserl calls an explosion of the noema.

26. The Perceiving Subject

More subtly, the notion of cross-identification by acquaintance is related to Husserl's ideas of *Ichbeziehung* and of the possible role of *Körperlichkeit* in our acts.[58] Since cross-identification by acquaintance depends on the point of view (in a literal or almost literal sense) of the person whose acts are being considered, for him such cross-identification is inevitably ego-bound. It presupposes an ego which is located in space and time (at least in perceptual space and time). However, this does not yet presuppose bodily existence in any very strong sense. And it is of course restricted to individuation by acquaintance, especially to perceptual individuation.

27. Inexistence Reconstructed

Most importantly, we can now see in what sense the object of an intentional act can be said to 'inexist' in it. Since such an act involves several possible worlds, the object must be the same (well-defined) in all these worlds. In other words, a correct exhaustive description of the act involves a specification of what it means for this object to exist in all these different worlds. It seems perfectly appropriate to interpret 'intentional inexistence' as amounting to this kind of involvement of its identificatory conditions in an act.

28. Conclusion

In fact, possible-worlds semantics yields even such further pieces of evidence for its promise as a general theory of intentionality as I cannot present here. Even without them, it seems to me that we can conclude that recent developments in possible-worlds semantics have proved to have profound implications for the notion of intentionality itself, and not just for a number of particular intentional concepts.

Academy of Finland and Stanford University

NOTES

[1] International Colloquium on Explanation and Understanding, Helsinki, 25–26 January, 1974. The present essay is a considerably enlarged version of the paper read at that Colloquium. In expanding and rewriting it, I have greatly profited from discussions with Professor Dagfinn Føllesdal and from his unpublished (as well as published) writings. He nevertheless is not responsible for my errors, the more so as he in so many words disagrees with some of my conclusions concerning Husserl. I have also profited from comments by Professor Yrjö Reenpää and from the unpublished writings of Professors Ronald McIntyre and David Smith.

[2] Franz Brentano, *Psychologie vom empirischen Standpunkt*, Ducker und Humblot, Leipzig, 1874, Vol. I, Book 2, Chapter 1, sec. v., p. 85; translated in Roderick M. Chisholm (ed.), *Realism and the Background of Phenomenology*, The Free Press, Glencoe, Illinois, 1960, p. 50.

[3] In the sequel, I shall nevertheless argue that the sense-impressions involved in normal perception are not of this kind.

[4] Cf. Brentano, op. cit., sec. iii.

[5] Ludwig Wittgenstein, *Philosophical Investigations*, Blackwell's, Oxford, 1953, pp. 89–103 and passim.

[6] Edmund Husserl, *Ideen zu einer reinen Phänomenologie und Phänomenologischen Philosophie*, sec. 85 (Husserliana ed., Martinus Nijhoff, The Hague, 1950, pp. 207–208; first ed., 1913, p. 171; Boyce Gibson translation, p. 226).

[7] Husserl, op. cit., sec. 124 (Husserliana ed., p. 305; first ed., p. 257; Boyce Gibson translation, p. 320).

[8] Cf. my books *Knowledge and Belief*, Cornell University Press, Ithaca, N.Y., 1962; *Models for Modalities*, D. Reidel, Dordrecht, 1969.

[9] Cf. Herbert Spiegelberg, 'Der Begriff der Intentionalität in der Scholastik, bei Brentano und bei Husserl', *Philosophische Hefte* (ed. by M. Beck) **5** (1936), 72–91.
[10] Cf. L. J. Savage, *The Foundations of Statistics*, John Wiley, New York, 1954, pp. 9, 82–87.
[11] What I have in mind in this qualification is seen from my paper, 'Surface Semantics', in *Truth, Syntax, and Modality*, Hugues Leblanc (ed.), North-Holland Publ. Co., Amsterdam, 1973, pp. 128–147.
[12] This seems to have been an instance of a deeper (and wider) tendency to think of all rational activities in goal-directed concepts. Cf. the first two chapters of my book, *Knowledge and the Known*, D. Reidel, Dordrecht and Boston, 1974.
[13] Cf. also the less striking formulations of the same point in *Picasso on Art: A Selection of Views*, Dore Ashton (ed.), The Documents of 20th-Century Art, The Viking Press, New York, 1972, pp. 27–31.
[14] Cf. my *Knowledge and the Known*, p. 36.
[15] Even more convincing evidence is obtained when it turns out that many problems concerning an artist's activity and its objects turn out to be but special cases of general problems in the semantics of intensional concepts. The problem of the identity of the object of an artist's creation (*e.g.*: Would he have created the same work of art if he had executed it differently in such-and-such respects?) is for example a special case of the more general problem of cross-identification. For a glimpse of this problem, see my 'Quine on Quantifying in a Dialog', in *The Intentions of Intentionality and Other New Models for Modalities*, D. Reidel, Dordrecht-Boston, 1975.
[16] *Critique of Pure Reason*, beginning of the Transcendental Aesthetic (A 19=B 33).
[17] See Aristotle, *Metaphysica*, IX, 6, 1048b, 23–35.
[18] W. V. Quine, *World and Object*, MIT Press, Cambridge, Mass., 1960, p. 9.
[19] Dagfinn Føllesdal, 'Phenomenology', Chapter 21 in *Handbook of Perception*, Vol. 1, ed. by E. C. Carterette and M. P. Friedman, Academic Press, New York, 1974.
[20] Cf. *Ideen*, sec. 85: "What forms the materials into intentional experiences and brings in the specific element of intentionality is ... *noesis*." The 'materials' in question are described by Husserl as 'sensory data', 'hyletic or material data', or in older terms, "*sensuelle*, wohl aber *sinnliche Stoffe*". (See pp. 173–174 of the original; p. 210 of the Husserliana edition; and p. 228 of the Boyce Gibson translation). Husserl's formulation clearly presupposes that it is *only* through a *noesis* that these 'materials' become intentional.
[21] Føllesdal, op. cit.
[22] Føllesdal, op. cit.
[23] Cf. Wolfgang Metzger, *Psychologie*: *Die Entwicklung ihrer Grundannahmen seit der Einführung des Experiments*, Dietrich Steinkopf, Darmstadt, 1954, p. 32. (Referred to and discussed by Eino Kaila, *Die perzeptuellen und konzeptuellen Komponenten der Alltagserfahrung*, Acta Philosophica Fennica, Vol. 13, Helsinki, 1963, pp. 65–69.) Another phenomenon which serves to illustrate my point here is the illusion of seeing three-dimensional Necker cube with the wrong orientation even when one's touch information gives the right orientation. See, e.g., R. C. Gregory, *The Intelligent Eye*, Weidenfeld & Nicholson, London, 1970, p. 40. Kaila discusses similar inversion phenomena; see op. cit., pp. 44–46.
[24] Cf. William H. Ittelson, *The Ames Demonstrations in Perception*, Princeton and London, 1952.
[25] See A. Michotte, *The Perception of Causality*, Methuen, London, 1963.
[26] As is well known, Husserl goes as far as to say that for noemata "*esse* consists ex-

clusively in its 'percipi'..." (*Ideen*, Husserliana ed., p. 246, first ed., p. 206, Boyce Gibson tr. p. 265).

27 Thus for instance Quine speaks of visual impressions as "colors disposed in a spatial manifold of two dimensions" (op. cit., p. 2). This quote also illustrates the fact that on the view I am criticizing spontaneous sense-impressions need not be devoid of structure by any means. What is at issue is whether they are already impressions of definite objects (are intentional in Husserl's sense) and whether they must be described in the same terms as these objects.

28 Cf. Kaila, op. cit., pp. 71–73.

29 Cf. Dagfinn Føllesdal, 'An Introduction to Phenomenology for Analytic Philosophers', in *Contemporary Philosophy in Scandinavia*, R. E. Olson and A. M. Paul (eds.), The Johns Hopkins Press, Baltimore, 1972, pp. 417–429, especially p. 423.

30 Cf. note 23 above.

31 See J. J. Gibson, *The Senses Considered as Perceptual Systems*, Houghton Mifflin, Boston, 1966. Most of David Katz' writings are also relevant here. There is a brief summary of some of his assumptions in Chapter 2 of his *Gestaltpsychologie*, Basel 1948.

32 See the proceedings of the 1973 Colloquium on Perception in Helsinki in *Ajatus*, Yearbook of the Philosophical Society of Finland, **36** (1974).

33 J. J. Gibson, op. cit., especially Chapter 1 and 13.

34 For instance, it is emphasized that colors are not normally seen just as colors as such, but as somehow connected with the objects of perception, that is to say, as colors *of* objects (surface colors), film colors, colors *of* transparent regions of space (volume colors), colors *of* light sources (luminous colors), colored illuminations *of* objects or *of* empty space, etc. (See Jacob Beck, *Surface Color Perception*, Cornell University Press, Ithaca, N.Y., 1972, especially Ch. 2, and David Katz, *The World of Colour*, Kegan Paul London, 1935.) Nor is this object-relatedness restricted to visual perception. It is perhaps even more remarkable in the sphere of touch. There are even analogues in the tactile-haptic area to several to the different modes of color perception. (See David Katz, *Der Aufbau der Tastwelt*, Barth, Leipzig, 1925, and J. J. Gibson, *The Senses Considered as Perceptual Systems*, Houghton Mifflin, Boston, 1966, Chapter 7.)

35 Cf. J. J. Gibson, op. cit., Ch. 13 and passim.

36 Cf. here Hintikka, 'Information, Causality, and the Logic of Perception' (note 32 above).

37 Cf. here Marx Wartofsky's contribution to the volume mentioned in note 32 above. The psychological literature on the experiential, conceptual, and cultural conditioning of perception is too vast to be surveyed here.

38 Further materials concerning Husserl's views on perception are contained especially in his *Phänomenologische Psychologie* (Husserliana, Vol. IX), *Analysen zur passiven Synthesis* (Husserliana, Vol. XI), *Vorlesungen zur Phänomenologie des inneren Zeitbewusstseins* (Husserliana, Vol. X), and *Ding und Raum*: *Vorlesungen 1907* (Husserliana, Vol. XVI). See also Elisabeth Ströker, 'Zur phänomenologischen Theorie der Wahrnehmung', forthcoming in the volume mentioned above in note 32, and Føllesdal's comments on Aron Gurwitsch in his 'Phenomenology' (note 19 above).

39 Translation modified from Føllesdal's; see *Husserliana*, Vol. XI, p. 363, lines 18–27. One problem with this passage is the Husserl is there speaking in so many words of the *filling component* of a perceptual noema not of sense data. Now clearly the two are related extremely closely to each other in Husserl. (Perhaps they are at bottom identical?) Yet it seems to be impossible to extract from Husserl any clear statement concerning their precise relationship. By any token, they nevertheless are sufficiently close to each

other in Husserl for us to rely on the quoted passage here, presupposing of course sufficient general caution in interpreting Husserl.

40 Føllesdal, 'Phenomenology' (note 19 above).

41 Cf. Michael Dummett's discussion of Frege's principle "the concept *horse* is not a concept" in Dummett, *Frege: Philosophy of Language*, Duckworth, London, 1973, especially pp. 210–212. Note also that Fregean senses could not operate as the references of our terms in opaque (oblique) contexts (as they do on Frege's doctrine) if they were unsaturated.

42 Frege, 'Über Sinn und Bedeutung', *Zeitschrift für Philosophie und philosophische Kritik* **100** (1892), pp. 25–50; see p. 26. Husserl's term '*Gegebenheitsweise*' is also highly suggestive, even though it refers only to a certain component of the noema.

43 Cf. my 'Quine on Quantifying in a Dialog' (note 15 above).

44 This point was emphasized particularly vigorously by Richard Montague; see *Formal Philosophy: Selected Papers of Richard Montague*, Richmond H. Thomason (ed.), Yale University Press, New Haven, 1974.

45 See my 'On the Logic of Perception', Ch. 8 of *Models for Modalities* (note 8 above).

46 The writings of Saul Kriple, Richard Montague, and Dana Scott offer good examples of this. Of them, Kripke has given the most sustained motivation for this view; see his 'Naming and Necessity', in *Semantics of Natural Language*, D. Davidson and G. Harman (eds.), D. Reidel, Dordrecht, 1972, pp. 253–355. Of some of the difficulties into which this treatment leads, cf. my paper, 'On the Proper Treatment of Quantifiers in Montague Semantics', in *Logical Theory and Semantic Analysis: Essays Dedicated to Stig Kanger*, Sören Stenlund (ed.), D. Reidel, Dordrecht, 1974, pp. 45–60.

47 Cf. my papers 'Quine on Quantifying In' (note 15 above) and 'The Semantics of Modal Notions and the Indeterminacy of Ontology', in *Semantics of Natural Language* (note 46 above), pp. 398–414, reprinted above as Chapter 2 of this volume.

48 David K. Lewis, 'Counterpart Theory and Quantified Modal Logic', *Journal of Philosophy* **65** (1968), 113–126.

49 Cf. my 'Quine on Quantifying In' (note 15 above).

50 The perfectly unproblematic phenomenon of seeing double gives us an example of splitting world lines, when perception is treated 'informationally' along the lines indicated earlier in this paper.

51 As one can easily see, merging is ruled out if and only if the following formula is valid: $(x)(y)$ (possibly $(x=y) \supset (x=y)$). In many treatments of different modal logics, it is not.

52 The principle says that from $F(a)$ and $a=b$ you can infer $F(b)$ for any sentence $F(x)$. Of its interpretation, see once again my 'Quine on Quantifying In' (note 15 above).

53 For the following, see my 'On the Logic of Perception' (note 45 above), 'Objects of Knowledge and Belief', *The Journal of Philosophy* **67** (1970), 869–883, and 'Knowledge by Acquaintance – Individuation by Acquaintance', in *Bertrand Russell: A Collection of Critical Essays*, David Pears (ed.), Doubleday, Garden City, N.J., 1972, pp. 52–79.

54 Cf. my *Models for Modalities* (note 8 above), pp. 97, 120–121, 141.

55 See also Ronald McIntyre, 'Intentionality and *de re* Modality' (preprint).

56 Cf. sec. 94 of the *Ideen*; also *Logische Untersuchungen* 5, *xvii* (Vol. 1, p. 402 of the first edition, p. 579 of the Findlay translation); 1, *xii* (p. 48 of the first ed., p. 288 of the translation); 5, *xxxvi* (pp. 472–3 of the first ed.; pp. 631–2 of the translation).

57 Cf. R. Sokolowski, *The Formation of Husserl's Concept of Constitution*, Phenomenologica, Vol. 18, Martinus Nijhoff, The Hague, 1970.

58 Cf., e.g., Ströker, op. cit. (note 38 above), sec. C, and the references given there.

J. N. FINDLAY

COMMENTS ON PROFESSOR HINTIKKA'S PAPER

I experience some difficulty in commenting on this paper, because Hintikka has approached intentional concepts in a manner so different from my own. He has, for instance, connected the directedness of a conscious intention with something purposive, since one of the common meanings of 'direction', as is the ordinary sense of 'intention', is undoubtedly connected with purpose. 'To what are all these acts and utterances directed?' often means 'What are you trying to achieve by these acts and utterances?' Now I would be infinitely far from denying that all conscious intentions are pervaded by purpose: to have something before us as the object of our awareness is certainly to try to keep it in consciousness, to try to envisage it from various aspects, to try to see it in varying contexts, some actual and possible, to compare it with other possibilities, and finally to imagine or perceive it rather than merely to think of it. But the purposive conation which inspires and sustains conscious intentionality is distinct from the conscious intentionality that it sustains: the former is consummated only in a *sequence* of conscious intentions, the latter is whole and complete in *each* conscious intention. Conscious intentionality is simply the 'thereness' or *Vorhandenheit* of something for consciousness: though it is sustained by purpose, it is in itself utterly disparate from anything purposive or conative. It is the unique givenness of something to consciousness, its appearance in the light of consciousness, and this is the same whether what is thus apparent is perceived or thought of. The possible criminality of the Empress Agrippina may be as much *vorhanden* to consciousness as the telephone on my desk. I do not know whether Hintikka clearly differentiates this kind of pure cognitive appearance in his experience: it has eluded many, as for instance Heidegger. I think, however, that it is basic to the understanding of intentionality.

Other things that Hintikka says lead me to think that he connects the directedness of intentionality with its direction to something individual, to a concrete *res* or thing that is to be found somewhere in the real world, which can then perhaps also be transferred in thought to other possible

Manninen and Tuomela (eds.), Essays on Explanation and Understanding, 111–116.

worlds. In default of there being such an individual, or in default of my knowing who or what it is, my intention can only be directed to a possible fact or truth, to a *dictum* rather than a *res*. Thus, my belief can, on Hintikka's view, only be directed to the present Prime Minister of Denmark if I have some individual specifically in mind, if I know there is such an individual, otherwise I do not have the present Prime Minister of Denmark as my intentional object, but only some general state of affairs which concerns some such person. It seems plain to me that whatever this is, it is not phenomenology. If I turn the light of my regard on the present Prime Minister of Denmark, the present Prime Minister of Denmark is my intentional object, what my cognitive beam is directed upon. The fact that I do not know of an individual who satisfies the description of being Prime Minister of Denmark, or the fact that there is in fact no such individual, are both totally irrelevant to my being intentionally concerned with something which is the one and only contemporary Prime Minister of Denmark, nor is it anything but a gross failure to understand what intentionality is to think that these facts constitute a difficulty for my thinking. The very defining character of intentional activity is that of its being indifferent to the reality of its object. It is as easy to refer to something one does not know or know of as something one does know or know of, and it is as easy to refer to something which does not exist as to something which does exist. I can intend the God Jupiter, the first God of the Romans, whom I do not know or know of, and who does not exist, as easily as I can intend Professor Hintikka. And while when I think of the God Jupiter, I am perhaps implicitly supposing that something is divine, rattles the thunder and gathers the clouds, etc., this is plainly *not* what I am thinking of. My concern is with a *res* or πρᾶγμα, not a *dictum* or λεκτόν. And if concern with the unknown or non-existent creates a difficulty, then there is just as much difficulty in my concern with an unknown or nonexistent fact as with an unknown or non-existent object.

I now come tho what seems to be the *pièce de resistance* of Hintikka's analysis: that intentionality always implies the presence of a concept involving the simultaneous consideration of several possible states of affairs or courses of events, or the consideration of several possible worlds. Such simultaneous consideration he tells us means logical parity and semantical explicability rather than anything overt; the explication may presumably be spread out in a 'logical specious present', whatever that is, rather than confined to a single illuminated instant, and the possible worlds that serve as the background of what we conceive or perceive, need not be total world-systems or histories, but minuscule parts of the universe, the alternative courses, for instance, that a single experiment might take. As the statement of a *condition* of intentionality I have no objection to what Hintikka says. If I take note of an object as being such-and-such, it is perfectly plain that I am not only ready to take note of the environing things which contrast with it and are *not* such-and-such: I am also ready to consider the possibility that *it* might not be such-and-such but something else. Plainly the point of anything we notice would vanish if there were no *contrast* between it and other things or facts and possibilities: everything we perceive or conceive we locate in that vast actual system we call the world, and we also locate it in that infinitely vaster system of what might be in the actual world though it does not actually happen to be so. What we are conscious of as an object is certainly always envisaged in the horizon of the great object we call Reality, and in the wider horizon of the object we call Possibility, and I should say in a still wider horizon which embraces as thinkables though not as believables the things which are *not* even possible. The theory of background horizons is a part of Husserlian phenomenology, as it is of certain forms of idealism. I certainly accept it. I do not go as far as the young men of California in thinking that possible worlds are to be treated as on a level with actual worlds – this view out-Meinong's Meinong – nor in thinking that they are really worlds at all; much less do I believe this in the case of impossible worlds. Possibilities or impossibilities of worlds are not possible or impossible worlds. But I do believe that to think of anything is to think of everything, and that that everything also includes what might or might not be as much as what is. But I find that Hintikka's Horizon-Theorem, as I shall call it, interesting and important as it is, is not definitory of intentionality, since it *presupposes* the concept of intentionality. For if I am simultaneously to *consider* various alternative possibilities in conceiving of any one fact or possibility, it is plain that I must be considering *each* of them in order to consider them all. There must be a conscious intention to a realized possibility A, that this is a big toadstool, even if this has to be accompanied by conscious intentions to other unrealized possibilities, that this for instance might have been a small mushroom instead of big toadstool. To be aware of what things are, I must be aware of what they might be or ought to be, but this is only to say that awareness is one of contrasting possibilities, in which an awareness of each contrasted possibility must however necessarily be included. I do not therefore think the Horizon-Theorem throws the least light on the notion of intentionality, since it presupposes it.

For the rest, however, I am in considerable agreement with what Hintikka says. I do believe that intentionality goes with intensionality or conceptuality; to be aware of anything is to be aware of it as such and such, even if this awareness takes the limiting form of being aware of it as something quite indefinite or even indeterminable. That there is something to *be* characterized or determined, but as yet totally uncharacterized or undetermined, is a definite case of intentionality, not infrequent in perception or thought, and involving a very interesting intentional object. $(\exists x)?x$. And I believe with Hintikka and Husserl that conceptuality is as much present in perceptual as in pure thought-confrontations. To think emptily of the viscosity of pitch and to perceive this viscosity in fulfilled fashion when I look at a pitch-lake in Trinidad, are experiences which differ profoundly in respect of fulfilment but not in respect of their intentional direction. To see viscous pitch is merely to live through an illustrated variant of a thought-reference to pitch as viscous.

I also agree with Hintikka in his criticisms of some of the features of sense-stuff or *Hyle*: it ought not to be thought of as an isolable element, but merely as the difference between a fulfilled, intuitive intention, on the one hand, and its corresponding emptied version, on the other. There is such a difference, and it is of all differences the most important, but it does not consist in the presence in intentional experience of a surd stuff which makes the difference. Husserl's conception of uninterpreted hyletic elements and interpretative *Auffassungen* which ensoul these, is absurd if treated literally: there is no trick in virtue of which non-intentional elements can be used to fulfil intentions. The only conceivable view is that uninterpreted sense-affections are possibilities of intuitive intentionality which are for some reason not actualized: I am affected appropriately to the intention of a red circle but I do not raise my affection to the intuition in question. I do not for this reason accept Hintikka's view that there are pure sense-affections, *e.g.* twinges of pain, which involve not even the intrinsic possibility of being used intentionally. It seems to me that one can, with sufficient detachment, perceive a toothache as one perceives the

Aurora Borealis: I in fact recommend such detached awareness of localized pains as a good way of diminishing their disagreeableness. Husserl's use of the concept of *Sinnesdaten* is of course thoroughly unsatisfactory: sometimes he means by them reduced intentional objects à la Moore, which are what remain when we remove all that is not intuited, sometimes on the other hand he means mere modifications of experience that can be used in intuitive fulfilment but are not actually being so used. On either view the hyletic elements are really hypostatized abstractions of the difference between a sensuously fulfilled and a not sensuously fulfilled or empty awareness. I do not know why Hintikka devotes so much space for arguing the intentional character of perception, since it is plain that even if a real object is acting on our senses and causing us to respond, it is only intentionally present if we *take* it to be present, and it is, even when real, much more than we intuitively see it as being, and, even when perceived as itself there, is not infrequently a pure illusion to which nothing in reality corresponds. Seeing is not some direct encounter infinitely superior to thinking and believing: it is merely an illustrated form of intentionality as much capable of purveying the false and the impossible as a thought or an utterance.

I am further in agreement with Hintikka that the sense in virtue of which we intend an object or state of affairs is not itself the object or state of affairs. It is not *id quod intelligimus*, but *id quo intelligimus* as St. Thomas would say. If I perceive you as a slim, young stripling or believe you to be a senator, I am not considering the detached meaning 'being a slim, young stripling' or 'being a senator', but objects which involve a great deal more than these abstracted contents. But I *can* make noematic senses or intentional objects into objects *simpliciter*, and they are then given in all their openness and incompleteness of structure. There is nothing wrong in objectifying senses or intentional objects: it is only wrong to identify them with the objects to which they introduce us, an identification which is, however, very tempting since they in a sense ordinarily vanish in such objects.

There are a great number of other points in Hintikka's paper that I cannot here criticize or approve. But I should like to say a few general things about his connection of 'intentionality' with a *t*, with 'intensionality' with an *s*. As a Platonist, I believe that intentionality with a *t* is only a special case of *intensionality* with an *s*, since I think that universals – not all of them but such as are genuine – are the only true, substantial entities, and that particulars only cling to being in so far as universals are instantiated in them. There are no thises, thats, nor heres and theres, nor nows and thens, nor mes and yous, only universals which are thisified and thatified, or herified and therified, or nowified and thenified, or meified and youified, etc. And universals always have an inherent reference to the extensional truths which concern the factual coinstantiation of universals, but all intensional truths concern their necessary or possible instantiation: the possible worlds and individuals of Professor Hintikka are all involved in the sense of certain universals. Intentionality with a *t* is merely a special status which occurs when certain universals are enjoyed by intelligent minds. It is the universals which intend their possible instances: minds only intend those instances by harbouring the corresponding universals. By harbouring *lapiditas* the mind acquires the intrinsic reference of *lapiditas* to *lapides*. All this is to be found in Aristotle and St. Thomas. I would like to see a complete intensional logic worked out in which intensions are the only true subjects of propositions, and all else merely concerns their actual and possible instantiation, and also the mental acts which through such intensions concern themselves with their possible instantiations. I am sure that, if Professor Hintikka can be induced to give up his belief in individuals, and will expunge them from his logic, he will be able to work out a better logic of both intensionality and intentionality than anyone else.

Boston University

JAAKKO HINTIKKA

REPLY TO J. N. FINDLAY

Professor Findlay's kind and generous comments show that I unfortunately have not been quite as clear in my paper as ought to have been. By 'directedness', I tried to mean directedness just in the sense Findlay so nicely describes. This intentional relation to what is present to one's consciousness need not be in the least purposive. References to purposiveness in my text were largely motivated by the character of the occasion on which the paper was originally presented, a symposium on explanation vs. understanding.

Nor need the *relatum* of the relation of intending something enjoy actual existence. The main problem with individuals in intentional (intensional) contexts is not their possible nonexistence. That is being taken for granted by most people, perhaps excluding one or two of Professor Findlay's "young men in California". I have never held that an intentional object could not be nonexisting, and I do not have the slightest intention to change my mind now. In this respect, Professor Findlay's criticisms just have to be given a new address.

Nor is the possible nonexistence of individuals the crucial issue in the so-called *de re-de dicto* distinction. (Admittely, on my formal reconstruction a *de re* reading does presuppose real existence, but this is surely due only to unfortunate limitations of the current notation of modal and epistemic logic.) Here, as in so many other important junctions in the theory of intentionality, the crucial question is uniqueness or well-definedness, not nonexistence. The possible nonexistence of the next Prime Minister of Denmark does not deprive him of his status as one of my intentional objects. But if I don't have any beliefs as to who he is or will be, and if I am perhaps even speculating as to who among several more or less well-known persons the P. M. might eventually be, my thought-act in focusing on the question of his identity is not directed to one well-defined intentional object, but on different people under different possible courses of events. Nor can I reify the future P.M. even in my intentional thinking into an additional entity. If I am asking myself which of my two

Manninen and Tuomela (eds.), Essays on Explanation and Understanding, 117–119.

old friends, *X* and *Y*, will be appointed, I have two intentional objects before my mind's eye, not three. I cannot help suspecting that Professor Findlay has missed the subtle difficulty which the directedness-theory of intentionality faces in dealing with such *de dicto* acts as, *e.g.*, those 'directed' to the next Prime Minister of Denmark. I do not see that his remarks on this point make Husserl's plight any less than I argued it to be.

As to Professor Findlay's main critical point, I am not averse to admitting that intentionality in his sense of directedness may have to be displayed toward each of the several possible worlds that one is considering. In this sense, directedness may indeed be prior to the multiplicity of possibilities. However, this is not the moot point. The real question is: Which one is it that makes an act intentional, its directedness or its involving a multiplicity of alternatives? I do not see that Findlay's observation helps at all answering this main question of my paper.

I am not sure, either, to what extent there is a real disagreement present here. I do not find Findlay's formulations false, but rather too metaphoric to my taste. Perhaps I can put my main thesis in the form of a prediction. When all metaphoric elements are removed from such locutions as "directedness upon an object or content", "*Vorhandenheit* of something for consciousness", and so on, it will be found (if I am right) that their underlying logical form is quite different from their apparent form (which is a two-place relation between a mind and an intentional object) and involves the kind of multiplicity of possibilities suggested in my paper. Because of the importance of this idea, I regret that Professor Findlay did not comment on its attempted partial realization in my paper in the form of a quasi-deduction of the main ideas of possible-worlds semantics from Frege's and Husserl's ideas about meanings and other intentional objects.

The distinction between possibilities of worlds and possible worlds may be without much real difference, I suspect. I shall be perfectly happy with the former, as long as they as it were do the work they have to do. Mainly, it has to be recognized that the question as to which individual would be which, were a certain possibility realized, is an important one, and sometimes even admits of several equally valid answers. As long as Findlay's "possibilities of worlds" allow for this, I can live happily with them.

I am enough of a Platonist to feel a mild temptation to take the course Professor Findlay recommends to me. However, I do not see that it solves all the problems of identity and quantification that have been the main testing-ground of intensional logicians in the last few decades. Nor can I but be painfully aware how much of our conceptual environment such an excision of all particulars would deprive us of.

PART II

CAUSALITY AND INTERVENTION

PETER WINCH

CAUSALITY AND ACTION

I want to raise some questions about Professor von Wright's treatment of the relationship between the causal explanation of events and the concept of human action. One of his main aims is to show the relation between the concept of 'Humean causation' (meaning by this no more than a causal relation in which "cause and effect are *logically independent* of one another" (cf. *Explanation and Understanding*, London 1971, p. 93) without commitment to Hume's regularity view) and the concept of human action. Two of the most important conclusions that he works towards are that human action cannot be fully explained in terms of Humean causation and that the concept of Humean causation itself depends on the possibility of human action, conceived in a teleological way.

Von Wright's discussion of causation is in terms of necessary and sufficient conditions and he develops a simple and perspicuous diagrammatic schema for the tracing and representation of various conditionship relations. Having done this, he raises an important question. Suppose that we have been examining a physical system and, by means of observation, tracing various conditionship relations between different states of that system. "... How do we know that the alternative possibilities of development, as familiar to us from the observations, really represent all the possibilities?" (p. 60). This question is closely related to the question of how we come to be in a position to assert counterfactual conditionals concerning the operation of causal systems, or to put it in another of the ways favoured by von Wright, what entitles us to say that '*nomic*' connections obtain between the states of a system.

Von Wright's answer to these questions is that such things are possible for us by virtue of our ability not merely to *observe* but also to *act* (in the form of experimental 'interference' with the system we are studying) in a way which cannot itself be understood in terms of Humean necessary and sufficient conditionship relations. His argument for this conclusion (and consequently the precise significance of the conclusion itself), however, is extremely difficult to follow and assess, partly because several of the con-

Manninen and Tuomela (eds.), Essays on Explanation and Understanding, 123–135. *All Rights Reserved.*

tentions on which the argument depends themselves depend in their turn on arguments which are not fully set out till later in his discussion of action and intentionality. I shall now devote some space to an attempt to disentangle the issues.

We are dealing with a certain physical system, *S*, the operation of which we have been systematically observing. "In the succession of occasions we have repeatedly noted the occurrence of a certain state *a*. It has always, in our experience, been followed by a state *b*, this again sometimes by c_1 sometimes by c_2, c_1, sometimes or always by ______, c_2 sometimes or always by ______, and so forth through a number of, say *n*, stages. In these successions of events we have, by the tools of causal analysis, been able to discern certain conditionship relations". (p. 60) What we wish to discover is whether these observed developments really represent all the system's possibilities. The initial state *a* has always, in our observation, originated from some immediately preceding state. Von Wright now asks us to assume "that there is a state α such that we feel confident, on the basis of past experience, that α *will not change* to the state *a* unless we change it to *a*" (*ibid.*). We then change α to *a* and observe the system to undergo one of the developments we have observed in the past. We are now, von Wright claims, in a position to draw "a very powerful logical conclusion", *viz.* "that neither α nor any state which occurred *anterior* to α can be a *sufficient* condition of the initial state of the system as instantiated" (p. 61). This is because any chain of successive sufficient conditions within the system "is interrupted at α. Because α, we assumed, will not change to *a* unless *we* change it." (*ibid.*)

When the argument is stated thus baldly it certainly looks as though it makes possible this "very powerful logical conclusion" only insofar as it involves a blatant *petitio*. The *conclusion* is that α is not a sufficient condition of *a*; a *premise* that α will not change to *a* without our intervention. A sufficient condition is explained by von Wright as follows: *p* is a sufficient condition of *q* if, "whenever *p* is, *q* will be there too; the presence (occurrence) of *p* suffices to ensure the presence (occurrence) of *q*" (p. 38). If I say that α will not change to *a* without my intervention, I am saying that the presence of α does not suffice to ensure the presence of *a* (something else is required in addition: my intervention). So am I not saying that α is not a sufficient condition of *a*? If this is so, then it is hard to see what role my actual intervention plays in the investigation; I seem to be in a position to make the assertion represented here as a conclusion, whether I change α to *a* or not. What difference does this intervention make? The fact that I succeed in effecting the change does nothing to show that the change could not also have come about in some other way. That possibility is only excluded because I am already, according to von Wright, assuming its exclusion. It is not even clear that reference to the *possibility* of my intervention is relevant. If I can assume that α will not change to *a* without my intervention, I can also assume that α will not change to *a* as long as the system is in its present state; and I can assert this, apparently, without bringing in mention of any possible human actions at all.

This is one's natural first reaction to the argument. However, von Wright seems to wish to forestall such a criticism with his claim that the experimenter's assumption is "not an assumption about a causal conditionship relation" (p. 61). This claim is plainly crucial to his case and would have deserved a detailed defence at this point in the argument. But all von Wright says (for the present) is: (1) "It is not assumed that the state α is a sufficient condition of not-*a*." (2) "Nor do we assume that changing α to *a* requires knowledge of sufficient conditions of *a*. Sometimes knowledge of such conditions plays an important part in our changing of the situation. But this is not always the case." (p. 61).

Point (2) seems perfectly acceptable, but not relevant to the issue presently under discussion. What about point (1)? One thing one is tempted to say is that *of course* it is not being assumed by the experimenter that α is a sufficient condition of not-*a*, since he is assuming that α, along with an appropriate action of his, will result in *a*. Von Wright would probably disallow this move, since it seems to be an assumption of his argument that the system *S* must be regarded as independent of the experimenter and his actions. Experimental manipulation involves an 'interruption' of the chain of sufficient conditions. This assumption is connected with certain claims about the logical structure of human action and its difference from the logical structure of physical systems, which are not developed until the next chapter. But if the results of action are left out of account, how is it *not* being assumed that α is a sufficient condition of not-*a*? It *is*, as von Wright says, being assumed that if *S* is in state α, it will, if left to itself, not change to state *a*. That is, whenever α is, not-*a* will be there too; that is, according to the definition of 'sufficient condition', α is a sufficient condition of not-*a*. At the very least it seems to be being assumed that α is not a sufficient condition of *a* and this, of course, is precisely the 'conclusion' which experimental intervention of the sort von Wright describes is supposed to enable us to draw.

In introducing his assumptions, von Wright admits that they involve "grave problems for a philosopher" (p. 60) such as how we know that α will not change to *a* independently of our action or that we can change α to *a*. But he dismisses these difficulties with the remark: "But we must also admit the empirical fact that situations of the kind just described are familiar to us. I know (feel sure) that the window in front of me will not open 'of itself', but that I can open it. I *may* be mistaken, of course. Surprising things happen in nature and unexpected disabilities sometimes befall a man. But on the whole such knowledge is reliable. If it were not, action would not be (commonly) possible – and, *a fortiori*, neither would the activity which we call scientific experimentation."

It is of some importance to be clear about the nature of the 'empirical fact' which we are being invited to admit here. It is *not*, as might at first appear, the fact that there are situations in which we can make something happen which would not otherwise happen. It is rather the fact that there are situations in which we feel sure that this is the case. Von Wright does not perhaps emphasize this sufficiently in the present context, but he does so in the important footnote 39 to Chapter III (p. 199). It is possible – though this is only conjecture on my part – that he feels that his argument escapes difficulties of the sort I have been discussing for this reason. He is not, so he might want to argue, assuming the point which the experimental procedure he describes is supposed to establish, but a different point about the state of mind of the experimenter. But whether this is von Wright's argument or not, it seems clear that it will not work for two reasons. In the first place it is, after all, the *experimenter* who is supposed to be enabled to draw the "powerful logical conclusion" and it is equally the experimenter who is supposed to initiate his experiment holding the assumption that things will not change unless he makes them do so, and this is the very conclusion which he is supposed to be enabled to draw by the experiment. In the second place von Wright seems to regard "feel sure" as replaceable by "know" in the statement of his assumption and he also talks of such knowledge as being "reliable" (p. 61) and as having an "experiential basis". (p. 199) We have to ask what this experiential basis is supposed to be. If it is simply *observation* of sequences, then von Wright would seem to be admitting that, after all, part of the required counterfactual element does not depend on actual performing of experiments. If, on the other hand, the experiential basis includes the past performance of experiments, then those experiments must again have involved the very assumption the basis of which is in question. And it will still not have been made clear how the experimenter becomes entitled to claim that a certain change would *not* have come about but for his intervention by the fact that it *has* come about with his intervention.

We are still only at the beginning of the procedure by which von Wright claims that we can establish "closedness for the 'interior' of the system". What we still have to do in other words is to exclude the possibility that some feature p of a state of S should occur of which α is a sufficient condition. We do this, von Wright says, "by refraining from the action of changing α to a and by observing what happens then. We *let the world change* independently of our interference – which may of course mean that it does not change at all, but remains in a state identical with α. If, when this 'untouched' world has passed through the ... stages, corresponding (in time) to the stages from a to the end-state of our system, it does *not* exhibit the feature p, we can be sure that α is not a sufficient condition of the occurrence of p in the end-state of our system. If again it exhibits the feature, then we have to reckon with the possibility that α actually is such a condition and that the system is therefore not closed". (p. 62)

One relatively minor puzzle about this passage is that what von Wright says "may" happen – that α persists unchanged – is precisely what, according to him, the experimenter assumes when he actively changes α to a; it seems therefore that he will also assume this when refraining from initiating the change. But the more important difficulty is precisely how what an investigator can discover by 'refraining' from setting a system in motion, which on other occasions he does set in motion, differs from what he can discover if he simply passively observes the system without ever actively intervening at all. If he is limited to passive observation, he can still observe whether or not p occurs in any of the end-states of S, given α as a starting point. Indeed, what he observes under these conditions will be precisely the same as what he observes on an occasion when he has *refrained* from changing α to a. What the experimenter discovers by initiating the change from α to a is that thereby he can bring about the occurrence of p in the end-state of S. He can also discover that when he does not initiate the change from α to a, p does not occur in the end-state of S. Very well. But why could he not discover this latter if he had *never* initiated the change from α to a? Von Wright clearly thinks that he has given an answer to this question, but I have not succeeded in discovering what this is.

I do not of course want to deny that without the possibility of experimental intervention, investigators would *in practice* be impossibly constricted in the provision of examples of the operation of systems given alternative initial conditions. But von Wright seems to want to make a much stronger *logical* point than this.

Let me return to von Wright's claim that when an experimenter changes α to a any "chain of sufficient conditions" leading up to α is "interrupted". As I understand him such an 'interruption' is characteristic of human actions involving an intervention in the course of nature; this conception is plainly important to von Wright's views about the dependence of the notion of causation on that of human action and also to much else in this book and I will now discuss it.

The 'result' of a human action is some state of affairs; for instance, if I fire a gun, the result of my action is that the gun is fired. The connection between an action and its result is an internal, logical one and therefore cannot be regarded as a form of Humean causation. If the gun does not fire, then, whatever I did, it will not count as having fired the gun. Von Wright insists: "It is a *bad* mistake to think of the act(ion) itself as a cause of its result". (p. 68) My action may also have consequences: *i.e.*, events which occur as a consequence of what I do; alternatively, events which I bring about *by* performing the action in question. Thus, if someone dies as a result of my firing the gun, his death is a consequence of (the result of) my firing the gun; it is something which I have brought about by firing the gun. The relation between (the result of) my action and such a consequence is one of Humean cause and effect.

Let us apply this distinction to the performance of a scientific experiment. In the previously discussed schema the change of state from α to a is the *result* of the experimenter's action of changing α to a. The changing of α to a is not *caused*, it is *done*. On the other hand, the subsequent changes of state in S are caused by a. To say this is to be prepared to assert certain counterfactual conditionals. For instance, if I say that *a* was a sufficient condition of the change from *b* to *c*, I must be willing to say that if *b* had not changed to *c*, *a* would not have been present. Now von Wright maintains that such a causal counterfactual conditional 'rests' on another counterfactual conditional, *viz.* the one that says that "... [*a*] ... would not have been there had we not produced it". (p. 72) He claims further that this latter counterfactual is "not a statement of a conditionship relation nor of a causal connection". Von Wright is not at this stage very explicit about his reasons for this important contention, but as far as I can see, there are two strands in his overall argument that he might wish to bring to bear. The first starts with the point that the relation between my action and its result is not one of causal conditionship. To this must be added his further claim that to the extent to which an agent thinks that a certain state of affairs would have come about anyway independently of his action, to that extent will he refrain from describing his action in terms which would make that state of affairs the 'result' of that action. So, putting these points together, we can say that the relationship of α to *a* is not a causal one and that it implies that α would not have changed to *a* if it had not intentionally been so changed. Now it is not clear to me that these two points are sufficient to justify the assertion that the counterfactual under discussion is not a causal one. It is true that the assertion of how *a* did come about (as a result of my action) is not the assertion of a causal connection. But does that mean that the assertion that *a* would not have come about but for my action is not a statement of (the absence of) causal conditionship? I cannot see that it does. And perhaps von Wright is confessing to similar misgivings when he says, towards the end of his discussion of causality: "One way of disputing my position would be to maintain that action cannot be understood unless causation is already intelligible. I shall not deny that this view too could be sustained by weighty arguments" (p. 74).

The other strand in the argument has to do with the question whether an action can itself be said to be caused. Von Wright allows that actions can be "brought about" *e.g.* by commands, threats or acts of persuasion, but denies that such "bringings about" involve causal connections: "It is a motivational mechanism and, as such, not causal but teleological" (p. 69). Now this point is certainly relevant to the status of certain kinds of counterfactuals involved in action claims: for instance, such a one as:

"I should not have set the system in motion if my boss had not threatened to fire me if I didn't". But these seem quite different from the counterfactuals which are here under discussion, which would be, for instance: "The system would not have come into motion if I had not set it into motion". So once again a real ground for von Wright's contention has not emerged.

It is also obscure how precisely the counterfactual conditional involved in a nomic, causal relationship is supposed to "rest" (von Wright's scare quotes) on the supposedly non-causal counterfactual involved in the statement that some state of affairs was the result of my action. Von Wright says: "It is *established* that there is a causal connection between *p* and *q* when we have satisfied ourselves that, by manipulating the one factor, we can achieve or bring it about that the other is, or is not, there". (p. 72) The connection involved in "If *p* had not occurred, *q* would not have occurred" seems logically quite independent of that involved in "If I had not performed the action of which *p* was the result, *p* would not have occurred". Of course *both* of these are involved in the claim that *q* came about as a consequence of my manipulation of *p*; but not, so far as I am able to see, in any sense which would justify us in saying that the one 'rested' on the other.

Von Wright's analysis leads him to the view that the *meaning* of '*p* is the cause of *q*' is 'I could bring about *q*, if I could do (so that) *p*.' (p. 74) I shall not discuss that view further here, though there is certainly more to be said about it. I shall also not discuss von Wright's interesting attempt (based on this analysis of causation) to describe a case in which an effect could be said to precede its cause. This is a case in which I perform some ('basic': see below) action, such as raising my arm and *thereby* bring about the occurrence of certain neural events in my brain necessary and sufficient for the movement of my arm and occurring before that movement. I shall, however, attempt some remarks about his treatment of the question whether intentional human action is compatible with full (Humean) causal explanation of all bodily movements. Von Wright's position on this question is as follows. He does not want to deny that there might be causally sufficient conditions, operating in my nervous system, for all movements of my body. Nor does he want to say that this, if it were the case, would make what we now call intentional action an illusion. He does claim however that my ability to act intentionally presupposes that I am not aware of the operation of these causes, if such there be. It may be instructive to compare von Wright's position with Spinoza's. Spinoza thought that my uncritical commonsense notions of my ability to act freely were due to my ignorance of the causes of my action and that if I were to become aware of those causes I should cease to suppose that I could act freely. Now in the first place von Wright differs from Spinoza in distinguishing between "my action" and "the bodily movements which on a given occasion constitute my action" and in thinking that the notion of a "cause" of my action analogous to the cause of some natural event is incoherent. Secondly, again unlike Spinoza, von Wright does not want to say that I am under any illusion in thinking about my actions in the terms in which I ordinarily do; he wants to say rather that it is a necessary condition of my thus thinking (and therefore of my acting intentionally at all) that I should be unaware of any causes there may be which are sufficient conditions of my relevant bodily movements. Perhaps I could put the point like this: if I were to become fully aware of the causally sufficient conditions of my bodily movements (and incidentally, von Wright does not commit himself to the existence of such sufficient conditions), I should no longer be able to act intentionally; but the discovery of such conditions would not entail the discovery that I never had really acted intentionally in the way in which, in my state of ignorance, I supposed I did. A third point which is a difference of emphasis as between Spinoza and von Wright, is also relevant here. By "ignorance" of causes von Wright means *unawareness* of causes: *i.e.*, lack of awareness of their operation in the particular situation in which I act; he does not seem to mean ignorance *in general* of the causes which operate in such cases (if they do).

It will be evident that this is a somewhat delicately poised position and I shall try and test its stability by examining von Wright's arguments for it. I noted earlier the distinction between the 'result' of an action, which is internally connected with that action, and the 'consequence' of an action, which is something *caused* by (the result of) what I do. The consequence is therefore something which I bring about by means of doing something else. This distinction is relative to the particular terms in which I describe my action (and hence too its result), but there are some things which I do, von Wright holds, which I cannot describe as things which I bring about by means of doing something else. He appropriates the term "basic actions" for these. Raising my arm in some of its instantiations, would be an example.

Now in many cases at least of non-basic action, something which I bring about by means of doing something else is also something which could be brought about by other means, perhaps by the operation of Humean causes in which no human agency is involved. It may also be that I discover that something which I thought I was *doing* was in fact the outcome of independent Humean causes. I thought I was opening the door, but then discover that the real cause of the door's opening was some mechanism of which I had been unaware. In such a case I am shown to be mistaken in accepting the counterfactual conditional which I have to accept in order to act ('the door would not open if I did not open it') and I therefore retract my original description of my action. I no longer say that I 'opened' the door, but perhaps that I 'pushed' the door. On the basis of this analysis von Wright concludes that I cannot (logically) combine acting under a certain description with observing the operation of causes of the result of my action so described.

Let us accept this argument, as it is applied to non-basic actions such as the above example. Is it still acceptable if applied to basic actions and to the putative neural conditions causing the bodily movements corresponding to those actions (which in the case of basic actions will be identical with the results of those actions)? Von Wright thinks that it is. "What is excluded... is that at one and the same time I raise my arm and *observe* the operation of the cause. For, observing the cause operate entails letting *it* lift my arm ('under my watching eyes'), and leaving it to the cause to do this is incompatible with lifting my arm myself. This is a logical ('grammatical') point. When I observe, I *let* things happen. When I act, I *make* them happen. It is a contradiction in terms both to let and to make the same thing happen on the same occasion. Therefore no man can observe the causes of the results of his own basic actions." (p. 130)

I find myself unable to decide whether or not von Wright is right about this. I do think, though, that there are difficulties about the way he states the case. For one thing, though this is certainly not a conclusive objection, it makes a difference that we are now dealing with basic actions. In the previous example, when I observed that there was a mechanism opening the door, I concluded that the action I was performing was not one of opening the door but simply one, say, of pushing the door. But certainly I did not conclude that I was not performing any action at all. That, however, is what I am being invited to conclude in the case in which I observe a neural mechanism lifting my arm. That is, in the first case I retract my original action description and substitute another, less ambitious description. In the present case, because we are concerned with a basic action, there is no less ambitious action description to substitute for that of 'raising my arm'; so here the retraction is being held to consist in the withdrawal of any claim to be acting at all. And this seems a move of a logically altogether different character from the original one.

My next point is connected with this. What I discover about the door is that the mechanism is going to open the door "in any case", that is, independently of any intentions of mine (which means here "in the absence of any intentional *actions* of mine"). Von Wright says just the same thing about the movement of my arm which is brought about by a neuromuscular mechanism. I intend to raise my arm and then someone points out to me a neural state in my brain which is a sufficient condition of my arm's rising. "Well," I am supposed to say, "I see my arm would have risen in any case". This is also supposed to mean "independently of any intentions of mine", but we must notice that in this case such a phrase can*not* be interpreted as "in the absence of any intentional *actions* of mine", precisely because we are here dealing with a 'basic' action. What is more, whereas the discovery about the door-opening mechanism was certainly a discovery about what was going to happen quite independently of any intentions of mine, it is not altogether clear that this is true in the 'basic action' case. We need here something which von Wright does not offer, a discussion of the precise relation we are assuming to hold between the neural mechanism and my intentions. It surely cannot be *assumed* that these are utterly independent in the way in which the door mechanism is independent of my intentions. Can I not suppose – and might I not have reason for supposing – that the operation of such Humean causes will only be observed on those occasions on which I am also acting intentionally? This question leads us naturally into a thicket of difficult questions in the philosophy of mind, which it would be out of place to try to discuss in detail here. For instance, the central thesis of 'central state materialism' is one view which would not allow the Humean cause of my movements to be thought of as operating independently of my intentions (since they would be thought of as 'contingently identical' with those intentions). I certainly do not mean to advocate this (to me unintelligible) view. It does not seem to be the only tenable alternative to a view like von Wright's. It might be supposed, for example, that a given intentional action, involving given bodily movements, is always accompanied by given neural processes (though I do not want to give the impression that I suppose we have any good reason at all for thinking this true). I do not see that von Wright has provided any argument against this possibility. His claim that we cannot think of ourselves as acting intentionally in cases where we think that the movements involved in the relevant action would be produced "anyway" by Humean causes will not serve, because on this supposition we do not think such a thing. We think rather that the causes producing those movements on occasions when we act intentionally will *only* be found to operate on occasions when we act intentionally under the relevant description. Nor need such a supposition entail any assumption about my intentions being *caused* by the neural processes – or indeed *vice versa*. And if someone does insist on postulating a causal relationship here, what would be the ground for assigning it one direction rather than the other? Indeed, on von Wright's own account of the notion of cause as *dependent* on the notion of human action, there would apparently be good reason for insisting that it is the intentional action which causes the neural processes. This is what he himself does in the claim, which I have alluded to already, that the neural process may be produced by me simply by intentionally acting in a certain way.

I am, therefore, not convinced that von Wright has succeeded in making out his case for saying that my awareness of the operation of Humean causes of my bodily movements is incompatible with my acting intentionally in a way which involves those movements. One might ask what becomes of the argument if one substitutes the word 'muscular' for 'neural' in it. Imagine that the muscles of my arm are laid bare in such a way that I can observe the varying degrees of their tension which make my arm rise. I (intentionally) raise my arm and, as I do so, watch the muscular causes of my arm's rising. Is there any logical defect in that last sentence? I doubt if von Wright (or anyone else) would want to claim that there is. But if that is so, what difference is made if we substitute back 'neural processes in my brain' for 'muscular processes in my arm'? If it is claimed that an important difference is made, then this must be due to some *special* significance which neural processes in the brain are supposed to have in this connection. And this in itself seems to show the insufficiency of an argument, like von Wright's, which relies *solely* on claims about the relation between Humean causes *as such* and intentional action.

King's College,
University of London

ALASDAIR MACINTYRE

CAUSALITY AND HISTORY

Whatever else may have afflicted our culture, we have in the past four hundred years greatly improved our knowledge of the workings of nature; and our knowledge of nature is such that our knowledge of particular natural causes and our knowledge of the truth of certain law-like physical or chemical generalisations are very often closely connected. If I do in fact know that on a given occasion what caused the change in the pressure exerted by a given volume of a particular gas was a change in its temperature, I do so partly in virtue of also knowing the truth of the gas law equations and perhaps also of some of the further law-like generalisations that comprise the kinetic theory. If I do in fact know the generalisations of geometrical optics to be true, it is because I and others have been able to produce specific visual effects on given occasions by making use of them. Why then should we be surprised if philosophers have tended to offer accounts of the notion of causality which link it in the most conceptually intimate ways to that of a law-like generalisation, some going so far as to make it part of the meaning of the words 'cause' and 'effect' and its cognates and translations that a cause and its effect are always members of classes which are linked by some law-like generalisation?

The attempt to show that causes and laws are definitionally and conceptually linked has as a matter of historical fact flourished especially in those intellectual contexts, such as those of eighteenth century empiricism and nineteenth century utilitarianism, where natural science was taken to be the exemplar of all knowledge. Historians or social scientists of our own time who have supposed that by and large the results of enquiry in their own disciplines would in the end exhibit the same logical and conceptual structure as the results of enquiry in the natural sciences have usually borrowed or adapted versions of Hume's or of Mill's account of causality from those earlier episodes. Consequently they have been forced to suggest some key role for law-like generalisations within the structure of their own disciplines. To such scholars it must be a source of discomfort to realize that what the social sciences as a matter of record

Manninen and Tuomela (eds.), Essays on Explanation and Understanding, 137–158.

do achieve by their enquiries is so rarely, if at all, the formulation of a warranted law-like generalisation. Some economic generalisations, such as those that underlie the truth that there is no possibility in a capitalist economy of full employment without inflation or those that connect changes in the money supply with inflationary or deflationary tendencies, are perhaps both universal in logic form and warranted. But even their logical status is far from clear, and the predictive power of the theories which embody them is, to say the least, notably less than that of the law-like generalisations of physics and chemistry. Some other social scientific generalisations are certainly framed so as to be universal and law-like; consider for example G. C. Homans' generalisations in *The Human Group*, either in their original version, such as "If the interactions between the members of a group are frequent in the external system, sentiments of liking will grow up between them, and these sentiments will in turn lead to further interactions, over and above the interactions of the external system", or in H. A. Simon's "formalized" version in *Models of Man* $dA(t)/dt = c_1[F(t) - yA(t)] + c_2[E(t) - A(t)]$. The only problem is that, as Stanislav Andreski has forcibly pointed out, they are unquestionably false and no one but a professional social scientist would even have been tempted to believe them.

The result is that most writers on the philosophy of social science appear to be analyzing the structures of a highly rigorous and organized, but unfortunately entirely imaginary, as yet to be invented set of disciplines rather than the messy actuality of the human sciences as they now exist. Confronted with this gap between analytical ideal and scientific reality, some other writers have adapted one of two alternative strategies. They have either denied that the social sciences and history yield genuine causal knowledge at all or else have claimed that the concept of causality employed in the social sciences and history is somewhow radically different from that employed in the natural sciences. Both these strategies ought to be rejected. It is clear on the one hand that the social sciences do yield some causal knowledge: a rapid decrease in the monetary supply in a modern capitalist economy after a period of undue expansion will generally cause a substantial increase in unemployment; other relevant factors being unchanged, increasing the detection rate of modern urban crime tends to decrease the crime rate; Robespierre was executed and his execution produced the events of Thermidor, *because* enough votes for his execution were available *because* a fear that Robespierre himself was planning a Thermidorean reaction had spread. It is clear on the other hand that it is quite implausible to give two different accounts of causality, one for nature and one for society. This is not just because on general semantic grounds we ought to be reluctant to multiply meanings of the word 'cause'; it is because so often we have to enquire as to the cause of a given phenomenon, while not yet knowing whether the cause is natural or social, and so often we have to frame hypotheses which invoke natural causes in such a way as to exclude social causes, and vice versa. This is sufficient to show that our presumption ought to be that any adequate account of the concept of causality will have to apply both to natural and to social enquiry; for otherwise we could make no sense of many questions which we not only ask, but even answer.

If then we have to reject both these strategies, why did they fail? The proponents of both took the view – as I shall argue, rightly – that our best understanding of human events does not characteristically depend upon any knowledge of law-like generalisations; but they also unfortunately shared the view – as I shall argue, mistakenly – that the only scientific concept of causality is one that must be explicated in terms of law-like generalisations. So all-pervasive indeed has this view come to be that the more philosophically minded practitioners of those disciplines where we should hopefully expect some discomfort with the conventional philosophical orthodoxy, such as history, often turn out themselves to have been educated into the belief that the first word on causality was said by Hume and the second, if not the last, by J. S. Mill. They thereby allow their historical work to be informed by their philosophical views rather than bring those latter views to judgment in the light of their experience of history. As a result even concrete historical debate comes to be informed by conceptual commitment of a distorting kind. Consider how concrete historical debates do in fact proceed.

What caused the American Civil War? The abolitionist answer was: slavery. The Southern answer was: the violation of states' rights by the federal government. Lincoln's answer was: the attempt to destroy the Union. To these partisan answers the professional academic historian is apt to retort, with D. H. Fischer, that "there is no such thing as *the* cause ...." What occurred – and what on the view of such academic historians always seems to occur – is that a number of antecedent conditions coincided and produced an outcome for the production of which each alone would have been insufficient. This view that "there is no such thing as *the* cause", has an honorable ancestry. It is to be found in Marc Bloch's assertion that "the monism of cause can be, for history, only an impediment". But even Marc Bloch does not raise the question of whether this view was genuinely derived from studying the explanations given by historians rather than borrowed from a type of philosophy whose pre-occupations are alien to those of the historian.

The view that there is and always must be a plurality of causes I shall call causal pluralism. Its advocates see themselves as opposing not merely the enthusiastic oversimplifications of partisans, but also the oversimplifications of any view which on theoretical grounds attempts to identify some single ultimate causal agency. Marxism-Lenism has been the paradigmatic example of what causal pluralists have cited as a distorting and oversimplifying causal monism. In other periods Hegelianism or the theological history of Christianity or Islam might equally well have provided them with their favorite target. But since they have tended to concentrate for reasons that are not obscure upon one variant of Marxism I shall follow them in their choice of example.

Consider the contrast between partisan explanations, pluralist explanations and Marxist explanations of the First World War. The partisans cited, depending upon their point of view; the arrogant expansionism of imperial German foreign policy after the fall of Bismarck; the Slav menace to European culture (as envisaged in the drawing rooms of Berlin in 1912); the crimes of Serbian nationalism; Sir Edward Grey's secret diplomacy; Germany's unprovoked aggression against Belgium; and so on (each of these being taken to exclude all or most of the others as causes or as *the* cause). Pluralists have tended to keep all these items – J. M. K. Vyvyan's chapter entitled 'The Approach of the War of 1914' in Volume XII of *The Cambridge Modern History* (1960), which uses such words as 'doctrinaire' and 'dogma' of those of whom causal pluralists disapproved – is a splendid example of this. But some causal pluralists of course have added other items such as the rigidity of the German railway time-table (which, according to A. J. P. Taylor, played such a key part in the German plans for a swift mobilisation, and the consequent defeat of France within sixty days, that at a crucial moment the choices facing the German government were severely limited). Against both partisans and pluralists Marxists-Leninists have argued that the cause of the First World War lay in the nature of imperialism, highest and final stage of capitalism.

Pluralist historians attempt to treat the issue between themselves on the one hand and partisans and Marxists on the other as though it is a relatively simple empirical issue to be settled by a straightforward recourse to the facts. They picture themselves as those who have tried to consider *all* the evidence and their opponents as those who have focussed narrowly and arbitrarily upon one issue at the expense of others, ignoring the variety of causes which empirical considerations thrust upon our attention. The vast majority of Anglo-American academic historians seem to be both in their practice and in their relatively rare moments of theory causal pluralists and the history of history over the past fifty years has shown them to be endlessly hospitable to new types of candidate for causal efficacy. To the political has been added the economic, to the economic the social, to the social the intellectual and most recently of all the psychohistorical. The invocation of some new type of cause is rarely at the expense of the existing array of causes. Luther's toilet training easily coexists with capitalism as a cause of the Reformation in pluralist versions.

What I am going to argue is that the conclusions of causal pluralist historians are in key part determined not by the evidence, but by the concept of causality which informs their work and which is in no way derived from that work – and that to that degree their position is *a priori.* Both partisans and Marxists are divided from pluralists by their use of a different and, as I shall argue, more adequate concept of causality, as well as by their empirical findings. I do not of course want therefore to assert that any particular Marxist or partisan explanation is true and I do not want to deny that many pluralist explanations are correct. But when this is so it is in spite of, and not because of, the *a priori* framework of causal pluralism.

The concept of causality which informs the theoretical standpoint of pluralism is that which I have already identified, derived in part from the accounts of causality given by Hume and Mill and in part from the patchwork of emendations which has emerged from attempts to deal with the more obvious inadequacies of their accounts. It has three central features. First, it treats causality as a relationship primarily between types of event and state of affairs rather than between individual events or states of affairs. Every singular causal statement is held to exemplify a law-like generalisation of the form: 'Wherever an event or state of affairs of such-and-such a type occurs, then an event or state of affairs of so-and-so type occurs.'

Secondly, an event or state of affairs which is a cause must on this view always satisfy either a necessary condition or a sufficient condition or both for the occurrence of that event or state of affairs which is its effect. The notions of necessity and sufficiency are interdefinable. To say that the occurrence of an event or state of affairs of type *A* is a necessary condition for the occurrence of an event or state of affairs of type *B* is to say that if the former occurrence does not take place then the latter occurrence will not take place either; and to say that the occurrence of *A* (similarly understood) is a sufficient condition for the occurrence of *B*, is to say that if *A* does occur, *B* will also occur. It follows that if the occurrence of *A* is a necessary condition for the occurrence of *B*, then the non-occurrence of *B* is a sufficient condition for the non-occurrence of *A*; and that if the occurrence of *A* is a sufficient condition for the occurrence of *B*, then the non-occurrence of *B* is a necessary condition for the non-occurrence of *A*.

Thirdly, as is evident from the preceding points, causality on this view is taken to be essentially a dyadic relation, at one level between particular events or states of affairs, at another level between types of events or state of affairs. So much is this taken for granted that although the nature of the relation between the two terms has been as major a topic for discussion as any in philosophy, the question of whether causality is indeed a *dyadic* relationship has received almost no discussion at all.

At the level of explicit philosophical theorising there are a number of accounts of causality which embody these three contentions. The most cogent are those which have an additional feature. To identify causes and effects is not just to identify actual or hypothetical successions of events; it is to identify points at which human agency did or could have made some effective intervention in the natural or social world. It is because of this that we learn what causes have what effects much more by our successful and unsuccessful interference with the environment than just by passive observation. This insight led D. A. T. Gasking to understand causal explanations as symmetrical with recipes for the production of effects – the recipe is prospective, the explanation retrospective. Professor G. H. von Wright was led to formulate a similar account of causality by pursuing a different line of argument. Because the relations which hold between types of events defined only in terms of necessity and sufficiency are symmetrical, but causal relations are asymmetrical, von Wright supplements an account in terms of necessity and sufficiency by one in terms of the notions of action and intervention. So he concludes that: "*p* is a cause relative to *q*, and *q* an effect relative to *p*, if and only if by doing *p* we could bring about *q* or by suppressing *p* we could remove *q* or prevent it from happening". (*Explanation and Understanding*, p. 10.) '*p*' and '*q*' are of course in this account names of types of state or event, not names of particular states or events.

It is going to be my contention that something very like von Wright's account is presupposed by both the practice and the theory of pluralist historians. But in order to make this claim a necessary preliminary is to show that an alternative account of causality is possible. I shall try to show this by arguing that an alternative account is necessary since, so I shall argue, any account which embodies the features that von Wright's embodies is bound to break down.

Consider first of all the difference between two kinds of link that may occur in causal chains, either of which might be invoked to explain the production of a particular sample of mercuric oxide. On the one hand there will be those links constituted by events in the sequence of sub-atomic relationships and interactions which constitute the coming together under certain conditions of a particular sample of mercury and a particular sample of oxygen. Each of these links will be specified with reference to some out of a body of law-like generalisations (although I ought to note that later in this argument the role of such generalisations even in this kind of account will have to be reexamined). Consider by contrast however a quite different type of answer to the question: "What causes this particular sample of mercuric oxide to come into being?" The professor of chemistry quarreled with his wife at breakfast; as a result of prolonging the quarrel he arrived too late at the laboratory to correct the laboratory technician's misunderstanding of the instructions the professor had given the night before. So in error the technician manufactured a sample of mercuric oxide. None of the links in this chain is specified with reference to any law-like generalisation at all. But our knowledge of these links is no less well-founded than our knowledge of subatomic sequences in chemistry. To know that the professor was late because he quarreled with his wife or that the technician manufactured the mercuric oxide because he misunderstood the professor's instructions we require no knowledge of any relevant law-like generalisations, and if we did acquire such knowledge our knowledge of the particular causal relations would be no better warranted. This is not because we are dealing here with events in the human rather than in the subatomic world; we often in *both* the natural and social world identify and understand particular causal relationships without invoking law-like generalisations. Philosophers who have clearly recognized this, such as Donald Davidson, have however still insisted that when we assert that a particular causal relationship holds we imply the truth of some generalisation even if we cannot formulate it. But what grounds could there be for maintaining this? They seem to be something like this: if two different occasions in apparently precisely similar environments what was apparently precisely the same causal agency produced two different effects, then we should have to look for some actual difference in the two apparently similar situations. But to say this is to state one often useful rule of method for finding out causes, not an implication of all assertive uses of the word 'cause' and its cognates. Moreover, it is a question of contingent empirical fact whether this rule of method will on a particular occasion yield results or not. We do not know *a priori* that it will and we therefore do not know – what Davidson's view requires us to know – that for every particular causal connection which we identify there is some law-like generalisation waiting to be advanced.

This detachment of our knowledge of particular causes from our knowledge of generalisations is of crucial importance in those areas such as history and the social sciences where many particular causal connections are open to view, but few, if any, genuine law-like generalisations. Very often when we try to improve our knowledge of particular causes in the area of human agency we do have to formulate and rely on generalisations, but a very different type of generalisation from that of the law-like generalisation of natural science, what we may call a 'By and large and for the most part...' type of generalisation which picks out characteristic rather than universal ways of behaving. This is the kind of generalisation about human behaviour on which Theodore Zeldin relies when he argues that the varying sense of community in different French villages in the nineteenth century was the effect of different ways of working the land and of different local traditions. (*France 1848–1965*, pp. 135–140). Zeldin does not need to appeal to any alleged law-like generalisation to establish his claims about particular causal chains in particular villages. Nor is it easy to see how discovery of any such would improve his or our understanding of what he has to teach us about the particularities of French peasant behaviour.

A second point at which the von Wright account of causality becomes questionable is when we realize that to identify a cause is not usually to identify either a necessary or a sufficient antecedent condition. Consider the example of a court trying to determine the causes of an automobile accident. They will consider as putative causes such factors as the degree of skill possessed by the driver, the patch of oil on the road, the bad light at that time of day, and so on. Now clearly whatever the court is doing it is not compiling a list of necessary conditions. For were it to do such it would have to include in its list such items as the invention of the automobile and the fact that the driver's parents met at least once.

The usual response to such considerations is to suggest that the court is engaged in making a selection from the total set of necessary conditions. But the crucial question to ask is: are we here dealing with necessary conditions at all? To say that the patch of oil on the road did cause an uncontrollable skid and so the accident, is not to say that *that* particular accident – that car coming off that road at that time and place and being damaged to just that degree – would not have happened but for that particular patch of oil. The importance of this point can be brought out be considering Sir Edward Grey's secret treaty commitments to the French as a cause not of the First World War, but of Britain's embroilment in it. It is quite's clear that the Union of Democratic Control were right to claim this as *a*, and some radical members of the U.D.C. may even have been right to claim it as *the* cause of this and much more than this. (The U.D.C. thesis was that the secret British commitment made the French excessively ready to go to war which in turn made the Russians excessively ready to go to war which in turn...). But had the secret commitments carried through by Grey not been made – if Grey had died for example before making them – would Britain not have backed up the French in 1914? It is quite clear that if we withdraw from history all Sir Edward Grey's actions a number of other paths could have led and were in 1905 – before any commitments had been made – equally likely to lead, on any mode of reasoning about probabilities, to the same outcome in 1914. It follows that in the case of such causes – a class which includes historical actions and events – we are not asserting when we claim that one event was the cause of another that the former satisfied a necessary condition for the occurrence of the latter.

To my use of this example a proponent of the conventional view might retort that if Sir Edward Grey's diplomatic action did not satisfy a necessary condition for British involvement in 1914, then that merely shows that British involvement was overdetermined – that Sir Edward Grey's activities did not satisfy a necessary condition for involvement, not because causes are not necessary conditions, but because other occurrences satisfied these conditions. But this retort would be based on a misunderstanding of the intended force of the example. If this were indeed a case of overdetermination, the contention would be that if we removed Sir Edward Grey's diplomatic activities from history, everything subsequent – which we now take to be the outcome of Grey's activities – would remain as it was; whereas this wildly implausibe contention is not what I am putting forward. What I am advancing is the view that if Sir Edward Grey's diplomatic activities had not occurred then some other events – which did not in fact occur – *might* well have led to British intervention in 1914. But if this is so, then Sir Edward Grey's diplomatic activity, although certainly in actual fact a cause of what occurred in 1914, equally certainly did not satisfy any necessary condition for the outbreak of war. Hence the notion of causality has application independently of the notion of a necessary condition.

But clearly if the notion of cause is independent of the notion of necessary condition, it cannot be elucidated in terms of it; and equally clearly if the notion of cause is independent of the notion of necessary condition then it is independent of the notion of sufficient condition too, if only because necessity and sufficiency are interdefinable. But it is worth underlining this latter point so far as historical events or actions are concerned.

Consider the causal chain of events leading from the decision by the Black Hand Organisation in Belgrade that Archduke Franz Ferdinand should be assassinated during his tour of Bosnia to the actual assassination by Gavrilo Princip at Sarajevo on June 28, 1914. Given the first event, in no sense did the second have to follow. Up to the moment that Princip pressed the trigger, we have no grounds for treating those events that were causes as satisfying a sufficient condition for the occurrence of the assassination. Notice too that the same is true of many chains of natural, non-human occurrences. Up to the moment at which the avalance kills the mountain cat which has strayed across its path, there has occurred no event which satisfies a sufficient condition for the occurrence of the mountain cat's death in just that way.

We are now in a position to understand why the alleged symmetry between causal explanations and causal recipes, embodied in the views of Gasking and von Wright, is misleading. From the fact that a particular cause produced a particular effect nothing whatsoever follows about how in general that effect can be produced. Moreover a recipe is normally disjunctive in form: to produce that effect, bring about either this cause or this cause or this one. But causal explanations embody nothing that corresponds to this disjunctive form. Of course in cases where our knowledge of particular causes producing particular effects is accompanied by a knowledge of causal laws, there is indeed a symmetry between laws and recipes. But to emphasize this is to emphasize the gap between our knowledge of causes and our knowledge of laws.

It is perhaps because causes in nature and society have been seen as the mere instantiation of laws that the elucidation of causality in terms of generalisations and of necessity and of sufficiency has prevailed for so long. It is perhaps for the same reason that causality has been envisaged as a dyadic relationship. That it is not can be understood from the following considerations. We never in citing a cause simply seek to explain why a particular revolution or famine or war happened; we seek to explain why that revolution or that famine or that war happened rather than something else. Any historian who adheres overrigidly to the canon firmly proclaimed by a number of modern historians, that the historian ought to take no interest in what would have happened if what did in fact happen had not happened, has in fact abjured the identification of causes altogether. But to discover what difference a putative cause made, if any, involves identifying that to which it made a difference, that ongoing state of affairs which, but for the intervention of that particular causal agency, would have produced some alternative outcome. To give a causal explanation we therefore need at least four related terms: there is first of all that which intervenes, secondly that state of affairs which is interfered with by the intervention, thirdly the actual effect of the intervention and fourthly the outcome that would have prevailed but for the intervention. Causality is a relationship between at least four items, not two.

Obviously my account so far has much in common with Hart and Honoré's view in *Causation and the Law*, and one way to develop my account further is to examine the points of agreement and disagreement with Hart and Honoré. What is most valuable in their account is something that they share with Gasking and von Wright, the attempt to understand causility primarily in terms of human activity. Even although there is not the necessary symmetry between explanations and recipes supposed by Gasking and von Wright, there is a crucial analogy between causal knowledge and practical knowledge. An agent looking forward always has to distinguish three features in his situation: that which he wishes to change, to abolish or to make, that which provides the relatively unalterable and invariant context for his projected actions and that which affords him a means for effecting the desired change. So he distinguishes in retrospect the cause of the change which he effected from the mere conditions which provided the context for his causal transaction. This distinction between causes and conditions is as crucial retrospectively as it is prospectively, and it is this distinction which Hart and Honoré are able to mark by their contrast between the ongoing background situation and the intervening cause. Insofar as they thereby make central to their account the difference between what the cause did bring about and what would have happened otherwise their view is one to which I am substantially indebted. Where their viewpoint is more open to question is on their specific characterisations of the background situation.

They characterize this in terms of a contrast between what is normal and expected and what is abnormal and unexpected. Since they are specifically concerned with causation as regarded by the law, this characterisation is in some ways unsurprising. For legal systems always do include some normative characterisation of normality: if we wish to identify as the cause of a railway accident the failure of a signalman to perform his ordinarily assigned duties, we do so by identifying this as the only relevant breach in the normal and to be expected workings of the railway system. But neither the natural scientist nor the historian is able to rely on some uncontested definition of normality in the way that the interpreter of the law is. This is of course so far not yet to say that the Hart and Honoré analysis is mistaken, but only that its application is restricted. However, there is a way in which it seems to be simply mistaken.

Hart and Honoré understand the nature of a cause in terms of an interference with ongoing regularities of some background. But it is clear that the regularities or trends with which such causes interfere are themselves sometimes constituted by non-causal sequences – such as the sequences of chess-playing – and sometimes by causal sequences – such as the sequential positions of our planetary system. We cannot therefore elucidate the nature of causality by refering to the interfering agency alone. We need an account of causality which will allow both that which is interfered with and the interfering agency to be understood as causes.

What then are the requirements that a more adequate account of causality must meet? First it must enable us to distinguish between causes and conditions; it must do so by exhibiting a cause as that which makes *this* happen rather than *that* which would otherwise have happened. A cause is what makes a difference. Secondly we must preserve that particularity of causality for which I argued earlier and avoid presenting causal connections as mirror images of causal laws. Any account of causality must of course be compatible with our understanding of causal laws, but it must preserve the asymmetry of recipes for change and causal explanations. Thirdly we must reformulate the insight contained in both the Gasking-von Wright view and the Hart and Honoré view, that the concept of causality is intimately linked to that of human agency. In the light of these requirements the following partial and incomplete account of causality, agency and laws suggests itself.

When an agent performs the same action on two different occasions or when two agents perform the same action, the same event does not always take place in the social or natural world. You bake bread and it rises; I bake bread and it does not. From the standpoint of all the intentional descriptions of the actions available we did precisely the same things; but what happened in the world were two different things. The primitive notion of a cause is the notion of what makes that difference between two events where there is no difference in the corresponding actions. The *primitive* discovery of causality is not, as Hume's account, let alone Gasking's or von Wright's, would suggest, the discovery of the powers and possibilities of human agency, but the discovery of the limitations upon these powers.

It does not follow that our account cannot transcend the limitations of that primitive discovery. More particularly it does not follow that human agents themselves cannot be causally efficacious, if only because what makes the difference in what happens when two agents do the same thing can often be traced either to the actions of a third agent or to the other actions or properties of one of the two agents. Moreover what interferes with success, what limits human powers, can itself be interfered with. A cause is what makes *any* outcome different from what it would otherwise have been. Such an outcome is always the product of the conjunction of the causal agencies already at work and some intervening cause or causes.

This account preserves from the Hart and Honoré view the crucial distinction between what was going to happen if some particular cause had not intervened and the causal intervention. It is of course not a complete analysis of the concept of causality. But it offers a starting point for developing the nature of the contrast between intervening causes on the one hand and what I called just now "causal agencies already at work" on the other. Agents intervening in the world – and for that matter gas clouds moving into planetary systems or rainwater impinging on granite – do not just encounter causal chains, they also encounter causal orders. By a causal order I mean an interrelated set of items, such as a planetary system, or a crystalline structure, or some types of educational system, where the relations between the items can only be formulated in terms of some type of generalisation. What these generalisations specify is a bond between the items of a certain type. It is important that bonds themselves function as causes, providing a background order in interfering with which intervening causes may be effective. Consider two examples.

A gas cloud enters a planetary system and instead of distorting the planet's trajectories is dispersed on its encounter with them. The gravitational bond between the planets was too strong. The precise relationship between the mass of the planets and the gravitational attraction was the cause of the gas cloud failing to alter the planets' courses. Or two children of different social class and of equal abilities enter the same school system with the same degree of motivation and parental encouragement. It is a school system at a time when and a place where the generalisation 'Social class determines educational opportunity' holds. The working-class child leaves school at fourteen; the middle-class child goes to a university. What caused the difference in outcomes was the precise relationship, the bond, between social class and educational opportunity.

It follows from this view that laws do not specify causal relationships – relationships, that, is between particular causes and particular effects – at all; they specify rather certain types of relationship in the world which can themselves be causally efficacious. Those relationships or bonds are of three types. There is first the type of cause where the bond is unbreakable. Examples of this include the key relationships in relativistic physics, but also the relationships between pressures, temperatures, and volumes specified in the gas law equations. There are not in a crucial sense, two events, one the change in temperature of a particular sample of a particular gas at a particular time, the other the change in pressure-cum-volume states. The relationship between the two aspects of the situation is not merely contingent, in this sense, that there is no way of raising the temperature, but not changing its pressure-cum-volume state, no way of interpolating some factor to make temperature changes independent of pressure-cum-volume changes.

This means that the generalisations of the gas law equations are very different from those of Newtonian mechanics. For these latter link antecedent and consequent contingently, so that if an event satisfying a particular antecedent occurs – two bodies approacing each other in certain specific ways – the outcome predicted in the consequent can always be prevented by introducing a third moving body into the situation. Yet the set of generalisations that constitute Newtonian mechanics specify not merely the links between antecedent and consequent events, but the precise nature of the intervention that would have to occur to alter a predicted outcome to some other specific outcome. Given the occurrence of a Newtonian antecedent event or state on a particular occasion the standard predictable Newtonian consequent event or state is not inevitable; but will only *not* occur if some other Newtonian antecedent appears on the scene. We find therefore both non-contingent bonds and contingent bonds of this kind in nature; but we also discover a type of bond where we can, over a certain unspecified range, link properties of events or states of affairs by generalisations, but without being able to specify the nature of the intervention necessary to breach the link. This type of bond, whose strength or fragility in respect of a variety of types of intervention is unknown to us, except as the result of our actually intervening, is exemplified in such generalisations about particular stretches of time and place as 'Social class determines educational opportunity'. The bondings of the social world are in general of this third type. Nor ought this to be surprising. The relation between gas pressure and gas temperature allows of no intervention; planetary systems and other large scale mechanical systems have a very high degree of contingent immutability; but human social orders are enormously mutable and appear vulnerable to many different types of intervening cause.

If this is correct laws do not specify causal connections and the Hume-Mill account of causality – with its numerous progeny – depends upon not distinguishing adequately the role of law-specified relationships, of what I have called bonds, as causes in the world from the role of laws in theories. That hallowed formula 'Whenever an event or state of affairs of type *A* occurs, then an event of state of affairs of type *B* occurs' *never* by itself specifies any possible causal relationship. It is indeed already at fault in not distinguishing the gas law equations from the generalisation of mechanics; but worse still, except in the gas law equation type of case, it yields no predictions whatsoever even when to it as major premise, such a minor premise as 'An event of type *A* occurred' is added. For we need to know what other events are occurring in the space-time neighborhood to make any predictions. For similar reasons the two premises cannot specify a causal relationship, a fact often implicitly conceded, as is their lack of predictive power, by the ritual invocation of the phrase 'in standard conditions'. But if all this is so, then orthodox philosophy of science from J. S. Mill to the Vienna Circle, is nothing but a series of unfortunate fictions.

It follows from my arguments that theories may be deductive, but explanations are not, for explanations are concerned with particular causes and particular effects. This obviously has important implications for the philosophy of physics and chemistry, but the concern of my present argument is after all with the understanding of human causation, of causation in history and the social sciences. In all historical sciences – and this point can be made as aptly about physical cosmology or geology or parts of medicine as about human history – we have to make the already emphasized distinction between causal interventions and the pre-existing causal order into which intervention is made, successfully or unsuccessfully, if we are to make any progress with problems of constancy and change that are central to such sciences.

Consider for example the explanation of a particular patient's recovery from pneumonia by the use of antibiotics. We explain the cure by first giving an account of the standard development of pneumonia as exemplified in this patient, using this account also to specify what would have occurred if the antibiotics had not been administered. The standard development of pneumonia constitutes a causal order, the bonds between certain properties being causally efficacious in producing each succeeding state. The antibiotic dissolves the bond and the causal order is overthrown.

Just as the cure is explained, so is the onset. The healthy lung in its succession of states constitutes a causal order on which the virus intervenes. Hence if we are to explain the patient's state at any given moment we shall find ourselves involved in an hierarchical form of explanation in which the cured but damaged lung is the outcome of the impact of the intervening cause, the antibiotic, on the causal order of pneumonia and the pneumonia in turn is the outcome of the impact of the intervening cause, the virus, on the causal order of the healthy lung. Explanations in historical sciences are thus hierarchical and non-deductive; and I take this form to be characteristic.

In this form of explanation the notion of a causal order whose strength or fragility is a matter of the causal efficacy of the bonds between the items that compose it is a crucial one. It is only where we can confidently identify a continuing causal order that we can say with any confidence what would have happened if the intervening cause had not in fact intervened. Otherwise uncertainty as to the alternative outcome must have its counterpart uncertainty as to how far, or even whether, we should rate the intervening cause as a cause at all.

It is important to notice at this point that of course what impacts upon and alters the course of the members of one causal order may be the members of another. A planetary system and a gas cloud are both causal orders; the social structure of Ashanti in the eighteenth century and the institutions of the slave trade are both causal orders. Which we take to be the intervening cause and which the ongoing regularities of the background will depend on some other criterion. Such a criterion is provided in all historical sciences by what we take to be the subject of our historical narrative. If we treat the development of Western trade as the subject of our narrative, then it is the discovery of West Africa which impacts on an existing situation; if we treat the development of civilization in Africa as our subject, then it is the slave trade which interferes with the *status quo*. In assigning causes in history therefore – and that is to say, in human affairs – we need an even more complex scheme than I have hitherto suggested.

Historians themselves are apt to treat the question of the choice of subject of narrative as being a matter or the preferences of the individual historian. One may choose to write the history of international trade, or part of it; another may choose to write the history of African civilization, or part of it. But one choice, so the orthodox academic view runs, is in no way superior to the other. Everything depends on what the individual historian happens to be interested in. This academic anarchism has as its consequence a reinforcement of causal pluralism. Marc Bloch quoted with approval Simiand's thesis that "For a doctor the cause of an epidemic would be the multiplication of a microbe and its conditions the dirt and ill health occasioned by poverty; for the sociologist and the philantropist, poverty would be the cause, and the biological factors, the condition". Bloch added: "This is in all honesty to acknowledge the subordination of the perspective to the peculiar angle of the enquiry."

It is beyond the scope of the argument so far to ask whether there are objective criteria for judging that of different types of narrative some are to be subordinated to others. But we ought at least to notice that if this were so, then the professional academic historian's conventional view would be one more source of error for academic history. But to develop such an argument would require a further consideration of the nature of narrative.

For the moment let me instead return to the debate about the First World War. The causal pluralists all represent the causal relationship as dyadic and causes themselves as all classifiable as necessary or sufficient. By this initial error they are forced to treat each particular relevant antecedent condition which they have identified as itself necessary, but not by itself sufficient, and so they enter into the process which I described earlier of adding cause to cause. Where they disagree most often with each other is in what they speak of as the weight to be assigned to different causes. The implicit metaphor is clear: it is that of a spring balance, to one side of which pieces of metal of varying weight are being added until the other side is raised. This additive view at once entails a multiplicity of causes, rules out any hierarchical structure for causal explanation and therefore any identification of 'the' cause. Given that some analogue of the Gasking-von Wright view of the concept of cause is implicit in these historians' causal pluralism, it is not surprising that what appeared to them and to philosophers who shared their views to be the central problem of the philosophy of history was the problem of the role of law-like generalisations. Hence the whole Hempel-Popper-Gardiner-etc. discussion. For it appeared to all these writers to be the case that *either* we could identify non-trivial generalisations *or* we could not offer historical explanations of any substance. But this consequence depended on the view now seen to be erroneous that explanations are deductive or quasi-deductive in the way that theories are.

If then causal pluralism approaches the First World War, like any other event, within a framework derived from this philosophical view, it is not surprising that pluralist historians come to the conclusions that they do. Consider by contrast the procedures of both partisans and Marxists. I choose as examples of partisan views those of Weber and Durkheim. Weber's view, revealed in letters even more than in public utterances, was that the order of European civilization, a fragile but important causal order in which Germany played the key part, was threatened by the intervening cause of expansionist Slav barbarism. Durkheim's view, expressed in a pamphlet designed for neutral consumption as well as in private letters, was that expansionist, aggressive German policies were the intervening cause threatening the fragile causal order of modern European civilization in which France played the key part. Neither Weber's nor Durkheim's partisanship are in retrospect reputable. (One Jaurès is worth a hundred Durkheims; one Liebknecht a hundred Webers.) But the logical structure of their arguments remains important. For both make it clear that how we identify the intervening cause depends upon how we identify the pre-existing causal order. The contrast with causal pluralism is complete; for causal pluralism has no concept of a preexisting causal order and therefore no possible concept of *the* intervening cause. The contrast with Marxism also becomes clear.

The Marxist-Leninist thesis against which causal pluralists have so inveighed can now be understood as follows. The proximate problem in the approach to the First World War is *not*: what were the causes of the war? It is: why did all the attempts to stave off a great European war have to fail in the end? The Marxist does not see the war caused by some external aggressive intervention upsetting a pre-existing stable order; he sees a pre-existing causal order moving by its own dynamics towards war in such a way that no peacemaking cause can intervene to prevent it. There may well be a problem on the Marxist-Leninist view in explaining why it took so long for a great European war to break out; but to look for the causes of that war in terms of a set of individual items such as diplomatic episodes or assassinations is to look in the wrong place.

The Marxist-Leninist thesis thus does have a criterion for identifying 'the' cause. But it is not of course clear that the Marxist view is true. What the historical debate would have to focus on next would be whether Marxism-Leninism did or did not correctly identify the pre-existing causal order. If it did, then we can enquire further about the production of this particular causal order, moving a step further in the hierarchy of explanation. But whether it did or not depends on an empirical investigation of the Marxist-Leninist claims about the bonds which hold together the phenomena of imperialism and capitalism.

I have followed the pluralist historians in using a Marxist-Leninist thesis as an example in this argument. But Marxism is of course only one out of a whole family of historical methodologies whose practice runs counter to that of causal pluralism. The virtue of the Marxist-Leninist thesis about the origins of the First World War as an example for my argument lay in its clarity; its disadvantages spring from the fact that it is perhaps too historically contentious on other grounds. But at least from its use we may learn this: that if Marxism fails, it fails as history and not just as philosophy, and any approach that seeks to refute it, or to replace it, or to claim its inheritance must also succeed as history and not just as philosophy.

Nonetheless it will be instructive finally to consider a rather different type of example. I have used the expressions 'intervening agency' or 'intervening cause' in order to contrast what these characterize with what I have called 'the ongoing regularities of the background' or 'causal orders... already at work'. This vocabulary is very much at odds, at first sight, with that used for instance in his account of the causes of the English civil war of the seventeenth century by Lawrence Stone. Stone distinguishes three types of cause. There are *pre-conditions*; an example is "structural weaknesses of government finance in England" in the seventeenth centure. There are *precipitating causes*; Stone's example here is "the extra-legal taxation of the 1630s". And there are *triggers*; Stone's example is "the financial collapse of 1640", which led to the summoning of the Long Parliament. Stone's classification of causes thus moves from the long-term and the general to the short-term and the particular; so much so that in his review of Stone's book Professor H. G. Koenigsberger (*Journal of Modern History*, March 1974) was able to suggest that no more was achieved by Stone's classification than a more familiar – we might add, banal – two-fold classification into 'long-term' and 'short-term'. Stone's reply to Koenigsberger is argumentatively weak, but illuminating, for Stone makes it clear that he believes that his three-fold classification was a translation and an expansion of the causal vocabulary of that most eminent of French historians, Fernand Braudel, who distinguishes *structures* from *conjunctures*. What I want to contend is that Stone himself has misunderstood and mistranslated Braudel, and so laid himself open unnecessarily to criticism.

In moving from *structures* to *conjunctures* we may of course in fact move from the long-term to the short-term (hence presumable Stone's misunderstanding); but what is crucial in Braudel's conceptual scheme is the movement from the structure of some causal order already in being to the point at which some intervening agency impacts on it so that the conjunction of background and intervening agency produces an outcome other than that which would otherwise have occurred. Stone's classification has the result that in his theorizing, although not of course in his excellent narrative, he misses the connection between the causal factors which supply the answer to the question of why the structure of government finance in England collapsed at the time and in the way that it did rather than at some other time and in some other way, a connection which Braudel's vocabulary is well designed to place in the foreground.

If we look at the practice of Braudel, and at that of his French colleagues who have more recently been associated with *Annales*, we shall find that – whatever their theory – their practice exemplifies in a high degree the treatment of causality as singular for which I have argued.

The product of their work could not be justly represented in terms of any set of law-like generalisations. On the contrary, if there is a danger, it is that we are all too apt to be submerged in their writings by a mass of singularities and particularities which sustain only cautious and limited comparisons. If we are to escape the danger of being so submerged, then the related work of a Marxist historian such as Jean Vilar – one of the relatively few Marxists who has repudiated in his practice the generalizing scientism which infects Marxists as different as Engels and Althusser – perhaps provides the kind of guide that we need. For these writers focus our attention on the relation of causal explanation in history to the writing of a certain kind of narrative. To carry the argument further, we would have to examine in depth the nature of historical narrative.

Boston University

ALEKSANDAR KRON

AN ANALYSIS OF CAUSALITY

1. Introduction

The basic aim of this paper is to give a brief sketch of a formal theory of causal relations. More precisely, our aim is (a) to show how the language of the first-order predicate logic can be applied to an analysis of this kind and (b) to discuss some both technical and philosophical aspects of such an analysis.

Before we develop the formal machinery, let us describe the motives for introducing some technical details to be given later.

If 'causation' is a name of a relation between 'cause' and 'effect' at all, then we must ask what are the individuals which such a relation can hold for. We suppose that causal relations are defined for (1) states of affairs and (2) for changes of states of affairs. Let s_1, s_2, s_3, and s_4 denote states of affairs. Then we can say that s_1 is a cause of s_2 and that s_2 is an effect of s_1; also, we can say that the change from s_1 to s_2 is a cause of the change of s_3 to s_4 and that the change of s_3 to s_4 is an effect of the change of s_1 to s_2. From (1) and (2) some other important meanings of 'cause' and 'effect' can be derived. We mention here only two of them. From (1), as it will be seen later, we can derive a meaning of the phrase 'The property P is a cause of the property Q'. From (2) we derive the meaning of the phrase 'The change c_1 of the individual a is a cause of the change c_2 of a'. It can be seen that other meanings of 'cause' and 'effect' can be derived as well, but we do not claim that our analysis exhaust the whole field of such possibilities.

The two main problems in our analysis are how to understand a (possible) state of affairs and how to represent it within first-order logic or model theory, and how to define that a state of affairs has changed to another state of affairs.

The first of our problems can be solved in various ways. Whevener we talk of a state of affairs, we do that only with respect to some specified relational structure $\mathfrak{A}$. Then a solution of our problem along Carnap's

Manninen and Tuomela (eds.), Essays on Explanation and Understanding, 159–182.

line would be the following one. Let L be our first-order language containing an individual constant for each individual in $\mathfrak{A}$, and a relational symbol for each relation in $\mathfrak{A}$. Then the diagram $D(\mathfrak{A})$ of $\mathfrak{A}$ could be taken as a description of the actual state of affairs with respect to $\mathfrak{A}$. (We use the notation of Bell and Slomson's *Models and Ultraproducts*.)

There are many difficulties involved in such a definition. For example, in this case L might be uncountable. Furthermore, in scientific and ordinary discourse we neither require that a description of a possible state of affairs contains only atomic sentences nor that it contains all sentences of $D(\mathfrak{A})$. Hence, in order to simplify the matter, we shall suppose that any set K of first-order sentences defined in $\mathfrak{A}$ such that $\mathfrak{A} \models K$ partially describes a state of affairs with respect to $\mathfrak{A}$. For example, if $\mathfrak{A} \models \exists v_1 A$, then $\exists v_1 A$ 'says' that there is an individual in $\mathfrak{A}$ such that A is true for it. But this is a 'fact' about $\mathfrak{A}$, isn't it? Consequently, $\mathfrak{A}$ itself may be considered as a complex state of affairs and any set K of sentences such that $\mathfrak{A} \models K$ as a partial description of it. Thus, if $\mathfrak{A} \models K$, K partially describes a sub-structure $\mathfrak{A}'$ of $\mathfrak{A}$. Namely, if $\mathfrak{A} = \langle \alpha, R \rangle$, where α is a non-empty set and R a set of relations defined on α, then $\mathfrak{A}' = \langle \alpha, R' \rangle$, where R' is the set of relations 'mentioned' in K. Again, $\mathfrak{A}'$ is a state of affairs partially described by K and 'contained' in the state of affairs $\mathfrak{A}$ partially described by K as well.

What means that $\mathfrak{A}_1$ has changed to $\mathfrak{A}_2$? The consideration of the concept of change is not the main issue of this analysis. But the fact that $\mathfrak{A}_1$ has changed to $\mathfrak{A}_2$ can be represented by a sequence $\langle \mathfrak{A}_1, t_1 \rangle$, $\langle \mathfrak{A}_2, t_2 \rangle$, where t_1 and t_2 are moments of time. Thus, in t_1 there is a state of affairs $\mathfrak{A}_1$ and in t_2 a state of affairs $\mathfrak{A}_2$. If $\mathfrak{A}_1 = \langle \alpha_1, R_1 \rangle$, $\mathfrak{A}_2 = \langle \alpha_2, R_2 \rangle$ and $\alpha_1 = \alpha_2$, we can say that the relations between individuals of α_1 have changed, provided that $R_1 \neq R_2$; if $R_1 = R_2$ and $\alpha_1 \neq \alpha_2$, we have the same relations in t_1 and t_2, but they hold for different individuals.

Consider now a denumerable sequence of ordered pairs of this kind. Such a sequence can be viewed as a sequence of states of a process or a sequence of values of a two-place vector.

Our definition of partial descriptions of states of affairs may seem to be weak: every consistent set of sentences is a description of a possible state of affairs, *i.e.*, it holds in a relational structure. In our further considerations we shall define descriptions of a very special kind which will play important roles in our analysis of causal relations.

Let us ask what means that a state of affairs $\mathfrak{A}_1$ is a cause of a state of affairs $\mathfrak{A}_2$. Suppose that K_1 and K_2 are partial descriptions of $\mathfrak{A}_1$ and $\mathfrak{A}_2$ respectively. Then our question can be answered by considering K_1 and K_2. Namely, $\mathfrak{A}_1$ is a cause of $\mathfrak{A}_2$ if there is a causal explanation of $\mathfrak{A}_2$ in terms of $\mathfrak{A}_1$. This is only an incorrect way of expressing the fact that K_2 can be causally explained in terms of K_1. Of course, such an explanation takes place within a theory T. Hence, the investigation of causal relations must concentrate on T, K_1 and K_2 first. To such an investigation the next part of our paper is devoted.

2. Causal ordering

In a system of linear nonhomogeneous equations H. A. Simon defined an asymmetrical relation interpreted by him as causal dependence. He has developed an appropriate conceptual apparatus and proved a number of interesting theorems concerning the connection between the causal ordering of such a system and the identifiability concept [1].

Following the basic ideas of Simon, we develop an analogous apparatus for an arbitrary set of formulas (wffs) with free individual variables. This apparatus will be used in the definition of a causal explanation.

Let L be a first-order language with individual variables

$$v_1, v_2, \ldots$$

and an arbitrary set of individual constants; if this set is denumerable, we write for its elements

$$\bar{a}_1, \bar{a}_2, \ldots; \quad \bar{b}_1, \bar{b}_2, \ldots.$$

Let T be a closed first-order theory with equality (the axioms of T contain no free variables) such that

$$* \quad T \vdash \forall v_m \exists v_n v_m \neq v_n$$

and such that for every wff $\exists v_n A$ there is an individual constant $\bar{a}_n$ and

$$** \quad T \vdash \exists v_n A \Leftrightarrow v_n = \bar{a}_n.$$

Thus, if $\mathfrak{A}$ is a model of T, then $\mathfrak{A}$ contains at least two individuals. We introduced ** in order to simplify the exposition; it can be omitted if we use a purely model-theoretic language.

DEFINITION 2.1. Let $A(v_1, \ldots, v_{n-1}, v_n)$ be a wff with free variables $v_1, \ldots, v_{n-1}, v_n$; v_n is *real* in A iff

$$T \vdash \forall v_1 \ldots \forall v_{n-1} \exists v_n A \quad \text{and} \quad T \vdash \forall v_1 \ldots \forall v_{n-1} \exists v_n \neg A .$$

If v_n is real in A, then it 'matters' to A. In other words, if T is consistent, then neither

$$T \vdash \forall v_1 \ldots \forall v_{n-1} \forall v_n A \quad \text{nor} \quad T \vdash \neg \exists v_1 \ldots \exists v_{n-1} \exists v_n A .$$

Let K be a non-empty set of wffs containing free variables, let K' be a finite non-empty subset of K, let $V' = \{v_1, \ldots, v_n\}$ be the set of real variables in K', let mK' and mV' be the numbers of elements in K' and V' respectively. Then we give

DEFINITION 2.2. K is *linear* iff for every K'

(a) for every $A \in K'$ there is a real variable in A;

(b) $mV' \geqslant mK'$;

(c) let $k = mV' - mK'$, let $\wedge K'$ be the conjuction of all elements of K', let $\bar{K}'$ be the set of universal closures of elements of K' with respect to $v_{j_1}, \ldots, v_{j_k}$, $1 \leqslant j_l \leqslant n$ and $1 \leqslant l \leqslant k$; if $\bar{K}'$ satisfies (a) and (b), then

$$T \vdash \forall v_{j_1} \ldots \forall v_{j_k} \exists ! v_{j_{k+1}} \ldots \exists ! v_{j_n} \wedge K' .$$

We shall say that K is *strongly linear* iff it is linear and

(d) if $T \vdash \wedge K' (\bar{a}_{j_1}, \ldots, \bar{a}_{j_k}, \bar{c}_{j_{k+1}}, \ldots, \bar{c}_{j_n})$,
$T \vdash \wedge K' (\bar{b}_{j_1}, \ldots, \bar{b}_{j_k}, \bar{d}_{j_{k+1}}, \ldots, \bar{d}_{j_n})$
and $T \vdash \bar{a}_{j_l} \neq \bar{b}_{j_l}$ for some $1 \leqslant l \leqslant k$, then $T \vdash \bar{c}_{j_r} \neq \bar{d}_{j_r}$ for some $k+1 \leqslant \leqslant r \leqslant n$.

(a) and (b) need no special comment; to see what (b) 'says', it is sufficient to note that if K contains $A(v_1)$ and $B(v_1)$, where v_1 is the only real variable in both $A(v_1)$ and $B(v_1)$, then K is not linear.

(c) means that for any choice of constants $\bar{a}_{j_1}, \ldots, \bar{a}_{j_k}$ we can define unique values $\bar{a}_{j_{k+1}}, \ldots, \bar{a}_{j_n}$ of $v_{j_{k+1}}, \ldots, v_{j_n}$ respectively such that $T \vdash \wedge K' (\bar{a}_{j_1}, \ldots, \bar{a}_{j_k}, \bar{a}_{j_{k+1}}, \ldots, \bar{a}_{j_n})$. The restriction that $\bar{K}'$ satisfies (a) and (b) is necessary. Consider, for example,

$$K' = \{A(v_1), B(v_1, v_2, v_3), C(v_1, v_2, v_3, v_4)\} ,$$

where v_1 is the only real variable in $A(v_1)$. Then $k = mV' - mK' = 1$. Now, if we apply Definition 2.2. without restriction, then we can choose a value for any of the variables $v_1, \ldots, v_n$ and the values of the remaining variables will be determined uniquely. But if we apply Definition 2.2. to $K''= =\{A(v_1)\}$, then there is a unique value of v_1, say $\bar{a}_1$, such that $T \vdash A(\bar{a}_1)$. If in K' we choose $\bar{a}_1'$ for v_1, where $T \vdash \bar{a}_1 \neq \bar{a}_1'$, then there are unique $\bar{a}_2, \bar{a}_3, \bar{a}_4$ such that

$$T \vdash B(\bar{a}_1, \bar{a}_2, \bar{a}_3) \wedge C(\bar{a}_1', \bar{a}_2, \bar{a}_3, \bar{a}_4),$$

but it is not the case that

$$T \vdash A(\bar{a}_1') \wedge B(\bar{a}_1', \bar{a}_2, \bar{a}_3) \wedge C(\bar{a}_1', \bar{a}_2, \bar{a}_3, \bar{a}_4),$$

provided T is consistent. The restriction that $\bar{K}'$ must satisfy (a) and (b) prevents us from choosing a value for v_1 in K', for

$$\{\forall v_1 A(v_1), \forall v_1 B(v_1, v_2, v_3), \forall v_1 C(v_1, v_2, v_3, v_4)\}$$

contains a formula with no real variable.

(c) is the most important point of Definition 2.2. It defines what it means (as we shall see soon) that a set of real variables depends on another set of real variables. In our example, v_2, v_3 and v_4 depend on v_1, but v_1 does not depend on v_2, v_3 or v_4. This is the main ingredient of the subsequent definition of Simon's causal dependence.

(d) does not exist in Simon's definition of a linear set, but we find strongly linear sets to be important in some contexts and theorems.

There are interesting consequences of Definition 2.1. and Definition 2.2. and we shall prove some of them. In the sequel K will always denote a linear set, $K', K'', \ldots, K_0, K_1, \ldots$ will denote finite non-empty subsets of K, $V, V', V'', \ldots, V_0, V_1, \ldots,$ will denote the sets of real variables in $K, K', K'', \ldots, K_0, K_1, \ldots$ respectively. This notation will be extended in obvious ways.

THEOREM 2.1. *If T is consistent, then every linear set is consistent.*

Proof. Let K be such a set. Using Definition 2.2. (c) and predicate logic, we obtain

$$T \vdash \exists v_{j_{k+1}} \ldots \exists v_{j_n} \wedge K',$$

for every K'. If K is inconsistent, there is a finite inconsistent K'. Hence, $\vdash \neg \wedge K'$ and $\vdash \neg \exists v_{j_{k+1}} \ldots \exists v_{j_n} \wedge K'$. Therefore, T is inconsistent; this proves the theorem.

THEOREM 2.2. *Let $A(v_1, \ldots, v_n) \in K$, where $v_1, \ldots, v_n$ are the only free variables in A; then $v_1, \ldots, v_1$ are real in A.*

Proof. $\{A(v_1, \ldots, v_n)\}$ is a finite subset of K. Hence, by Definition 2.2. (c),

$$T \vdash \forall v_{j_1} \ldots \forall v_{j_{n-1}} \exists! v_{j_n} A(v_1, \ldots, v_n),$$
$$v_{j_l} \in \{v_1, \ldots, v_n\}, \quad 1 \leqslant l \leqslant n.$$

Now, we easily derive

$$T \vdash \forall v_{j_1} \ldots \forall v_{j_{n-1}} \exists v_{j_n} A(v_1, \ldots, v_n)$$

and

$$T \vdash \forall v_{j_1} \ldots \forall v_{j_{n-1}} \exists v_{j_n} \forall v_k (A(v_{j_1}, \ldots, v_{j_{n-1}}, v_k) \Rightarrow v_{j_n} = v_k).$$

From the latter, using *, it follows that

$$T \vdash \forall v_{j_1} \ldots \forall v_{j_{n-1}} \exists v_{j_n} \neg A(v_1, \ldots, v_n).$$

This shows that v_{j_n} is real in $A(v_1, \ldots, v_n)$.

THEOREM 2.3. *For any $A \in K$, it is not the case that*

$$T \cup K - \{A\} \vdash A,$$

provided that T is consistent.

Proof. It is sufficient to prove the theorem for any finite subset K' of K. For, suppose that for every finite $K' \subseteq K$ and any $A \in K'$ it is not the case that $T \cup K' - \{A\} \vdash A$, but that $T \cup K - \{A\} \vdash A$. Then there is a finite $K'' \subseteq K - \{A\}$ such that $T \cup K'' \vdash A$. Obviously, then $T \cup K''' - \{A\} \vdash A$, where $K''' = K'' \cup \{A\}$, contrary to our hypothesis.

Let $mK' = 1$; then $K' - \{A\} = \emptyset$. Since A contains a real variable, it is not the case that $T \vdash A$ (remember that T is a closed theory).

Let $mK' = n > 1$, $A \in K'$, let $mV' = k$ and let V'' be the set of free variables in $K' - \{A\}$. We shall distinguish two cases: (I) $mV' = mV''$ and (II) $mV' > mV''$.

Case (I). Let $K'' = K' - \{A\}$ and let $V' = V'' = \{v_1, \ldots, v_k\}$. Suppose that

$$T \cup K'' \vdash A;$$

then by predicate logic

$$T \vdash \wedge K'' \Leftrightarrow \wedge K' \quad \text{and} \quad T \vdash \wedge K''(v'_{j_k}) \Leftrightarrow \wedge K'(v'_{j_k}),$$

where $\wedge K''(v'_{j_k})$ and $\wedge K'(v'_{j_k})$ are obtained from $\wedge K''$ and $\wedge K'$ respectively, by substitution of v'_{j_k} for v_{j_k}. Since

$$T \vdash \forall v_{j_1} \ldots \forall v_{j_{k-n}} \forall v_{j_{k-n+1}} \exists!\, v_{j_{k-n+2}} \ldots \exists!\, v_{j_k} \wedge K''$$

and

$$T \vdash \exists!\, v_{j_{k-n+2}} \ldots \exists!\, v_{j_{k-1}} \exists\, v_{j_k} (\wedge K'' \wedge \forall v'_{j_k} (\wedge K'' (v'_{j_k}) \Rightarrow \Rightarrow v_{j_k} = v'_{j_k})),$$

by the substitutivity of equivalence we obtain

$$T \vdash \exists!\, v_{j_{k-n+2}} \ldots \exists!\, v_{j_{k-1}} \exists v_{j_k} (\wedge K' \wedge \forall v'_{j_k} (\wedge K' (v'_{j_k}) \Rightarrow \Rightarrow v_{j_k} = v'_{j_k}))$$

and

$$T \vdash \forall v_{j_1} \ldots \forall v_{j_{k-n}} \forall v_{j_{k-n+1}} \exists!\, v_{j_{k-n+2}} \ldots \exists!\, v_{j_k} \wedge K'.$$

On the other hand, we obtain

$$T \vdash \exists!\, v_{j_{k-n+1}} \ldots \exists!\, v_{j_k} \wedge K'.$$

Let $B = \exists!\, v_{j_{k-n+2}} \ldots \exists!\, v_{j_k} \wedge K'$. Then

$$T \vdash \exists v_{j_{k-n+1}} (B \wedge \forall v'_{j_{k-n+1}} (B(v'_{j_{k-n+1}}) \Rightarrow v_{j_{k-n+1}} = v'_{j_{k-n+1}}))$$

and

$$T \vdash \exists v_{j_{k-n+1}} \forall v'_{j_{k-n+1}} (B(v'_{j_{k-n+1}}) \Rightarrow v_{j_{k-n+1}} = v'_{j_{k-n+1}}),$$

where $B(v'_{j_{k-n+1}})$ is obtained by substitution of $v'_{j_{k-n+1}}$ for $v_{j_{k-n+1}}$ in B ($v'_{j_{k-n+1}}$ does not occur in B). By predicate logic,

$$T \vdash \forall v_{j_{k-n+1}} \exists v'_{j_{k-n+1}} v_{j_{k-n+1}} \neq v'_{j_{k-n+1}} \Rightarrow \Rightarrow \exists v'_{j_{k-n+1}} \neg B(v'_{j_{k-n+1}})$$

since $v_{j_{k-n+1}}$ is not free in $\neg B(v'_{j_{k-n+1}})$. Using *, we derive

$$T \vdash \neg \forall v_{j_{k-n+1}} \exists!\, v_{j_{k-n+2}} \ldots \exists!\, v_{j_k} \wedge K'$$

and thus T is inconsistent, contrary to the hypothesis of the theorem. Hence, it is not the case that $T \cup K'' \vdash A$.

Case (II). Let $K'' = K' - \{A\}$. Since $mV'' < mV'$, there are variables in K' which do not occur in K''. Let

$$V'' = \{v_1, \ldots, v_k, \ldots, v_m\}$$

and let

$$v_k, v_{k+1}, \ldots, v_m, v_{m+1}, \ldots, v_{m+n}$$

be the variables of A, $k \leqslant m$ and $n \geqslant 1$. Suppose that

$$T \cup K'' \vdash A.$$

Then, using predicate logic,

$$T \vdash \exists v_1 \ldots \exists v_m \wedge K'' \Rightarrow \exists v_k \ldots \exists v_m \ldots \exists v_{m+n-1} \forall v_{m+n} A .$$

By Definition 2.2. (c),

$$T \vdash \exists v_1 \ldots \exists v_m \wedge K'' ;$$

hence,

$$T \vdash \exists v_k \ldots \exists v_m \ldots \exists v_{m+n-1} \forall v_{m+n} A .$$

By Definition 2.2. (c) and predicate logic,

$$T \vdash \forall v_k \ldots \forall v_m \ldots \forall v_{m+n-1} \neg \forall v_{m+n} A .$$

Thus, T is again inconsistent, contrary to the hypothesis of the theorem. This completes the proof.

DEFINITION 2.3. A finite non-empty $K' \subseteq K$ is *self-contained* iff $mV' = mK'$; K' is *sectional* iff $mV' > mK'$.

As a trivial consequence of Definition 2.3. we have

THEOREM 2.4. *If K' is self-contained and $V' = \{v_1, \ldots, v_n\}$, then* $T \vdash \exists! v_1 \ldots \exists! v_n \wedge K'$.

THEOREM 2.5. *If K' and K'' are self-contained, then $K' \cap K''$ is self-contained.*

Proof. $mK' + mK'' - m(K' \cap K'') = m(K' \cup K'')$ and $mV' + mV'' - m(V' \cap V'') = m(V' \cup V'')$. But $mK' = mV'$ and $mK'' = mV''$. Hence, $mV' + mV'' - m(K' \cap K'') = m(K' \cup K'')$. On the other hand, $m(K' \cup K'') \leqslant m(V' \cup V'') = mV_{K' \cup K''}$, where $V_{K' \cup K''}$ is the set of variables in $K' \cup K''$. We get $m(K' \cap K'') \geqslant m(V' \cap V'')$. But $K' \cap K'' \subseteq K$; hence, $mV_{K' \cap K''} \geqslant m(K' \cap K'')$, where $V_{K' \cap K''}$ is the set of free variables in $K' \cap K''$. Obviously, $V_{K' \cap K''} \subseteq V' \cap V''$ and $mV_{K' \cap K''} \leqslant m(V' \cap V'')$. Therefore, $mV_{K' \cap K''} = m(V' \cap V'') = m(K' \cap K'')$. This completes the proof.

From the proof of the preceding theorem we have

THEOREM 2.6. *If K' and K'' are self-contained, then $K' \cup K''$ is self-contained.*

DEFINITION 2.4. A self-contained $K' \subseteq K$ is *minimal* iff it contains no self-contained proper subset.

THEOREM 2.7. *If* K' *and* K'' *are minimal, then* $K' \cap K'' = \emptyset$ *and* $V' \cap V'' = \emptyset$.

Proof. $K' \cap K'' = \emptyset$ follows from Theorem 2.6. and Definition 2.4. Now, $m(K' \cup K'') = mK' + mK''$ and $m(V' \cup V'') = mV' + mV'' - m(V' \cap V'')$. Since $mV' = mK'$ and $mV'' = mK''$, we have that $m(V' \cap V'') > 0$. This implies $m(K' \cup K'') > m(V' \cup V'') = mV_{K' \cup K''}$, contrary to the hypothesis that K is linear.

DEFINITION 2.5. Let M be the union of all minimal subsets of K. We say that K is *causally ordered* iff

(a) $M \neq K$,

(b) for every non-empty $K' \subseteq K$ there is a self-contained $K'' \subseteq K$ such that $K' \subseteq K''$.

That a linear K may be the union of minimal subsets is clear if

$$K = \{A(v_1), B(v_2), C(v_3), \ldots\}.$$

That a linear K may have no self-contained subset is clear if we take

$$K = \{A(v_1, v_2), B(v_1, v_2, v_3), C(v_1, v_2, v_3, v_4), \ldots\}.$$

The importance of (a) and (b) will be seen from several theorems about causally ordered K, which we prove below. The reader familiar with Simon's original definition of a causally ordered system will notice that Definition 2.5. coincides with Simon's definition if K is finite.

THEOREM 2.8. *If* K *is causally ordered, then* $M \neq \emptyset$.

Proof. Since K is non-empty, it contains a non-empty finite subset K'; by Definition 2.5. (b), there is a self-contained K'' such that $K' \subseteq K''$. K'' is either minimal or it contains a minimal subset.

THEOREM 2.9. *If* $K' \subseteq K - M$, *then* K' *is sectional.*

Proof. If K' is self-contained, then K' contains a minimal subset. Hence, it is not the case that $K' \subseteq K - M$.

THEOREM 2.10. *If* K' *and* K'' *are sectional and* $K' \cup K''$ *is self-contained, then either* $K' \cap M \neq \emptyset$ *or* $K'' \cap M \neq \emptyset$.

Proof. Since $K' \cup K''$ is self-contained, it is not the case that $K' \cup K'' \subseteq$ $\subseteq K - M$. Hence, $(K' \cup K'') \cap M = (K' \cap M) \cup (K' \cap M) \neq \emptyset$.

THEOREM 2.11. *If $K' \subset K$ is minimal, then there is a self-contained $K'' \supset K'$ such that $K'' \cap (K-M) \neq \emptyset$.*

Proof. Take any sectional K'''; if $K' \cup K'''$ is self-contained, then the theorem is proved; if $K' \cup K'''$ is sectional, by Definition 2.5. (b) there is a self-contained K'', $K' \cup K''' \subset K''$. This proves the theorem.

THEOREM 2.12. *Let V_M be the set of free variables in M and let V_{K-M} be the set of free variables in $K-M$. Then $V_M \cap V_{K-M} \neq \emptyset$.*

Proof. By Definition 2.5. (b), Theorem 2.8. and Theorem 2.11., it follows that there are a self-contained K'' and minimal $K_1, \ldots, K_n$, $n \geqslant 1$, such that

$$K_1 \cup \cdots \cup K_n \subset K'', K' = K'' - (K_1 \cup \cdots \cup K_n)$$

contains no minimal subset and $K'' \cap (K-M) \neq \emptyset$. Let $V', V_1, \ldots, V_n$ be the sets of real variables in $K', K_1, \ldots, K_n$ respectively. We have

$$\begin{aligned} mK'' &= m(K_1 \cup \cdots \cup K_n \cup K') = mK_1 + \cdots + mK_n + mK' = \\ &= mV'' = m(V_1 \cup \cdots \cup V_n \cup V') = \\ &= mV_1 + \cdots + mV_n + mV' - m((V_1 \cup \cdots \cup V_n) \cap V'), \end{aligned}$$

since $(K_1 \cup \cdots \cup K_n) \cap K' = \emptyset$ and $K_i \cap K_j = V_i \cap V_j = \emptyset$ by Theorem 2.5., $1 \leqslant i, j \leqslant n$, $i \neq j$. (Note that $(V_1 \cup \cdots \cup V_n) \cap V'$ is not necessarily empty, for V' is not necessarily $V'' - (V_1 \cup \cdots \cup V_n)$). Since $mK_i = mV_i$, we have

$$mK' = mV' - m((V_1 \cup \cdots \cup V_n) \cap V').$$

If $mK' = mV'$, then K' is self-contained, but not minimal. Hence, it contains a minimal subset, which is a contradiction. Thus, K' is not self-contained. Hence, $mV' > mK'$ and thus

$$m((V_1 \cup \cdots \cup V_n) \cap V') > 0.$$

But $(V_1 \cup \cdots \cup V_n) \cap V' \subseteq V_M \cap V_{K-M}$; this concludes the proof.

THEOREM 2.13. *Let K' be sectional, let $K_1, \ldots, K_n$ be minimal, let $K_0 = K_1 \cup \cdots \cup K_n$ and let $K_0 \cup K'$ be self-contained. Then*

(a) *every $A \in K'$ contains a $v_k \notin V_0$ and at least two free variables;*

(b) $V' \cap V_0 \neq \emptyset$;

(c) if $V_0 \subset V'$, *then* $mV' - mK' = mV_0$;

(d) $V' \neq V_0$.

Proof. $\{A\} \subseteq K'$ is sectional; hence, $mV_A > 1$ and A contains at least two free variables. On the other hand, if $V_A \subseteq V_0$, then $m(K_0 \cup \{A\}) > > mV_0$, which is impossible. This proves (a).

(b) Analogous to the proof of Theorem 2.12.

(c) Since $K_0 \cup K'$ is self-contained, $m(K_0 \cup K') = mV_{K_0 \cup K'} = mK_0 + mK'$. But $mV_{K_0 \cup K'} = mV_0 + mV' - m(V_0 \cap V')$. Since $V_0 \subset V'$, $mV_{K_0 \cup K'} = mV'$. Hence, $mV' - mK' = mV_{K_0 \cup K'} - mK' = mK_0$.

(d) From (c), since $K' \neq \emptyset$.

THEOREM 2.14. *Let K' be sectional and let*

$$\{v_1, \ldots, v_n\} = V_M \cap V' ;$$

choose individual constants $\bar{a}_1, \ldots, \bar{a}_n$, substitute them for $v_1, \ldots, v_n$ respectively in K' and let K'_c be obtained in this way. Then K'_c satisfies (a), (b) *and* (c) *of Definition* 2.2.

Proof. By Definition 2.5. (b) there is a self-contained $K'' \supset K'$. Since K'' is not minimal, it contains a proper minimal subset. Let K_0 be the union of all minimal subsets of K''. Suppose that there is a $v_k \in V'' \cap \cap (V_M - V_0)$; then there is a minimal K_1 such that $v_k \in V_1$. We shall show that $K_1 \cap K'' = \emptyset$.

Suppose that $K_1 \cap K'' \neq \emptyset$. $K_1 \cap K''$ is self-contained, by Theorem 2.5.; hence, it is minimal or it contains a minimal K_2. By Theorem 2.7., either $K_1 \cap K'' \cap K_0 = \emptyset$ or $K_2 \cap K_0 = \emptyset$. Thus, either $K_1 \cap K'' \subseteq K'' - K_0$ or $K_2 \subseteq K'' - K_0$, and it follows that K_0 is not the union of all minimal subsets of K''. Hence, $K_1 \cap K'' = \emptyset$.

Now, by Theorem 2.6. we have $V_1 \cap V'' = \emptyset$. This is contrary to our hypothesis that there is a $v_k \in V_1 \cap V''$. Hence, $V' \cap V_M \subseteq V_0$.

This shows that for every sectional K' there is a sectional $K''' \supseteq K'$ and a union K_0 of minimal subsets such that $K''' \cup K_0$ is self-contained and that Theorem 2.13. can be applied. By (a) of 2.13., every $A \in K'_c$ contains at least one free variable. The condition (a) of Definition 2.2. is satisfied.

We have to show that $mK'_c \leqslant mV'_c$. Now, $mK'_c = mK'$ and $V' \cap V_M \subseteq V_0$; hence, $mK'_c + mK_0 \leqslant mV' + mV_0 - m(V' \cap V_0)$ and $mK'_c \leqslant mV' - m (V' \cap V_M) = mV'_c$. Thus, K. satisfies (b) of Definition 2.2.

In order to show that K'_c satisfies (c) of Definition 2.2., suppose that

$$V'_c = \{v_{n+1}, \ldots, v_{n+m}\}$$

and that the set $\bar{K}'_c$ of universal closures with respect to

$$v_{j_1}, \ldots, v_{j_k}, k = mV'_c - mK'_c,$$

satisfies (a) and (b) of Definition 2.2., where $v_{j_i} \in V'_c$, $1 \leqslant i \leqslant k$. Obviously,

$$V' = V'_c \cup (V' \cap V_M) = \{v_1, \ldots, v_n, v_{n+1}, \ldots, v_{n+m}\}.$$

Moreover, the set $\bar{K}'$ of universal closures of elements of K', with respect to

$$v_1, \ldots, v_n, v_{j_1}, \ldots, v_{j_k}$$

satisfies (a) and (b) of Definition 2.2. We also have

$$mV' = mV'_c + m(V' \cap V_M),$$

since $V'_c \cap V_M = \emptyset$; hence,

$$mV' - mK' = mV'_c - mK'_c + m(V' \cap V_M) = k + n.$$

By Definition 2.2. (c)

$$T \vdash \forall v_1 \ldots \forall v_n \forall v_{j_1} \ldots \forall v_{j_k} \exists! v_{j_{k+1}} \ldots \exists! v_{j_m} \wedge K'.$$

Using predicate logic, we easily obtain

$$T \vdash \forall v_{j_1} \ldots \forall v_{j_k} \exists! v_{j_{k+1}} \ldots \exists! v_{j_m} \wedge K'_c.$$

Therefore, K'_c satisfies (c) of Definition 2.2.

COROLLARY. *Let K be causally ordered and let* $(K-M)_c$ *be the set of results of substitution of constants for variables of* $V_M \cap V_{K-M}$ *in the wffs of* $K-M$. *Then* $(K-M)_c$ *is linear.*

The values of free variables in the wffs of $(K-M)_c$ depend on the values of free variables in the wffs in M, but not vice versa. Of course, we need individual constants in L in order to substitute them for the variables in the wffs of M. This could be avoided in a purely model-theoretic approach, but we do not insist on this point.

It is clear that $(K-M)_c$ contains at least one minimal subset (this follows easily by Theorem 2.13. (c), since K is causally ordered). Let us define K^1 as $(K-M)_c$ and let M^1 be the union of all minimal subsets of K^1. K^1 is not necessarily causally ordered. If it is, we can find the values of the free variables in M^1, introduce individual constants for them and substitute the constants for the elements of $V_{M^1} \cap V_{K^1-M^1}$ in the wffs of $K^1 - M^1$ etc. We give a simple example. Let

$$K = \{A(v_1), B(v_1, v_2), C(v_2, v_3), D(v_1, v_2, v_3, v_4). \ldots\}.$$

Then

$$M = \{A(v_1)\}.$$

Suppose that we have $T \vdash A(\bar{a}_1)$. Then

$$K^1 = \{B(\bar{a}_1, v_2), C(v_2, v_3), D(\bar{a}_1, v_2, v_3, v_4), \ldots\}$$

and

$$M^1 = \{B(\bar{a}_1, v_2)\}.$$

Suppose that

$$T \vdash B(\bar{a}_1, \bar{a}_2);$$

then

$$K^2 = \{C(\bar{a}_2, v_3), D(\bar{a}_1, \bar{a}_2, v_3, v_4), \ldots\}$$

and

$$M^2 = \{C(\bar{a}_2, v_3)\}$$

etc.

If K^r is causally ordered, it is not necessary that K^{r+1} is causally ordered. Hence, it is possible that our procedure of substituting the values of variables (constants) obtained from the minimal subsets has to be stopped (in case that $M^{r+1} = K^{r+1}$) or that we have to substitute some arbitrary constants for some variables of K^{r+1} (in case that for some sectional subset of $K^r - M^r$ there are only sectional supersets).

Now we define a finite or countable sequence of causally ordered sets $K^0, K^1, \ldots, K^r, \ldots$, starting with a causally ordered set K.

DEFINITION 2.6. Let K be causally ordered. Then $K^0 = K$; suppose that K^r is causally ordered and let M^r be the union of minimal subsets of K^r; let

$$V_{M^r} = \{v_1, \ldots, v_n, \ldots\}$$

and let

$$C_{M^r} = \{\bar{c}_1, \ldots, \bar{c}_n, \ldots\}$$

be the set of individual constants such that for every minimal K_r' of K^r, where

$$V_r' = \{v_{j_1}, \ldots, v_{j_k}\},$$

$$T \cup K'_r \vdash \wedge K'_r(\bar{c}_{j_1}, \ldots, \bar{c}_{j_k}),$$

where $\bar{c}_{j_i}$ is substituted for v_{j_i} in $\wedge K'_r$, $1 \leqslant i \leqslant k$ and $\bar{c}_{j_1}, \ldots, \bar{c}_{j_k} \in C_{M^r}$. Then K^{r+1} is obtained from $K^r - M^r$ by substitution of $\bar{c}_n$ for v_n in the wffs of $K^r - M^r$, for all n. (For the sake of simplicity, we supposed that K was countable.) K^r is the *derived set of order r*.

We shall now prove an important theorem concerning a sequence of causally ordered sets.

THEOREM 2.15. Let K^0 be causally ordered, let M^0 be the union of all minimal subsets of K^0, let $K^1, K^2, \ldots, K^r, \ldots$ be the (finite or infinite) sequence of derived causally ordered sets furthermore, let $M^1, M^2, \ldots$ $M^r, \ldots$ be the unions of all minimal subsets of order $1, 2, \ldots, r, \ldots$ respectively; finally, let $M^\omega = \bigcup_r M^r$. Then

(a) $\quad T \cup K^0 \vdash B$

for all $B \in M^\omega$ and

(b) $\quad T \cup M^\omega \vdash B'$

for all $B' \in K^0$.

Proof. Let us prove (a) first. We obviously have

$$T \cup K^0 \vdash B$$

for all $B \in K^0$. Suppose that

$$T \cup K^0 \vdash B$$

for all $B \in K^r$ (induction hypothesis). Let K'_r be a minimal subset of order r. We have

$$T \vdash \exists! v_1 \ldots \exists! v_n \wedge K'_r,$$

where $V'_r = \{v_1, \ldots, v_n\}$. Hence, by **

$$T \vdash v_1 = \bar{a}_1 \wedge v_2 = \bar{a}_2 \wedge \cdots \wedge v_n = \bar{a}_n.$$

In this way we obtain

$$T \vdash v_n = \bar{a}_n$$

for all $v_n \in V_{M^r}$. Note that the minimal subsets of K^r are disjoint as well as the sets of their free variables and that $\bar{a}_n$ is unique. By induction hypothesis, for all

$$B(v_{j_1}, \ldots, v_{j_k}, v_{j_{k+1}}, \ldots, v_{j_{k+m}}) \in K^r - M^r,$$

where

$$v_{j_1}, \ldots, v_{j_k} \in V_{M^r}, \quad k \geqslant 0,$$

we have

$$T \cup K^0 \vdash B(v_{j_1}, \ldots, v_{j_k}, v_{j_{k+1}}, \ldots, v_{j_{k+m}}).$$

Therefore,

$$T \cup K^0 \vdash B(\bar{a}_{j_1}, \ldots, \bar{a}_{j_k}, v_{j_{k+1}}, \ldots, v_{j_{k+m}}),$$

and this proves (a).

(b) Let $B' \in K^0$; then there is a finite self-contained subset $K_0' \subseteq K^0$ such that $B' \in K_0'$. Let M_0' be the union of all minimal subsets of K_0'; obviously, $M_0' \subseteq M^0$. Now, $K_0' - M_0'$ may be empty. If it is not empty, $(K_0' - M_0') \subseteq K^0 - M^0$ and after substituting constants from C_{M^0} for variables of $V_{M^0 \cap (K^0 - M^0)}$ in the wffs of $K^0 - M^0$, we obtain a finite self-contained $K_1' \subseteq K^1$, where K_1' is obtained from $K_0' - M_0'$ by this substitution. Again, $K_1' - M_1'$ may be empty, where M_1' is the union of all minimal subsets of K_1'. If it is not, then there is a finite self-contained $K_2' \subseteq K^2$ obtained from $K_1' - M_1'$ by substution of constants from C_{M^1} for variables from $V_{M^1 \cap (K^1 - M^1)}$ in the wffs of $K^1 - M^1$, and so on. Briefly, since K_0' is finite, there is a finite sequence $K_0', K_1', \ldots, K_r'$ of self-contained subsets of $K^0, K^1, \ldots, K^r$ respectively such that K_{s+1}', $0 \leqslant s < r$, is obtained from $K_s' - M_s'$ by the same substitution by which K^{s+1} is obtained from $K^s - M^s$. It is clear that there is always a K_r' such that $K_r' = M_r'$. Hence,

$$T \cup M^\omega \vdash \wedge K_r'.$$

Suppose that

$$T \cup M^\omega \vdash \wedge K_s',$$

$0 < s \leqslant r$ (induction hypothesis). We also have

$$T \cup M^\omega \vdash \wedge M_{s-1}',$$

where M_{s-1}' is the union of all minimal subsets of K_{s-1}'. Let

$$V_{s-1}' = \{v_1, \ldots, v_n\}.$$

Obviously,

$$T \vdash \exists! v_1 \ldots \exists! v_n \wedge M'_{s-1}$$

and using **

$$T \vdash v_1 = \bar{a}_1 \wedge \cdots \wedge v_n = \bar{a}_n .$$

Now, $\wedge K'_s = \wedge K'_{s-1}(\bar{a}_1, \ldots, \bar{a}_n)$, where $\bar{a}_1, \ldots, \bar{a}_n$ are substituted for $v_1, \ldots, v_n$ respectively in $\wedge K'_{s-1}$. We have therefore

$$T \cup M^{\omega} \vdash \wedge K'_{s-1} .$$

In this way we obtain K'_0. This procedure can be applied to every $B' \in K^0$ and this proves (b).

There is an interesting consequence of Theorem 2.15. Let $\mathfrak{A}$ be any relational structure such that $\mathfrak{A} \models_x T \cup K^0$, where K^r is causally ordered for every $r = 0, 1, \ldots, x$ is a valuation and all variables of L are in V^0. Then x is unique. For, by Theorem 2.15., we easily obtain

$$\mathfrak{A} \models_x T \cup K^0 \quad \text{iff} \quad \mathfrak{A} \models_x T \cup M^{\omega} .$$

It is clear that there is one and only one x such that $\mathfrak{A} \models_x T \cup M^{\omega}$.

DEFINITION 2.7. Let K'_r be a minimal subset of a derived K^r, let $v_k \in V'_r$ and let $v_k \notin V''_s$ for all K''_s, $s < r$, where K''_s is a minimal subset of K^s; then v_k is *endogenous* to K'_r; if $\bar{a}_k$ has been substituted for v_k in K'_r, then v_k is *exogenous* to K'_r.

DEFINITION 2.8. Let V'_r and V''_s be sets of endogeonous variables in K'_r and K''_s respectively, where K'_r and K''_s are minimal and of order r and s respectively; then V''_s is *directly causally dependent* on V'_r iff there is at least one $v_k \in V'_r$ exogenous to V''_s, and we write $V'_r \to V''_s$. V''_s is *indirectly causally dependent* on V'_r iff there are $V'_1, \ldots, V'_n$ such that

$$V'_r \to V'_1, V'_1 \to V'_2, \ldots, V'_n \to V''_s .$$

In this case we write again $V'_r \to V''_s$.

Now we can define what means that a state of affairs is a cause of another state of affairs.

DEFINITION 2.9. Let K^0 be causally ordered, let K'_r and K''_s be derived minimal subsets of order r and s respectively, $r<s$, let

$$V'_r = \{v_1, \ldots, v_k\},\ V''_s = \{v_{k+1}, \ldots, v_{k+m}\} \quad \text{and} \quad V'_r \to V''_s\ ;$$

let $\mathfrak{A}_1$ and $\mathfrak{A}_2$ be relational structures such that the relation symbols of K'_r and K''_s are defined in $\mathfrak{A}_1$ and $\mathfrak{A}_2$ respectively. Then $\mathfrak{A}_1$ is a cause of $\mathfrak{A}_2$ and $\mathfrak{A}_2$ is an effect of $\mathfrak{A}_1$ iff

$$\mathfrak{A}_1 \models \wedge K'_r(\bar{a}_1, \ldots, \bar{a}_k) \quad \text{and} \quad \mathfrak{A}_2 \models \wedge K''_s(\bar{a}_{k+1}, \ldots, \bar{a}_{k+m}).$$

Obviously, in Definition 2.9. $\mathfrak{A}_1$ and $\mathfrak{A}_2$ are states of affairs, in accordance with our introductroy considerations.

From Definition 2.9. a meaning of the phrase "The property F is a cause of the property G" can be derived. First of all, the relation $\to$ between sets of endogenous variables can be defined to hold between sets of wffs of minimal subsets. Let us write $K'_r \to K''_s$ iff $V'_r \to V''_s$, where, for example,

$$V'_r = \{v_1, \ldots, v_k\} \quad \text{and} \quad V''_s = \{v_{k+1}, \ldots, v_{k+m}\}.$$

Suppose that a property (relation, predicate) $F(v_1, \ldots, v_k)$ is defined as $\wedge K'_r$ and that a property $G(v_{k+1}, \ldots, v_{k+m})$ is defined as $\wedge K''_s$. Then we can say that $F(v_1, \ldots, v_{k+m})$ is a cause of $G(v_{k+1}, \ldots, v_{k+m})$ if there are a causally ordered K^0 and derived minimal subsets K'_r and K''_s such that $K'_r \to K''_s$.

3. Change

In this section we shall consider the concept of change. Let J^0 and K^0 be causally ordered and let $J^1, \ldots, J^r, \ldots$ and $K^1, \ldots, K^r, \ldots$ be the sequences of causally ordered sets derived from J^0 and K^0 respectively; let V^0, $V^1, \ldots, V^r, \ldots$ and $W^0, W^1, \ldots, W^r, \ldots$ be the corresponding sets of free variables and let $M^0, M^1, \ldots, M^r, \ldots$ and $N^0, N^1, \ldots, N^r, \ldots$ be the corresponding unions of minimal subsets. Suppose that $V^0 = W^0$ and that there is a mapping f from the set of all minimal subsets derived from J^0 (including those in J^0) to the analogous set of minimal subsets derived from K^0 (including those in K^0) and satisfying the following conditions:

(1) f is $1-1$ (obviously, f: $M^\omega \to N^\omega$).

(2) if $f(J'_r) = K'_s$, then $V'_r = W'_s$ and $r=s$.

(3) $J'_r \to J''_s$ iff $f(J'_r) \to f(J''_s)$.

Let $J'_r \subset J^r$ be a minimal subset such that

$$T \vdash \wedge J'_r \not\Leftrightarrow \wedge f(J'_r).$$

It is clear that although J^0 and K^0 have the same causal ordering, the relation symbols of J'_r differ from the relation symbols of $f(J'_r)$.

Now, let $\mathfrak{A}_1$ and $\mathfrak{A}_2$ be two relational structures such that (a) $\mathfrak{A}_1 \models_x J^0$ and (b) $\mathfrak{A}_2 \models_y f(J^0)$, where, obviously, if $\mathfrak{A}_1$ and $\mathfrak{A}_2$ have the same individuals, then $x \neq y$. $\mathfrak{A}_1$ and $\mathfrak{A}_2$ are two different sates of affairs with the same causal ordering. If we have (a) in a period t_1 of time and (b) in a period t_2 later than t_1, then we can say that $\mathfrak{A}_1$ has changed to $\mathfrak{A}_2$. There is no difficulty to think of the transition from $\mathfrak{A}_1$ to $\mathfrak{A}_2$ as of a change in a system S of individuals, where $\mathfrak{A}_1$ and $\mathfrak{A}_2$ are two consecutive states of S. Of course, the transition from $\mathfrak{A}_1$ to $\mathfrak{A}_2$ is here not explained, but simply introduced.

DEFINITION 3.1. Let J^0, K^0 etc. be defined as above and let $J'_r \subset J^r$ be minimal and such that $T \vdash \wedge J'_r \not\Leftrightarrow \wedge f(J'_r)$; if $K'_r = f(J'_r)$, then we write $X(J'_r, K'_r)$ and, in order to simplify the exposition, we say that J'_r has changed to K'_r; we write $X(J^0, K^0)$ iff $X(J'_r, K'_r)$ for some J'_r and some r.

Now we shall prove several theorems concerning X and $\rightarrow$.

THEOREM 3.1. *Let J^0 be causally ordered, let $J^1, \ldots, J^r, \ldots$ be causally ordered and derived from J^0; let $K^0, K^1, \ldots, K^r, \ldots$ be obtained from the preceding sequence by changing one and only one minimal subset J'_r of order r, i.e., $X(J'_r, K'_r)$. Then for all minimal J''_s such that not $J'_r \rightarrow J''_s$,*

$$T \cup K^0 \vdash \wedge J''_s .$$

Proof. Every $v_k \in V^0$ is endogenous to one and only one minimal subset J'_k. For, it is endogenous to no other minimal subset of order k, it is substituted in all minimal subsets of higher order and it does not occur in any minimal subset of order lower than k.

We shall distinguish two cases; (1) $s \leqslant r$ and (2) $s > r$.

Case (1). Let $v_k \in V'_r$; then v_k is neither endogenous nor exogenous to any minimal J''_s different from J'_r, $s \leqslant r$. Obviously, $T \vdash \wedge J''_s \Leftrightarrow \wedge K''_s$, since J''_s is K''_s (this follows by a trivial induction). Therefore,

$$T \cup K^0 \vdash \wedge J''_s ,$$

by Theorem 2.15.

Case (2). Suppose that not $J'_r \rightarrow J''_s$, $r<s$; then not $K'_r \rightarrow K''_s$. Let $p=r+1$, let K'_p be minimal and such that not $K'_r \rightarrow K'_p$ and let N_p be the union of all such K'_p. Then no $v_k \in W'_r$ is exogenous to any K'_p. From the proof of Theorem 2.15. (a) it follows that

$$T \cup K^0 \vdash B,$$

$B \in K^r - N^r$. But K'_p is obtained by substitution of constants for variables in some finite $K' \subset K^r - N^r$ and no $v_k \in W'_r$ occurs in the wffs of K'. The relation symbols of K' are identical with the relation symbols in the corresponding $J' \subset J^r - M^r$ and $W' = V'$. Hence,

$$T \cup K^0 \vdash \wedge J'.$$

The exogenous variables of J'_p occur in the wffs of J', as well as in a finite M'_r not containing J'_r such that by Case (1) (since $\wedge M'_r$ is $\wedge N'_r$)

$$T \cup K^0 \vdash \wedge M'_r.$$

But we have

$$T \cup M'_r \vdash \wedge J'_p,$$

by Definition 2.6. and this proves the theorem for $p=r+1$.

Consider J''_s; it is obtained by substitution of constants (say $\bar{a}_1, \ldots, \bar{a}_k$) for variables (say $v_1, \ldots, v_k$, the exogenous variables of J''_s) in a finite $J_{s-1} \subset J^{s-1} - M^{s-1}$. Let $r<p<s$, suppose that not $J'_r \rightarrow J'_p$ and let M_p be the union of all such J'_p, for all p. Then $v_1, \ldots, v_k \in V_p$ and, in particular, there is a finite union $M''_p \subset M_p$ of minimal subsets of different orders such that $v_1, \ldots, v_k \in V''_p$. Each J'_p is obtained by substitution from a finite $J_{p-1} \subset J^{p-1} - M^{p-1}$. Let our induction hypothesis be

$$T \cup K^0 \vdash \wedge J_p \quad \text{and} \quad T \cup K^0 \vdash \wedge M''_p,$$

for all, p, J_p and M''_p. Then we have

$$T \cup K^0 \vdash \wedge J_{s-1} \quad \text{and} \quad T \cup K^0 \vdash \wedge M''_{s-1}$$

for all J_{s-1} and all M''_{s-1}.

The exogenous variables of J''_s occur in the wffs of some J_{s-1} and in the wffs of the corresponding M''_{s-1} and do not causally depend on the variables of J'_r. For that particular M''_{s-1} we have

$$T \cup K^0 \vdash \wedge M''_{s-1}(\bar{a}_1, \ldots, \bar{a}_k),$$

by Definition 2.6. On the other hand, $\wedge J_s''$ is $\wedge J_{s-1}$ $(\bar{a}_1, \ldots, \bar{a}_k)$. This proves the theorem.

Theorem 3.1. can be stated differently, as follows. Suppose that the hypotheses of the theorem are satisfied. If

$$\mathfrak{A} \models_x J^0 \quad \text{and} \quad \mathfrak{A} \models_y K^0,$$

then x and y coincide at all places where the variables are not causally dependent on the variables of J_r'.

THEOREM 3.2. *Let the hypotheses of Theorem* 3.1 *be satisfied and let* $X(J_r', K_r')$ *consist in changing one and only one wff A of* J_r'. *More precisely,* $A \in J_r'$ *is replaced by B, where B contains the same free variables as A does, and* K_r' *is the result of this change. If T is consistent,*

$$V_r' = W_r' = \{v_1, \ldots, v_k\}$$

and both

$$T \vdash \wedge J_r'(\bar{a}_1, \ldots, \bar{a}_k) \quad \text{and} \quad T \vdash \wedge K_r'(\bar{b}_1, \ldots, \bar{b}_k),$$

then for no j, $T \vdash \bar{a}_j = \bar{b}_j$, *where* $1 \leqslant j \leqslant k$.

Proof. Let us first note that if $T \vdash \bar{a}_j = \bar{b}_j$ for all j, then from the hypotheses of the theorem we can derive a contradiction. For, then

$$T \vdash \wedge J_r'(\bar{a}_1, \ldots, \bar{a}_k) \Leftrightarrow \wedge K_r'(\bar{a}_1, \ldots, \bar{a}_k),$$

while, on the other hand, from $X(J_r', K_r')$ it follows that

$$T \vdash \wedge J_r'(\bar{a}_1, \ldots, \bar{a}_k) \not\Leftrightarrow \wedge K_r'(\bar{a}_1, \ldots, \bar{a}_k)$$

and T is inconsistent. Hence, under the hypotheses of the theorem, it is not the case that $T \vdash \bar{a}_j = \bar{b}_j$ for all j. Now we shall show that if $T \vdash \bar{a}_j = \bar{b}_j$ for at least one j, then $T \vdash \bar{a}_j = \bar{b}_j$ for all j.

Let V_A be the set of free variables in A and let V_1 be the set of free variables in $J_r'' = J_r' - \{A\}$. We shall show that $mV_1 - mJ_r'' = 1$. In order to do that, it is sufficient to prove $V_A \subseteq V_1$.

Suppose that the contrary is the case. Then for some j, $v_j \notin V_1$. Hence $V_A - V_A \cap V_1 \neq \emptyset$, *i.e.*, $V_A \cap V_1 \subset V_A$ and $mV_A - m(V_A \cap V_1) > 0$. But $mV_r' = mV_1 + mV_A - m(V_A \cap V_1)$ and it follows that $mV_r' > mV_1$ and $mV_r' - 1 \geqslant mV_1$. Since $mJ_r'' = mJ_r' - 1 = mV_r' - 1$, we obtain $mJ_r'' \geqslant mV_1$ and $mJ_r'' = mV_1$. Therefore, J_r' is not minimal and this proves $V_A \subseteq V_1$.

Now, $mV_r' = mV_1 = mJ_r'$. Hence, $mV_1 - mJ_r'' = 1$.

Since J'_r is minimal, $\bar{J}''_r$ with respect to any variable satisfies (a) and (b) of Definition 2.2. By Definition 2.2. (c),

$$(1)\qquad T \vdash \forall v_j \exists! v_1 \ldots \exists! v_{j-1} \exists! v_{j+1} \ldots \exists! v_k \wedge J''_r .$$

Since J''_r is K''_r, if $T \vdash \bar{a}_j = \bar{b}_j$ for at least one j and

$$(2)\qquad T \vdash \wedge J''_r(\bar{a}_1, \ldots, \bar{a}_k) \quad \text{and} \quad T \vdash \wedge K''_r(\bar{b}_1, \ldots, \bar{b}_k),$$

then $T \vdash \bar{a}_j = \bar{b}_j$ for all j. But (2) follows from the hypotheses of the theorem. Therefore, for no j, $T \vdash \bar{a}_j = \bar{b}_j$.

THEOREM 3.3. *Let J^0 be causally ordered, let $J^1, \ldots, J^r, \ldots$ be causally ordered and derived from J^0; let $K^0, K^1, \ldots, K^r, \ldots$ be obtained from the preceding sequence by $X(J'_r, K'_r)$ for exactly one J'_r, where $X(J'_r, K'_r)$ consists in replacing exactly one $A \in J'_r$ by B, $V_A = W_B$. Let*

$$V'_r = W'_r = \{v_1, \ldots, v_k\},$$
$$T \vdash \wedge J'_r(\bar{a}_1, \ldots, \bar{a}_k) \quad \textit{and} \quad T \vdash \wedge K'_r(\bar{b}_1, \ldots, \bar{b}_k).$$

Suppose that $V'_r \to V''_s$ directly, where

$$V''_s = W''_s = \{v_{k+1}, \ldots, v_{k+m}\},$$

and suppose that

$$T \vdash \wedge J''_s(\bar{a}_{k+1}, \ldots, \bar{a}_{k+m}) \quad \textit{and} \quad T \vdash \wedge K''_s(\bar{b}_{k+1}, \ldots, \bar{b}_{k+m}).$$

If T is consistent and there is exactly one $v_i \in V'_r$ exogenous to V''_s, then for no $1 \leqslant j \leqslant m$, $T \vdash \bar{a}_{k+j} = \bar{b}_{k+j}$.

Proof. J''_s and K''_s are obtained from finite $J_r \subset J^r - M^r$ and $K_r \subset K^r - M^r$ respectively, by substitution of constants for variables, where, obviously J_r is K_r.

Let $V_r = W_r = \{w_1, \ldots, w_n, v_{k+1}, \ldots, v_{k+m}\}$, $n \geqslant 1$. We know that if we choose arbitrarily values for some $n = mV_r - mK_r > 0$ variables, then the values of the remaining ones are determined uniquely. Suppose that

$$T \vdash \wedge J_r(\bar{c}_1, \ldots, \bar{c}_q, \bar{c}_{q+1}, \ldots, \bar{c}_n, \bar{a}_{k+1}, \ldots, \bar{a}_{k+l}, \bar{a}_{k+l+1}, \ldots, \bar{a}_{k+m})$$

and

$$T \vdash \wedge J_r(\bar{c}_1, \ldots, \bar{c}_q, \bar{d}_{q+1}, \ldots, \bar{d}_n, \bar{a}_{k+1}, \ldots, \bar{a}_{k+l}, \bar{b}_{k+l+1}, \ldots, \bar{b}_{k+m})$$

and that for no p, $T \vdash \bar{c}_p = \bar{d}_p$, $q+1 \leqslant p \leqslant n$, and for no u, $T \vdash \bar{a}_u = \bar{b}_u$, $k+l+1 \leqslant u \leqslant k+m$. We shall show that $q + l_x < n$.

If $q+l \geqslant n$, then we can choose arbitrarily values for n variables from the list $w_1, \dots, w_q, v_{k+1}, \dots, v_{k+l}$ such that the values of the remaining variables are not determined uniquely, contrary to Definition 2.2 (c). Hence, $q+l<n$.

From the hypotheses of the theorem, we have that J_s'' is

$$J_r(\bar{c}_1, \dots, \bar{c}_{n-1}, \bar{c}_n, v_{k+1}, \dots, v_{k+m})$$

and that K_s'' is

$$J_r(\bar{c}_1, \dots, \bar{c}_{n-1}, \bar{d}_n, v_{k+1}, \dots, v_{k+m}),$$

where $\bar{c}_n$ is $\bar{a}_i$ and $\bar{d}_n$ is $\bar{b}_i$ and $\bar{a}_i$ and $\bar{b}_i$ are substituted for $w_n = v_i \in V_r'$. By Theorem 3.2. it is not the case that $T \vdash \bar{c}_n = \bar{d}_n$. Moreover, we have

$$T \vdash \wedge J_r(\bar{c}_1, \dots, \bar{c}_{n-1}, \bar{a}_i, \bar{a}_{k+1}, \dots, \bar{a}_{k+m})$$

and

$$T \vdash \wedge J_r(\bar{c}_1, \dots, \bar{c}_{n-1}, \bar{b}_i, \bar{b}_{k+1}, \dots, \bar{b}_{k+m}).$$

By the preceding argument, there is no j such that $T \vdash \bar{a}_{k+j} = \bar{b}_{k+j}$ (otherwise we would have $l \geqslant 1$ and $n-1+l<n$). This completes the proof.

By an easy inductive argument Theorem 3.3. can be generalized to indirect causal dependence, *i.e.*, to the case where directly

$$J_r' \to J_{r+1}'', J_{r+1}'' \to J_{r+2}'', \dots, J_{r+s-1}'' \to J_{r+s}''$$

and for each J_{r+m+1}'' there is exactly one $v_i \in V_{r+m}''$ exogenous to V_{r+m+1}'', $0 \leqslant m \leqslant s-1$.

Suppose that the causal ordering is defined on a strongly linear set; then we call such an ordering S-causal ordering.

THEOREM 3.4. *Let J^0 be S-causally ordered, let $J^1, \dots, J^r, \dots$ be S-causally ordered and derived from J^0; let $K^0, K^1, \dots, K^r, \dots$ be obtained from the preceding sequence by $X(J_r', K_r')$ for one and only one J_r' and let this change be as in Theorem 3.2. Let*

$$T \vdash \wedge J_r'(\bar{a}_1, \dots, \bar{a}_k) \quad \textit{and} \quad T \vdash \wedge K_r'(\bar{b}_1, \dots, \bar{b}_k);$$

then for all J_s'' such that directly $J_r' \to J_s''$, if

$$T \vdash \wedge J_s''(\bar{a}_{k+1}, \dots, \bar{a}_{k+m}) \quad \textit{and} \quad T \vdash \wedge K_s''(\bar{b}_{k+1}, \dots, \bar{b}_{k+m}),$$

then it is not the case that $T \vdash \bar{a}_{k+j} = \bar{b}_{k+j}$, at least for one $1 \leqslant j \leqslant m$.

Proof. Let $V'_r = W'_r = \{v_1, \ldots, v_k\}$ and let

$$V''_s = W''_s = \{v_{k+1}, \ldots, v_{k+m}\}.$$

Now, for some $1 \leqslant i \leqslant k$, v_i is exogenous to both J''_s and K''_s; suppose that J''_s and K''_s are obtained from $J_r \subset J^r - M^r$ and $K_r \subset K^r - M^r$, respectively, by substitution of $\bar{a}_i$ and $\bar{b}_i$ for v_i, such that J''_s is

$$J_r(\bar{a}_1, \ldots, \bar{a}_k, v_{k+1}, \ldots, v_{k+m})$$

and K''_s is

$$K_r(\bar{b}_1, \ldots, \bar{b}_k, v_{k+1}, \ldots, v_{k+m}).$$

From the hypotheses of the theorem,

$$T \vdash \wedge J_r(\bar{a}_1, \ldots, \bar{a}_k, \bar{a}_{k+1}, \ldots, \bar{a}_{k+m})$$

and

$$T \vdash \wedge K_r(\bar{b}_1, \ldots, \bar{b}_k, \bar{b}_{k+1}, \ldots, \bar{b}_{k+m}).$$

By Definition 2.2. (d), if for all j $T \vdash \bar{a}_{k+j} = \bar{b}_{k+j}$, then for all i $T \vdash \bar{a}_i = \bar{b}_i$, contrary to Theorem 3.2. Therefore, at least for one j it is not the case that $T \vdash \bar{a}_{k+j} = \bar{b}_{k+j}$.

Let us now reconsider $X(J'_r, K'_r)$. In Definition 3.1. we have required only that $T \vdash \wedge J'_r \not\Leftrightarrow \wedge f(J'_r)$ and $V'_r = W'_r$. In Theorems 3.2., 3.3. and 3.4. a specific concept of change is used where a single $A \in J'_r$ is replaced by B. Let us write $X_1(J'_r, K'_r)$ for such a change. As a consequence of $X_1(J'_r, K'_r)$, in Theorems 3.3. and 3.4. we had that if $J'_r \to J''_s$, then it is not the case that $T \vdash \wedge J''_s \Leftrightarrow \wedge K''_s$. For, suppose $T \vdash \wedge J''_s \Leftrightarrow \wedge K''_s$; then, if

$$T \vdash \wedge J''_s(\bar{a}_{k+1}, \ldots, \bar{a}_{k+m})$$

and

$$T \vdash \wedge K''_s(\bar{b}_{k+1}, \ldots, \bar{b}_{k+m}),$$

we derive $T \vdash \bar{a}_{k+j} = \bar{b}_{k+j}$, for all $1 \leqslant j \leqslant m$, contrary to these theorems. If T is a complete theory in the sense that for all closed A either $T \vdash A$ or $T \vdash \neg A$, then Theorems 3.2., 3.3. and 3.4. can be strengthen such that 'not $T \vdash \bar{a}_j = \bar{b}_j$' (in 3.2.) and 'not $T \vdash \bar{a}_{k+j} = \bar{b}_{k+j}$' is replaced by '$T \vdash \bar{a}_j \neq \bar{b}_j$' and '$T \vdash \bar{a}_{k+j} \neq \bar{b}_{k+j}$' respectively. Moreover, in this case we would also have $T \vdash \wedge J''_s \not\Leftrightarrow \wedge K''_s$ and we would be allowed to write $X(J''_s, K''_s)$. Thus, if T is complete, as a consequence of Theorem 3.1. we have: if $X(J'_r, K'_r)$ and not $J'_r \to J''_s$, then not $X(J''_s, K''_s)$ (under the hypotheses of the theorem). As a consequence of Theorems 3.3. and 3.4. we have:

if $X(J'_r, K'_r)$ and $J'_r \to J''_s$ directly, then $X(J''_s, K''_s)$ (under the hypotheses of the corresponding theorem).

4. Concluding remarks

In this section we shall try to explain what we mean by the phrase 'The change of $\mathfrak{A}_1$ to $\mathfrak{A}_2$ is a cause of the change of $\mathfrak{A}_3$ to $\mathfrak{A}_4$'.

Let J^0 and K^0 have the same causal ordering, let (a) $\mathfrak{A}_1 \models \wedge J'_r(\bar{a}_1, \ldots, \bar{a}_k)$ in t_1, (b) $\mathfrak{A}_2 \models \wedge K'_r(\bar{a}'_1, \ldots, \bar{a}'_k)$ in t_2, (c) $X(J'_r, K'_r)$ and let t_2 be later than t_1. If (d) $J'_r \to J''_s$, (e) $\mathfrak{A}_3 \models \wedge J''_s(\bar{b}_1, \ldots, \bar{b}_n)$ in t_1 and (f) $\mathfrak{A}_4 \models \wedge K''_s(\bar{b}'_1, \ldots, \bar{b}'_n)$ in t_2, then we say that the change of $\mathfrak{A}_1$ to $\mathfrak{A}_2$, *i.e.*, $X(J'_r, K'_r)$ is a cause of the change of $\mathfrak{A}_3$ to $\mathfrak{A}_4$, *i.e.*, of $X(J''_s, K''_s)$.

Suppose that $\mathfrak{A}_1, \mathfrak{A}_2, \mathfrak{A}_3, \mathfrak{A}_4$ are such that $\bar{a}_i$ and $\bar{a}'_i$ denote the same individual a_i in $\mathfrak{A}_1$ and $\mathfrak{A}_2$ respectively, $1 \leqslant i \leqslant k$, and that $\bar{b}_j$ and $\bar{b}'_j$ denote the same individual b_j in $\mathfrak{A}_3$ and $\mathfrak{A}_4$ respectively, $1 \leqslant j \leqslant n$. If (a), (b), (e) and (f), we can say that the individuals $a_1, \ldots, a_k (b_1, \ldots, b_n)$ have changed, *i.e.*, that they are in different relations in t_1 and t_2. If, moreover, (d) holds, then we can say that the change $X(J'_r, K'_r)$ of individuals $a_1, \ldots, a_k$ is a cause of the change $X(J''_s, K''_s)$ of individuals $b_1, \ldots, b_n$. Finally, if $k = n$ and a_1 is $b_1, \ldots, a_k$ is b_k, then we say that the change $X(J'_r, K'_r)$ of $a_1, \ldots, a_k$ is a cause of the change $X(J''_s, K''_s)$ of $a_1, \ldots, a_k$.

This completes our analysis of causality.

University of Belgrade

BIBLIOGRAPHY

[1] H. A. Simon, 'Causal Ordering and Identifiability', in *Studies in Econometric Method*, ed. by W. C. Hood and T. C. Koopmans, John Wiley & Sons, New York, 1953.

RAIMO TUOMELA

EXPLANATION AND UNDERSTANDING OF HUMAN BEHAVIOR

I. THE ARISTOTELIAN AND GALILEAN TRADITIONS IN THE SOCIAL SCIENCES

In his interesting work *Explanation and Understanding* G. H. von Wright discusses the old, but recently much debated, controversy between the Aristotelian and the Galilean paradigms for research in the social sciences. The Aristotelian tradition emphasizes the purposive aspects and purposive explanation of human behavior while Galilean science looks for causal covering-law explanation of behavior.[1] Explanation of action by citing its reasons is often understood to represent the Aristotelian mode of explanation. Among the recent advocates of teleological explanation especially the Neowittgensteinians (*e.g.*, Anscombe, Melden) have to be mentioned. The views of von Wright also belong to this group. He claims that teleological explanations conforming to the pattern of (Aristotelian) *practical syllogism* are central to the social sciences:

"It is a tenet of the present work that the practical syllogism provides the sciences of man with something long missing from their methodology: an explanation model in its own right which is a definite alternative to the subsumption-theoretic covering-law model. Broadly speaking, what the subsumption-theoretic model is to causal explanation and explanation in the natural sciences, the practical syllogism is to teleological explanation and explanation in history and the social sciences" (von Wright, 1971, p. 27).

Even if nomological causal explanations of human behavior are not to be found in von Wright's opinion, he still thinks the notion of cause plays an important role in the social sciences. For he argues that there is an intimate connection between the concepts of cause and (human) action, although so that the notion of action conceptually and epistemologically precedes that of causation.

In what follows I shall discuss and criticize both von Wright's analysis of causation and his views on the role of practical syllogism within expla-

Manninen and Tuomela (eds.), Essays on Explanation and Understanding, 183–205.

nation. This paper does not attempt to be a review of von Wright's book, which abounds with interesting ideas and suggestions. I shall have to restrict myself to some major issues on which I disagree with von Wright and omit a discussion on many topics, especially most of the ideas to be found in the last chapter of the book.

II. CAUSALITY

1. G. H. von Wright's account of the notion of causality can be termed 'interventionist' or 'experimentalist'[2]: "I would maintain that we cannot understand causation, nor the distinction between nomic connections and accidental uniformities of nature, without resorting to ideas about doing things and intentionally interfering with the course of nature" (von Wright, 1971, pp. 65–66). According to this interventionist view, causation consists basically of this idea: by his intervention in a system an agent can bring about changes which would not otherwise have occurred (cf. von Wright, 1971, p. 60f).

To describe von Wright's idea, consider two *generic* events p and q. (I have understood that von Wright's *generic* events are what are more usually called event-*types* or *kinds* of events.) Now, if an agent can act so that one of these events becomes instantatiated and if this, for all instantiations, produces or brings about an instantiation of the other one, then according to the interventionist account the first (generic) event is a cause of the second.[3] (Singular causation is then obtained derivatively from generic causation.) For instance, as I can ventilate the room by opening a window which does not in this situation become open otherwise, the opening of the window is said to cause the ventilation of the room. Here the opening of the window is called the *result* of my action of opening the window, whereas the resulting ventilation of the room is called a *consequence* of my action. (Notice, however, that von Wright's above characterization of causality in terms *bringing about* or producing does not always hold good. For there are many cases in which an action can generate or produce another action without a causal relation holding between them. Thus, for instance, an institutional 'fact' such as signing a check by means of a signature on a slip of paper does not involve causal generation.)

More explicitly, von Wright offers the following *semantic* characterization of cause and effect (p. 70): "p is a cause relative to q, and q an effect relative to p, if and only if by doing p we could bring about q or by suppressing p we could remove q or prevent it from happening". In the first of the cases in the definiens the cause-factor is a sufficient, but in the second it is (merely) a necessary condition of the effect-factor. (In philosophical literature total cause-factor is often understood as a *sufficient* condition of the effect-factor; but I shall not here take up the sufficiency-necessity dispute nor shall I discuss the general difficulties connected with this kind of conditional analyses of causation; see the criticisms in Tuomela (1974a).)

In more complex and more realistic cases the causal factors will have to be relativized to an environment of other factors. This is because often or usually the cause-event itself is not a sufficient nor a necessary condition of the effect. Only when added to some other factors this cause-event may turn the whole constellation into a complex sufficient condition of the effect. But here we do not have to discuss these problems either, however essential they may be for other purposes.

The above characterization of causality must still be somewhat clarified. When a causal bond obtains between the generic events p and q in the manner of the above definition, p is assumed to be either a sufficient or necessary *nomic* condition of q. This feature of nomicity seems to be tantamount to *counterfactual support* in von Wright's analysis. We can thus assert: If p had occurred, when in fact it did not, then q would have occurred (cf. von Wright, 1971, p. 71). When we then add to this the requirement of the *potential or 'would-do' manipulability* of the cause-factor p as was done in the above semantic characterization we can say that the causal relation has become established and the following holds: "We can produce [bring about] q, *viz.* by producing [doing] p" (see von Wright, 1971, p. 72). But von Wright does not require the *actual* manipulability of p (over and above the above potential manipulability) for the *existence* of the causal bond. At most he requires actual manipulability for *testing* the causal bond empirically (cf. von Wright, 1971, p. 70).

In cases when the cause-factor p is not actually manipulable by action we then have to be satisfied with the following *meaning-giving* counterfactual conditional: If we could do p, when we in fact cannot, we could produce q. Let me point out immediately that the various theoretical and unobservable factors which often are cited as causes in science seem to fit von Wright's characterization very badly, if at all. At least the value of such a characterization as a philosophically illuminating explicans is rather small, at least for a 'non-experimentalist'. On the other hand his account clearly has great merits as a clarification of the old interventionist idea of causation, which is intimately connected with many important philosophical notions, such as freedom of will, and action.

My following critical remarks can be divided into those which are directed against von Wright's analysis as an explicate of the interventionist cause, and those which claim that in any case an irreducible notion of interventionist cause is not sufficient to exhaust the topic of scientific causation. I will start with analysis of backward or retroactive causation which belongs to the first group of critical remarks and then proceed with some more general remarks, most of which fall in the second group.

2. G. H. von Wright discusses in his book the following example which he takes to prove the existence of *retroactive causation* as well as to prove the independence of the cause-effect asymmetry of time considerations: "Suppose one could 'watch', one way or other, what happens in my brain and that one has been able to identify the neural event, or set of events, *N*, which must occur, we think, if my arm is to rise. I say to somebody: 'I can bring about the event *N* in my brain. Look'. Then I raise my arm and my interlocutor observes what happens in my brain. He sees *N* happen. But if he also observes what I do, he will find that this takes place a fraction of a second after *N*. Strictly speaking: what he will observe is that the result of my action, *i.e.*, my arm going up, materializes a little later than *N* occurs" (von Wright, 1971, p. 77).

Can an agent thus really bring about temporally prior neural events by means of *basic* actions such as raising one's arm? I claim that von Wright's argument is fallacious if it is taken to prove the existence of retroactive causation. To see why, let us reconstruct the argument as follows:

(1) I raise my arm at t_1.

(2) My raising my arm at t_1 brings about (causes) the neural event *N* to occur at t_0.

(3) The rising of my arm at t_1, brings about (causes) the occurrence of the neural event *N* at t_0.

Raising one's arm is (normally) a basic action. As basic actions (actions that can be done without doing anything else) are actions which an agent is able to perform under normal circumstances, we can perhaps see their close relation to von Wright's idea of causation, even if the requirement of manipulability by *basic* actions is not a part of von Wright's concept of cause. If an agent were (normally) unable to raise his arm, the rising of his arm would not be a cause (in von Wright's sense) of the neural event *N*.[4] We could only say that *N* is a contingently or factually necessary nomic condition of the rising of the arm.

As I find the whole idea of retroactive real or objective causation unacceptable by itself I would almost like to retort that any account of causation (at least outside microphysics) leading to a 'proof' of retroactive causation should be rejected merely on this ground (provided this proof really is valid). But even if von Wright's analysis of causation were basically acceptable a couple of other critical remarks against von Wright's analysis can still be made.

The first remark concerns the problem of the reach of retroactive causation. We might try to construe a *reductio ad absurdum* argument as follows. In the above example the neural event *N* is a necessary condition of the arm's rising. Similarly, for instance, the cooling of the earth a few billion years ago, or my being born, are both factually necessary conditions of my arm's rising at t_1. What is there in the above argument which prevents us from saying something like this: my rising my arm this morning caused, say, my birth as well as the cooling of the earth a few billion years ago? I think von Wright's argument does not clearly tell us that. He further claims, seemingly on *a priori* and not on scientific grounds, that retroactive causation has a very short reach: "It never stretches in time beyond the obtaining of the state which the agent himself *considers* as the initial state of his action, the state which he, in acting, transforms into the result of his action" (von Wright, 1971, p. 81, my italics). But von Wright gives us no convincing reasons for this more specific claim, either, and, I believe, none can be given. Perhaps he would argue that the relationship between my arm's rising and the neural event is nomic in the sense of belonging to one and the same 'closed system' and supporting the corresponding conditional statement, whereas this much cannot be said about my above examples. However, I am inclined to think the opposite. What is a closed causal system and what is not is in the final analysis a factual scientific problem. But in any case I propose that the following statement may be considered nomic in the required sense:

If I raise my arm, I must have been born for that to be possible.

It is, furthermore, of interest to notice that von Wright in the above passage characterizes the initial state of the causal system by means of what the agent considers to be the case (and what he is confident about) and thus makes the reach of retroactive causation depend on the beliefs of the agent. This additional agent-dependent or 'subjectivistic' element fits badly with von Wright's discussion in general, as he considers causal connections to obtain between *objective* extensionally describable events (also cf. his less subjectivistic definition of cause cited earlier).

Another general critical point that can be made here against von Wright's example of retroactive causation is this. There does not seem to be compelling reasons for accepting von Wright's neurophysiological 'theory' (which entails the nomological relationship between actions and neural events) on *a priori* grounds. I think future neurophysiological theories might very well falsify von Wright's speculative hypothesis, no matter how intuitively plausible we may or may not find it. Determinants of arm risings may be more complicated entities than suggested in the above example, and their relationship to behavior may be different from what is assumed here (cf. our remarks below, too).

The argument concerning the reach of retroactive causation can be criticized still from another angle. Von Wright's combining N with the raising of my arm to form a causal system but refusing to join the latter with the cooling of the earth seems to be connected with the role of intention (and its duration) over and above the mere physical event of the rising of the arm. But here von Wright is somewhat unclear (see e.g. von Wright, 1971, pp. 80–81). He does not explicitly tell us even whether psychological terms like 'intention', 'decision', 'want', etc. are referring terms at all, nor, if they are, whether they refer to behavior, brain states, some mental entities, or what.[5] In any case, I think it is somewhat dubious to think of a basic action as being a temporal event located at a specific discrete time point, such as t_1, or even an interval, but this is what von Wright does in connection with his example. Our everyday 'mental' discourse is not so closely related to the spatiotemporal physical framework that such exact criteria could be given. (However, the physical event of the rising of my arm may be considered locatable.) At least I would like to claim that if a basic action is thus locatable it will include t_0 (statement (2) then becomes false) and hence (presumably) the interval $t_0 - t_1$. If my claim is accepted then von Wright's example cannot be used to prove the existence of retroactive causation. We have only my raising of my arm at *about* $t_0 - t_1$, and this process has in some sense an equivalent physical description in which the neural event N (at t_0) is a factual necessary condition of the rising of my arm (at t_1). I would, furthermore, be willing to assert that the agent raises his arm (or gets his arm to go up) by making the neural event N happen, although this move is not needed for my present argument.

Some recent neurophysiological findings can, in fact, be interpreted to support my claim that the (bodily) action really must be taken to include the time point t_0. Thus, in an experiment subjects were asked to (voluntarily) push a rod in a tube at irregular intervals (see Becker *et al.*, 1973). Simultaneously EEG recordings were made from several positions of the head, including the mid-vertex position C_z, which is central for limb motion. The results show that there exists a very clear readiness potential about 0.5–0.8 seconds before the pushing movement (depending on the rapidity of the movement). This indicates that the action of pushing the rod starts a little before the pushing movement. If the pushing movement were blocked we would be left with this specific readiness potential together with its dramatic reduction right at the cerebral 'beginning' of the movement. It is not yet quite clear whether we have here a strict and specific necessary and sufficient cause of arm movement. In any case the readiness potential and its reduction seems *necessary* for arm movement.

It seems that we thus have evidence for saying that the *action* of rod pushing (or arm raising, if you like) does not take place strictly after the neurophysiological cause of the arm rising has taken place. If this is accepted, it again follows that von Wright's claim concerning retroactive causation is spurious.

3. Now we may go into some more general philosophical reasons which speak against accepting von Wright's analysis of causation. I shall first discuss and criticize the very idea of experimentalist causation. Then I proceed to some additional remarks on von Wright's analysis.

Let us recall that his notion of cause is (or may be called) *pragmatic* in the specific sense that it essentially relies on (the actual and potential manipulations of) human agents. (But, we may ask, what agents, and what about interindividual differences?). However, I do not find von Wright's otherwise interesting analysis very helpful as it becomes to depend in a very strong but insufficiently clarified sense on the troublesome notions of action and counterfactual conditional. What is more, I claim that, at least for scientific and theoretical purposes, causes should be analyzed in more objectivistic and agent-independent terms. To be sure, von Wright's analysis of cause has obvious *methodological* merits when applied to testing the existence of causal bonds, viz. for obtaining knowledge about causal relationships.

But, contrary to the explicit claim by von Wright (1971, p. 74), I claim that it does not very much clarify the *meaning* of the term 'cause' in the sense this term is significant in science (even in social science) and the philosophy of science. (However, compare also von Wright's own reservations concerning his interventionist or experimentalist notion of cause in von Wright, 1971, pp. 36–37.) Scientific theories in the 'pure' sciences (which aim at theoretical knowledge but not primarily at control, for instance) do not usually say anything – nor is that needed – about the actions of scientists (or other agents) who use the theory for the purposes of testing, explaining, predicting, etc. Indeed I want to argue that there is an important type of cause in science which cannot thus instrumentalistically be analyzed in terms of potential or would-do manipulability. To these cases an interventionist account of causation cannot be fruitfully applied. Let me briefly indicate what I have in mind.

Consider two typical examples of manipulative causality. The first example says that a cause of malaria is the bite of a mosquito. The second, somewhat hypothetical, example would be: a cause of cancer is the injection of a specified amount of, say, nicotine. These examples clearly give causes in the manipulative sense. But there is clearly a sense in which these are not 'genuine' or 'real' causes. In this sense the 'real' cause of malaria would be the operating of a certain amoeba transmitted by the mosquito. A more real cause of cancer again would be, not the injection of anything and perhaps not even the getting of a dose of nicotine into the organism, but rather some event or process in the organism requiring this dose to activate or evoke it. If my claim about the importance of this kind of theoretical, normally unobservable and nonmanipulable 'underlying' causal entities (e.g. viruses, electrons, black holes, latent wishes) is accepted, it follows that the idea of potentially manipulable causes is not very helpful even for epistemological and methodological purposes.

The experimentalist account of causality is, as I have already indicated, intellectually based on an instrumentalistic or 'control' image of science. This claim holds true no matter how much the account is elaborated, widened, or liberated. According to the control image the task of a science is to find out how the outputs of the system investigated depend upon its inputs (cf. Skinner's program). The inner states of the system cannot genuinely be accounted for as they are not allowed to destroy the lawful functional relationship output$=f$(input). As soon as it becomes 'obvious' that such relationships cannot be found, theoretical states (and theoretical causes) have to be taken into account, which again often entails the abandonment of strict (actual or even potential) manipulability as a semantical requirement. It is still clearer that strict manipulability cannot be accepted as a methodological requirement. (See Tuomela (1973a) and (1973b) for a discussion and elaboration of this kind of situation.)

Going back to the details of von Wright's analysis of causation, it seems to me that this account is to a great extent *circular*. This is immediately seen by recalling von Wright's definition of causation, and the discussion concerning it, earlier in this paper. For von Wright in effect equates the cause-effect relationship with the nomic bringing about or producing q by means of p. His analysis can certainly be regarded as a partial clarification of this bringing about relation. But the analysis leaves unclear what the real causal force gluing p and q together is. Recall that p and q are meant to be generic objective events. Thus the causal connection between them is exemplified by the connections between the instantiations of p and q. Obviously, we then need an account of causation which tells us something about the 'causal glue' both at the singular and at the generic level, so to speak. But von Wright does not give us very much in this respect.

As we saw, he seems to equate bringing about with *causal* bringing about (disregarding other notions of bringing about), causal bringing about with nomicity, nomicity with causal nomicity (disregarding *non-causal* nomic connections, of which there are plenty). Finally he equates (causal) nomic connections with connections sustaining counterfactual conditionals. This together with the would-do manipulability of the cause factor is then the core of von Wright's analysis of causation. But the only analysis it gives of the objective bringing about or producing relation is then that it should support the corresponding counterfactual conditional. This is very little as an analysis of a causal relationship. In particular it says absolutely nothing about the causal glue between an instantiation of p and an instantiation of q when p is a cause of q. Thus von Wright's experimentalist analysis of causation circularly relies on the *causal* notion of bringing about, which is left without sufficient clarification.

Next a few words about the asymmetry between causes and effects, which von Wright, too, wishes to establish. Do causes and effects become asymmetric in the analysis given by von Wright? It seems to me that no conclusive argument is to be found in his discussion. To review the matter briefly, if the cause-factor p were nothing but a sufficient or necessary nomic condition of the effect-factor q, the cause and the effect would not be asymmetric. As von Wright excludes time as a factor creating asymmetry there seems to be only one, but a more obvious, possibility left. It is that somehow the potential possibility to *directly* do p, but not q, is the decisive criterion. What von Wright seems to be suggesting is that if p is a basic action then q cannot be, whence asymmetry follows (cf. von Wright, 1971, p. 76). Even if this approach were otherwise acceptable it would have the undesirable consequence that then we could speak of causation only in the case p is a basic action. I think von Wright allows 'doing p' to include also cases where p is not a basic action. But as far as I can see, in von Wright's analysis even *symmetry* holds for cases where p and q are both non-basic actions. That is, it seems to follow from his analysis that, for any tokens of p and q, if causality holds in one direction it paradoxically also holds in the other direction.

The problem of asymmetry is thus still without an explicit solution within von Wright's experimentalist approach. Perhaps one might just try to work out a coherent account of asymmetry from the fact that the effect-factor q will always be *more* indirect (due to the bringing about relation) than p even in cases where *both* p and q are actions that our example agent can do (or could do if circumstances were appropriate). But even that would not of course be an acceptable way out for a 'non-experimentalist', who thinks that the fixation of causes and effects and the asymmetry of causation has to be analyzed non-anthropomorphically.

I cannot here try to sketch a full alternative and more satisfactory analysis of causality than the experimentalist account. But in any case I begin with an agent-independent or objectivistic 'entailment' characterization in terms of singular (nonrepeatable) events.[6]

Let thus p and q be singular events. Then we say that the statement 'p caused q' is true only if there are suitable statements P and Q describing

p and q, respectively, and if there is a nomic backing theory T such that T jointly with P (plus perhaps some context description) deductively explains Q. Deductive explanation can be explicated in terms of the model of explanation developed in Tuomela (1972). Theory T supposed to clarify the sense in which generic event of the kind P brings about or produces generic event Q. (I am here neglecting cases of indeterministic causation.) To make the above criterion for singular causal claims sufficient as well some additional conditions pertaining to the singular events p and q and some other conditions related to T have to be added (see Tuomela, 1974a and 1974b). Space does not here permit a fuller discussion of this version of an objectivistic entailment account, not is it possible to discuss here how this approach can be used to discuss the asymmetry and the direction of causation or other related problems.

Finally I would like to make a remark concerning the 'race' between agency (freedom of the will) and causality von Wright discusses in his book (see von Wright, 1971, pp. 81–82). According to him there is a close connection between the problem of freedom of the will and the problem of causation (and determinism) within the interventionist account, the result in this race being that agency wins causality. But in the alternative account I have begun to sketch no such connection is to be found (see Tuomela, 1974b). Furthermore, it is compatible that an agent acts freely (in an important sense) while his actions can be purposively caused by his effective intentions, wants or other effective pro-attitudes.

III. PRACTICAL SYLLOGISM AND TELEOLOGICAL EXPLANATION OF ACTION

Let us now consider teleological explanation of actions and especially explanation by means of the practical syllogism.

Before going to this topic let me point out that the description and explanation of actions is by no means the only and perhaps not even the most important task of the social sciences (including psychology). For instance, such diverse topics as the rise and development of an individual's cognitive system, the structural features of a society, etc. are certainly other theoretically respectable and important objects of study. However, following von Wright I shall here confine myself to discussing the explanation of actions only.

There is a vast philosophical literature on practical inference. I shall restrict myself to one of the schemes of practical inference von Wright employs to explicate teleological explanation of action. I will omit the discussion of its relationship to other proposed schemes (and to Aristotle's original view, whatever it strictly speaking is). The version of the practical syllogism which I will discuss and which suffices for our purposes is what von Wright calls his 'final formulation'. It can be stated as follows (cf. von Wright, 1971, p. 107):

(P1) From now on A intends to bring about p at time t.
(P2) From now on A considers that unless he does a no later than at time t', he cannot bring about p at time t.
(C) Therefore, no later than when he thinks time t' has arrived, A sets himself to do a, unless he forgets about the time or is prevented.

(Alternatively, we could omit the unless-clause from the conclusion (C) and add a premise saying that A does not forget about the time and is not prevented.) In the present form of the syllogism (P1) and (P2) are premises (constituting the explanans of a teleological explanation) from which the statement or explanandum (C) 'follows' as a kind of 'practical' conclusion.

What we have here is a scheme in which the symbols A, a, p, and t can be regarded as variables or placeholders. To get a specific practical inference (or teleological explanation) we have to substitute for them, respectively, names or descriptions for an agent, an action (i.e. action-type), a goal (an action-type or a state), and a time point. (P1) then becomes a contingent singular statement describing the agent's particular intention at a specified time. Similarly, (P2) becomes a contingent singular statement describing the agent's 'information' (belief, opinion, etc.) concerning the means (*i.e.*, the action a) considered necessary (normally: contingently necessary) for achieving the goal p. Finally, (C) yields a contingent singular statement describing the agent's action or, rather, his embarking on action. Whether an agent's setting himself to do a or his embarking on a amounts to much more than to his *decision* to do a, I cannot here discuss (see Tuomela, 1974b). If not, von Wright's practical syllogism above is only a scheme for intention transferral (or, alternatively, for explaining intentions).

I have here described only the third person form of the syllogism; the first person version involves some well known peculiarities which need not be taken up in the present context.

There is a number of difficult problems connected with practical syllogism. One major problem is in what sense (C) can be said to follow from the premises (P1) and (P2). Is the inference in some sense *logically conclusive*?

Before commenting on von Wright's answer to this question one remark is needed. I have above assumed in the manner of von Wright that all the sentences occurring in the practical syllogism are *descriptive* statements which are true or false. This basic assumption has been disputed by many authors, but I shall not here join their ranks.

Neowittgensteinians commonly claim that intentions and other 'pro-attitudes' are logically connected to the actions which 'flow' from them.[7] This claim is accepted by von Wright in a form, *i.e.*, in the sense that the premises of the syllogism cannot be verified (logically) independently of the verification of the conclusion: "The verification of the conclusion of a practical argument presupposes that we can verify a correlated set of premises which entail logically that the behavior observed to have occurred is intentional under the description given to it in the conclusion" (von Wright, 1971, p. 115).

It seems to me, that von Wright could as well have made his point without employing the (methodological) notion of verification. For his general idea is the conceptual one that, both in discussing the premises and the conclusion of the practical syllogism, we have to assume the 'teleological framework' or the 'standpoint of agency'. That is, we must assume *a priori* that we are dealing with something the agent *does* rather than with something that 'happens to him' (cf. von Wright, 1971, pp. 111–115). Thus it seems we can say that action (as opposed to 'mere' movement) and the (overall) intention connected with and expressed by it are intrinsically connected both in an ontic and in a semantic sense (and not only methodologically). To describe something as action *means*, we can say, that *there is* a "conduct plan" (*e.g.* a practical syllogism) which matches the action (cf. Tuomela, 1974b). – If this correctly describes von Wright's position, I agree with him on this problem (with some qualifications to be mentioned below).

If the logical connection argument is understood as above, one can, in addition, agree with von Wright in that "the premises of a practical inference do *not* with logical necessity entail behavior". "It is only when action is already there and a practical argument is constructed to explain or justify it that we have a logically conclusive argument. The necessity of the practical inference schema is, one could say, a necessity conceived *ex post actu*" (von Wright, 1971, p. 117).

Furthermore, we can also say the following. If one, as is usual, accepts that mental acts (propositional attitudes, acts of will) as (dispositional and intentional) *concepts* 'intrinsically' contain their objects in themselves, logical independence can still obtain in the following sense: "The logical dependence of the specific *character* of the will on the nature of its object is fully compatible with the logical independence of the *occurrence* of an act of will of this character from the realization of the object" (von Wright, 1971, p. 94).

So far I have agreed with von Wright's analysis.[8] Let us now investigate in more detail the alleged logical validity of practical syllogism. As I said above I am willing to accept the logical connection argument roughly in the sense of there being a conceptual connection between the premises and the conclusion of the practical syllogism. But I do it only with two kinds of qualifications. The first type of qualifications is related to the *ceteris paribus*-assumptions (cf. the unless-clause) of the practical syllogism and I think there is no real disagreement between von Wright's and my position here. The point is simply that, in my opinion, von Wright has not made the various factors involved here sufficiently clear and explicit. These problems have been discussed in the literature (cf. *e.g.*, Churchland, 1970), and I shall be brief in my comments.

We may say that the theory of the practical syllogism is concerned with the 'qualitative' philosophical foundations of rational deliberation and decision making (cf. its quantitative counterpart within statistical decision and game theory). However, viewed from this angle it is somewhat unsatisfactory. For it does not, at least explicitly, take into account the agent's *other* competing goals (intentions, wants, desires, etc.) than the one mentioned in premise (P1). Nor does it explicitly consider more than one means (action) for achieving the goal. These restrictions are dealt with by implicitly assuming that (1) A then had no other goal p' which he preferred to p; and (2) there was no other action a', also necessary for p, such that A preferred a' (or, rather, a' together with its various consequences) to a (with its consequences).

In any case, the practical syllogism then explicitly deals with only a small final fragment of rational deliberation.[9] What is a greater deficit is that it, in being so restricted, fails to take notice of the existence of the variety of (rational) decision-making principles. That is, it fails to recognize that there are several different but intuitively equally rational principles for combining goals and information to yield action (or a decision to act). For instance, A might slightly prefer p to p' but also strongly prefer a' to a. Whether he would do a or a' depends on his decision-principle, and what this is, is a contingent matter not to be decided on merely conceptual grounds. (Compare here the situation within decision-making under ignorance and under risk within statistical decision-making, where Milnor's paradox and other comparable puzzles show the existence of incompatible but in a sense equally rational decision-principles.)

With the above implicit assumptions concerning the agent's preferences over goals and actions, can we consider the practical syllogism logically conclusive? I think not yet. We still have to require (at least implicitly) that the agent is in another strong sense rational so that, *e.g.*, his present emotional state or his having an Oedipal complex or his miscalculations should not disturb his deliberation. Thus we are talking about rationality in two different senses here. Let us refer to the various notions (or senses) of rationality discussed in the previous paragraph by rationality$_1$, and to the present kind of rationality concerned with the absence of emotional and cognitive disturbances by rationality$_2$. Only after adding the rationality$_2$-assumption can the practical syllogism be accepted as expressing a conclusive semantic principle (and I shall discuss the price of this assumption later).

Furthermore, I think the syllogism can be formalized in a suitable language to show its logical form and to make it a case of 'theoretical' inference. In such a formalized version one makes use of a *general* principle (statement) concerning the combination of pro-attitudes and beliefs to yield action (see Tuomela, 1974b, for a formalization using event ontology). Using symbols $\bar{A}$, $\bar{p}$, $\bar{a}$, and $\bar{t}$ as quantifiable variables I thus propose that the following generalization may be used:

(L) For any agent $\bar{A}$, intention $\bar{p}$, action $\bar{a}$, and time $\bar{t}$, if $\bar{A}$ from now on intends to realize $\bar{p}$ at $\bar{t}$ and considers the doing of $\bar{a}$ no later than $\bar{t}'$ necessary for this, and if 'normal conditions' obtain between now and $\bar{t}'$, then $\bar{A}$ will do $\bar{a}$ no later than when he thinks the time $\bar{t}'$ has arrived.

I think (L) can be considered a valid semantic principle provided its 'normal conditions' are assumed to include all the qualifications discussed above plus the factors of the unless-clause of the conclusion (C) of the original version of the syllogism. I have here taken the consequent of (L) to be about the agent's doing *a* rather than his only setting himself to do *a*, for we may pack into the normal conditions everything that distinguishes setting oneself to do from doing. In order to make the inference logically valid we still need one obvious premise:

(P3) 'Normal conditions' obtain between now and $\bar{t}'$.

As our conclusion we now use:

(C) No later than he thinks time $\bar{t}'$ has arrived $\bar{A}$ does $\bar{a}$.

The result is that statement (C) logically follows from the conjunction of (P1), (P2), (P3), and (L). Let us call this version of the practical syllogism the *amended* version.

One interesting thing about the amended version is that it relies on the generalization (L), without which the syllogistic inference is not logically conclusive. This means that in a formal and Pickwickian sense we have here a kind of covering 'law' explanation of action. Only the law is non-contingent.

It may be debated whether the practical syllogism can always be turned into a theoretical inference by means of a generalization (L). For one might argue that there is no one single generalization (L) which suits all circumstances and that the schema of the practical syllogism is essentially 'open' in this respect. However, I cannot here elaborate on this point (see the remarks in Tuomela, 1974b).

What is essential about our amended version is that it deals only with an idealized (in a specified sense rational$_1$ and rational$_2$) agent. It concerns primarily the behavior of concepts in our common sense conceptual system (or rather in an idealized reconstruction of it). Borrowing a distinction from linguistics, we are, so to speak, here dealing only with (conceptual) '*competence*' and not with '*performance*'. Performance would here mean the actual behavior (and dispositions to behave) of blood-and-flesh agents. This actual behavior instantiates or exemplifies certain conceptual structures which are, so to speak, on the level of competence. Thus, for instance, an agent's actual deliberation process may instantiate the conceptual structure defined by our above schema of the practical syllogism. What a study of competence gives us is *understanding* of behavior. For recall that the model of practical syllogism presupposes that behavior is 'intentionalistically' understood as action. Furthermore, our idealized agent's action is explained by citing its reason or motive (or the like). This kind of explanation places the action in a larger culture-dependent setting of goals, values, norms and other standards together with some accompanying beliefs and opinions. All this is, of course, very important and central for explaining action in the sense of understanding it. But still it does not (fully) explain actual performance. As in linguistics, a theory of performance is of course needed and it has to rely on our knowledge of competence.

To put the above in a nutshell, a theory of competence or an intentionalistic theory of behavior assuming rationality is incomplete, if not vacuous, as a psychological theory (or theory of performance). This is because it assumes a strong kind of rationality (or intelligence) on the part of the behaving agent instead of explicating and explaining rationality and clarifying its conditions of applicability (which it should do). The assumption of rationality is nothing but a version of a *homunculus*-assumption.

If we now consider our amended version of practical syllogism from the above point of view of performance, it seems that we may have to replace at least the idealized requirement of rationality$_2$ in the list of normal conditions by something more informative. What can be offered instead? We have to require at least this: *A* must *know how* to do *a* and he must be physically and psychologically *able* to do *a* (von Wright, too, considers these two factors). There is no need here to discuss these factors in detail (for a reason to be given below). When we speak of a person's know-hows and abilities we may have to presuppose much about his psychological 'make-up' (*e.g.*, his general abilities, personality traits, skills, other wants and beliefs and his emotional life). This is a general feature about man's psychological aspects: these properties are strongly interconnected. In addition, these psychological features are related to the outer physical and social environment. For instance, environment-factors are crucial in the development of a man's cognitive system (his wants, values, beliefs, etc.). It affects both the latent and manifest, passive and active, long-term and short-term, etc., features of man's inner life.

I take my above remarks to be part and parcel of our accepted common sense psychology, and I shall not elaborate these points here. The general point I am making should be clear enough. It is that if the requirement of rationality$_2$ is dropped from the normal condition-assumptions, there is no finite (or even recursive) list of *specific* factual assumptions (of the above kind) about the agent and his environment which would make the generalization (L) a valid semantical principle. That is, any way of specifying the generalization by substituting a specific list of factual assumptions for the normal conditions-clause of (L) gives a *contingent* (synthetic) generalization (L′). The set of needed normal condition-assumptions is always open-ended and inexhaustible *ex ante*. Another reason why (L′) becomes synthetic is due to a kind of *indeterminacy* of the agent's 'subjective world' *vis-à-vis* the scientist's or, more generally, a fellow-man's 'world'. As the scientist starts investigating the agent from his outer point of view he has to connect or correlate the agent's world (beliefs, opinions, goals) with the 'objective' outer world. Therefore, the scientist's generalization (L′) will contain among its specific *ceteris paribus* assumptions *objective* factual statements which the acting agent may be unaware of and which he may even consider false.

As soon as we want to shift from *ex post actu* understanding of the behavior of a real agent to predicting behavior and to finding the actual operative reasons or motives (etc.) for action, the study of generalizations of the general form (L′) becomes a task of empirical scientific psychology. (Instead of speaking about intentions such generalizations may be about wants, desires, duties, and other related conative attitudes, which may generate intentions. Thus their degree of analyticity may be rather small.) Thus we have two empirical (or, rather, factual) elements in our amended practical syllogism: (1) The generalization (L′) has empirical (or at least factual) content. But it is still not, for conceptual reasons, a generalization with a high degree of lawlikeness (see Tuomela, 1974b, for an argument related to the agent's capability of 'free choice'). (2) Empirical information is needed for as certaining the truth of the other premises as well.

Let us see how far we have come. We are studying the behavior of people in some society. We have agreed to base our descriptions and explanations on the conceptual system internalized by that society, that is, on a certain system of concepts, general conceptual rules and some basic assumptions concerning the intentionality of behavior. (This kind of anthropological investigation is of course empirical and contingent in its nature, but this is of course a different matter than the contingent character of the principle (L′) internalized by that society.) Explaining actions by citing reasons and motives was then found to be a way of explanation which gives 'sense' or 'meaning' to them. This represents an aspect of explanation which can be called *understanding*. However, this is only one part of our explanation task. The other part is of course to explain actions by citing the real or actual reasons or motives for the agent's actions. Part of this problem already came up when discussing the amended form of practical syllogism when we noticed that any (for the scientist) 'concretely usable' version (L′) of (L) was in fact synthetic (within some community). This means that the question of what *specific* (teleological) conceptual frame a given society has internalized is not merely an *a priori* or a merely conceptual problem, after all.

After this important observation many possibilities stand open for us. A central part of the explanation-task has become a factual task of finding out which relevant (strict or probabilistic) generalizations hold true of people in that given society. For instance, a scientist may be more interested in finding generalizations whose consequents are action-descriptions rather than action-tryings. What he will find out cannot of course be dictated *a priori*. The only thing accepted *a priori* is a general and vague teleological framework, which among other things includes the general idea that a behavior theory should explain actions by citing want-belief-type reasons and motives (the preceding notions understood very widely). Our scientist may then find out how close actual agents are to rational agents (and which kind of rational$_1$ and rational$_2$ agents) and how well a certain practical syllogism actually represents their cognitive processes. He may find out whether really the action (of the practical syllogism model) a has to be believed *necessary* for the goal p. Whether the agent has to regard action a as necessary or sufficient or as having high probability, etc., for leading to the goal when he acts on a certain intention *cum* the relevant belief, cannot be dictated *a priori*. All we can say *a priori* is essentially what relates to the concept of intention, and that does not include the requirement that there be a means-action necessary for the intended goal.

According to von Wright, to teleologically explain an action is to construct a practical syllogism with true premises for it (von Wright, 1971, p. 99). But we have just noticed that this identification is unwarranted. An agent may act on or because of an intention (or because of a want), and thus act 'teleologically', without there being a relevant practical syllogism (of the kind von Wright requires) to match his action.

Not only is the model of the practical syllogism too *narrow* (in the above sense) but it is also too *wide* in the sense illustrated by the following example. Let us suppose an agent *A* collects autographs. Also assume *A* is fond of theater. *A* now learns that a famous person *X* will visit *A*'s home town and that *A*'s only chance to get *X*'s autograph is to go to certain theater next Friday night, as *X* will then attend a performance. *A* now forms the intention to get *X*'s autograph that Friday night. But he also forms an intention to see the play the theater runs (even if he might have been able to get the autograph during an intermission without seeing the play). In fact he has independently wanted to see the play, and it so happens that he will not be able to see the play on any other night than next Friday. Now consider the following two practical syllogisms, both with true premises and a true conclusion.

(PS1) *A* intends to see the play.
A considers that unless he goes to the theater on Friday night he will not see the play.
―――――
A goes to the theater on Friday night.

(PS2) *A* intends to get *X*'s autograph.
A considers that unless he goes to the theater on Friday night he will not get *X*'s autograph.
―――――
A goes to the theater on Friday night.

But now it is consistent with my above description of the situation to argue that (PS2) gives a satisfactory teleological explanation of *A*'s action but that (PS1) does not. *A* would not have gone to the theater on Friday night if he had not intended to get the autograph, we may argue. However, we can agree, it is not warranted to say that *A* would not have gone to the theater on Friday night if he had not intended to see the play. The intention to see the play ultimately became 'generated' only on the basis of *A*'s intention to get the autograph.

The difference in the explanatory power of (PS1) and (PS2), which is indicated by the respective truth and falsity of the above conditionals, is their difference in displaying *A*'s *operative reason.* I believe that the only way to distinguish explanatory practical syllogisms from non-explanatory ones is to inquire for the *causal* effects on action the premises of these practical syllogisms have. Thus I argue that a practical syllogism teleologically explains *A*'s action if and only if *A* acts *on* the practical syllogism. But to act on a practical syllogism is tantamount to the intention *cum* belief of the practical syllogism to *purposively cause* the agent's action.[10] What is more, it is part of the *concept* of an effective intention that it purposively causes (the bodily part of) the action. I have elsewhere defended this view and cannot here go into any details (see Tuomela, 1974b).

It should be noticed that a single intention may not always be the *full* purposive cause of the agent's action, but that only a combination of several intentions gives it. That does not, however, affect the main idea. It should also be noticed that we are here dealing with *singular* causal claims, which have to be backed by means of some *laws* (cf. Section I). If there are no laws (in a *strong* sense) in psychology, as may be claimed, these backing laws must be physical (in a broad sense). However, true generalizations of the type (L′), and related ones summarizing people's propensities to make practical inferences, may still be used in giving reasons for believing in the truth of some particular singular claims of purposive causation.

NOTES

[1] Chapter I of von Wright's book contains a brief historical survey of these traditions during this and the last century. Here I would like to emphasize the fact that von Wright seems to identify explanation in the natural sciences with explanation by means of Hempel's subsumptive model of explanation. However, as has frequently been pointed out in the literature, this identification is not warranted.

[2] This experimentalist notion of causality has a long history and also many contemporary advocates. See *e.g.*, Collingwood (1940) and Gasking (1955) for discussions of it.

[3] It seems obvious to me that an experimentalist or a manipulative notion of cause must be taken to hold between *generic* (rather than individual or singular) events, as manipulability entails repeatability.

For simplicity, I will assume in this paper, in accordance with von Wright, that *events* form the ontological category to which causes and effects belong. I know this assumption is controversial but I do not think my discussion depends upon the most critical issues.

4 Note here the following important fact which shows at least that von Wright wants to think of causes as non-intentionally characterizable events: "It is not what I decide or intend that matters to the occurrence of *N*, but the event, whether intentional or not, or my arm going up" (von Wright, 1971, p. 77.)

5 However, on p. 115 of his book we learn that intending and intentionality is not anything 'behind' or 'outside' behavior. Thus it is not a mental act or characteristic experience accompanying behavior but rather "behavior's intentionality is its place in a story about the agent". This seems to indicate that such psychological terms are not referring terms at all. However, this again is hard to consolidate with what von Wright later in Chapter III says about intentions to do something which are prior in time to doing that something (cf. von Wright, 1971, pp. 103–107; cf. also his remark on deciding on p. 74 and of "occurrences of acts of will" on p. 94). This again seems to indicate that 'intention' is a referring-term and that what is refers to is action. At least we have to assume that *statements* about intentions are true, or false, and presumably they have to be *about* actions.

6 This kind of objectivistic analysis is what von Wright (1971) opposes, *e.g.*, on pp. 70–71 but what he still seems to fall back upon, almost defeating himself, on pp. 72–74. He says there that his previous analysis "does not mean that whenever a cause can be truly said to operate some agent is involved. Causation operates throughout the universe – also in spatial and temporal regions forever inaccessible to man. Causes do their job whenever they happen, and whether they 'just happen' or we 'make them happen' is accidental to their nature as causes". Consequently, we may say that von Wright's idea can in these cases, which are just the normal ones in science, be applied only metaphorically, which is to say very little. The quoted strange passage makes one wonder what von Wright himself thinks to really have been accomplished by his interventionist account of causation. Also recall that von Wright's interventionist analysis of causality was argued to depend on a causal notion of bringing about. The causality in question must presumably be *objective* causality.

7 We still lack a general satisfactory account on the similarities and dissimilarities between the conceptual behavior of different psychological proattitudes (such as intentions, volitions, wants, desires, etc.). In any case I am willing to submit that they all intrinsically contain a disposition to behave towards a goal, and this in fact suffices for our present discussion.

What we would also need is a better systematic clarification on the semantic and ontological status on the differences between concurrent and preformed intentions, between such broad concepts as intentions, reasons, etc. and such more specific psychological concepts as desires, wishes, hopes etc. For our purposes it suffices to say that wants, desires, interests, obligations, etc. can form intentions and stand as (partial) reasons.

8 Strictly speaking, it seems that I am willing to accept a slightly stronger version of the logical connection argument. For I think that a breakdown in the logical connection must *always* be attributed to a failure of some of the normal condition assumptions to hold true (cf. below for the various normal condition assumptions). However, von Wright does not seem to think so, to judge from the example concerning a kind of 'akrasia' on p. 116 in von Wright (1971). I find this example troublesome for von Wright and in need of an explanation.

9 It should be noticed that von Wright's theory of the practical syllogism does not deal with an agent's plans for acting in any broad sense. In general such 'conduct plans' may

contain descriptions of wants, desires, wishes, etc. in addition to descriptions of the agent's intentions. Furthermore, such conduct plans may be concerned with long chains or series of actions instead of only one action (cf. Tuomela, 1974b).

Secondly, the practical syllogism does not specify how the agent's wants, desires, obligations, etc. generate his intentions or serve as the agent's basis in his intention formation.

Thirdly, the theory of the practical syllogism says nothing about what is going on either in the agent's body or 'in his mind' when his intention becomes effective and 'yields' action.

[10] Kim (1975) has investigated more carefully various forms of intention generation as well as the explanatory power of the practical syllogism. However, I do not think his remedies go to the heart of the matter. What is needed is an explication of the explanatory power of intentions which goes "underneath the conceptual surface" into ontological matters. This is what a study of purposive causation can be thought to give us.

REFERENCES

Becker, W., Iwase, K., Jürgens, R., and Kornhuber, H.: 1973, 'Brain Potentials Preceding Slow and Rapid Hand Movements', paper read at the 3rd International Congress on Event Related Slow Potentials of the Brain, held in Bristol, August 13–18, 1973.

Churchland, P. M.: 1970, 'The Logical Charater of Action-Explanations', *The Philosophical Review* **79**, pp. 214–236.

Collingwood, R. G.: 1940, *An Essay on Metaphysics*, Oxford University Press.

Gasking, D.: 1955, 'Causation and Recipes', *Mind* **54**, pp. 479–487.

Kim, J.: 1975, 'Intention and Practical Inference', this volume, pp. 249–269.

Tuomela, R.: 1973a, *Theoretical Concepts*. Library of Exact Philosophy, Vol. 10, Springer-Verlag.

Tuomela, R.: 1973b, 'Theoretical Concepts in Neobehavioristic Theories', in M. Bunge (ed.), *The Unity of Scientific Method*, Synthese Library, D. Reidel Publishing Company, pp. 123–152.

Tuomela, R.: 1974a, 'Causes and Deductive Explanation', forthcoming in PSA 1974, the proceedings of the Fourth Biennial Meeting of the Philosophy of Science Association, held in Notre Dame, Indiana, November 1–3, 1974.

Tuomela, R.: 1974b, *Human Action and Its Explanation*, Reports from the Institute of Philosophy, University of Helsinki, No. 2 1974, forthcoming in expanded form in Synthese Library, D. Reidel Publishing Company.

von Wright, G. H.: 1971, *Explanation and Understanding*, Cornell University Press.

PART III

HUMAN ACTION AND ITS EXPLANATION

ANTHONY KENNY

HUMAN ABILITIES AND DYNAMIC MODALITIES*

In his pioneering *Essay in Modal Logic*, G. H. von Wright, after systematising the uses of various modal words, said in a concluding note

> An important use of [the modal words] is connected with the notions of an ability and of a disposition and with the verb 'can'. For example: Jones can speak German (=it is possible for Jones to make himself understood in German); Jones cannot speak German (=it is impossible for Jones to make himself understood in German). We shall call the modal concepts which refer to abilities and dispositions dynamic modalities (I am indebted for the term to Mr. Geach)... The question whether the dynamic modalities, i.e. the logic of abilities and dispositions, is subject to exactly the same formal rules as the alethic modalities will have to be investigated separately. (*An Essay in Modal Logic*, Amsterdam, 1951, p. 54.)

So far as I know von Wright himself has never since directly investigated that question. In *An Essay in Deontic Logic and the General Theory of Action* (1968) he took a different approach to the notion of ability (pp. 47–57), which he employed also in *Explanation and Understanding*. In this paper I propose to investigate his original question, namely whether the dynamic modality of ability is subject to the same formal rules as the alethic modalities formalized by systems such as Lewis's S4 and S5 and von Wright's own M. I shall answer this question with a definite negative. I shall then ask whether the logic of ability, while not equivalent to any alethic modal logic, can be formalized in a system which resembles alethic systems in the way that current logics of obligation, knowledge and belief do. I shall conclude that unlike deontic, epistemic and doxastic logics, the logic of ability cannot be captured in a modal system with a possible world semantics of the kind familiar since the work of Kripke and Hintikka. To put the point paradoxically in the terms of von Wright: the thesis of my paper is that dynamic modality is not a modality.

I

A number of points in the passage from *An Essay in Modal Logic* call for comment. Von Wright mentions two types of dynamic modality: ability

Manninen and Tuomela (eds.), Essays on Explanation and Understanding, 209–232.

and disposition. I take it that the power of speaking German is given as an example of an ability. No example is suggested of a disposition: perhaps such properties as the brittleness of glass and the power of acqua regia to dissolve gold are meant. If we understand 'ability' and 'disposition' in this way, it must be at least initially an open question whether the same formal rules hold for abilities as hold for dispositions. Aristotle drew a sharp distinction between rational powers, such as the ability to speak Greek, and natural powers like the power of fire to burn. If all the necessary conditions for the exercise of a natural power were present, then, he maintained, the power was necessarily exercised: put the wood, appropriately dry, on the fire, and the fire will burn it; there are no two ways about it. Rational powers, however, are essentially, he argued, two-way powers, powers which can be exercised at will: a rational agent, presented with all the necessary external conditions for exercising a power, may choose not to do so. Jones may be a skilled Germanist and be at the podium of a hall filled with sharp-eared German speakers; for reasons of his own he may refuse to speak or launch into ancient Gaelic. (*Metaphysics*, *Theta*, 1046a–1048a.)

If Aristotle was right to draw such a distinction – and many recent philosophers have argued for a similar distinction, notably M. R. Ayers in *The Refutation of Determinism* – then the formal rules for ability will not be quite the same as those for disposition. A logic of Aristotelian ability would need to include a law corresponding to the medieval tag *eadem est potentia oppositorum* to express the two-way openness of rational powers. Henceforth I shall have little to say explicitly about natural powers or the dispositions of inanimate agents. I shall concentrate on the human skills and abilities of which the ability to speak a language furnishes a paradigm example.

It is noteworthy that von Wright found a certain difficulty in translating English sentences containing 'can' into a canonical form suitable for representation within his symbolic systems. He translated 'Jones can speak German' as 'It is possible for Jones to make himself understood in German'. It is not clear to me why he altered 'speak German' into 'make himself understood in German'; if such changes always need making then any modal logic to formalise 'can' will need a formidable appendix of translation rules. I think, however, that this is an accidental difficulty, and that the point would have been as well made by saying that 'Jones can speak German' means the same as 'It is possible for Jones to speak German'. What is more serious is that, whichever of the two forms we use, what follows 'it is possible that' is not a proposition, and therefore cannot be formalised by '*Mp*' if '*M*' is a proposition forming operator on propositions and '*p*' is a propositional variable.

There is, of course, nothing in the letter '*M*' which prevents it from being used as a proposition-forming operator on something other than a proposition. It might be an operator which forms propositions out of verbs, like the English expression 'I can...'. In an earlier part of his *Essay* von Wright presented a system Ml in which it was a condition of well-formedness that each (propositional) variable fell within the scope of exactly one modal operator. Clearly, there is no reason why we should not treat the variables in the system Ml as verb-variables and the operator '*M*' as interpretable as 'I can' (or, less egotistically, attach subscripts to it in such a way that 'M_a' means 'A can', 'M_b' means 'B can' and so on). The operator '*L*', which is equivalent to '*NMN*' might then plausibly be read as 'I can't help,' since it seems to mean 'It is not the case that I can not...'. If we allow ourselves to do this, then we can interpret formulae of the system Ml as laws of a logic of ability. Some of the results appear quite plausible: for instance

> *CLpMp*: If I can't help øing, then I can ø.
> *CAMpMqMApq*: If either I can do *p* or I can do *q*, then I can either do *p* or do *q*.

Both of these look *prima facie* like laws we would want in any formal logic of ability.

There would be two ways, however, in which such a system would have to remain radically incomplete as a system to formalize the logic of ability. In the system Ml formulae like '*CpMp*' and '*CLpp*' would be ill-formed, since one of the variables is not within the scope of a modal operator. This would be reflected in the suggested interpretation by the corresponding English expressions not being genuine sentences (e.g., 'If swim I can swim', 'If I can't help smoking smoking'). In such a system we would not be able to render symbolically sentences such as 'If I am swimming, then I can swim' or 'If I can't help smoking then I smoke' which, whether or not they are logical *truths*, should surely be capable of formulation in any system of dynamic modality. A system like Ml would be incapable of formalising any inference from what a man actually does to what he can do, or from what he can't help doing to what he will actually do. Indeed, despite what we suggested a moment ago, the system isn't strictly capable of handling the notion of not being able to help doing something. For if '*p*' is not a propositional variable, then a formula like '*NLNp*' contains an equivocal use of '*N*', once as an operator on propositions and once as an operator on verbs.

Secondly, the system Ml does not admit of the iteration of modal operators, since every variable must be within the scope of only one such operator: formulae such as '*MMp*' or '*LLp*' are illformed. Thus we could not express in this system – even if we waive the difficulty just mentioned – propositions such as 'If I can't help being able to ø, then I can ø' and 'If I can ø, then I can't help being able to ø' which seem to express a truth and a falsehood respectively. Many of our abilities seem to be abilities for being able: the ability to learn German, for instance, seems to be the ability to become able to speak German. It seems therefore that we want a logic of ability which will permit the iteration of the ability-operator in some way, just as we want a logic of ability which will permit the ability-operator to operate on propositions. Clearly an ability-operator which fulfils the latter of these requirements will also be able to fulfil the former, since a statement of ability is itself a proposition.

An ability-operator which fulfils these requirements is 'I can bring it about that...' or 'I can make it the case that...'. If we interpret the operator '*M*' in this way, and the propositional variables in the normal way, then we can interpret as English sentences not only the well-formed formulae of the system Ml, but also the much larger class of well-formed formulae of the system M, which permits the iteration of the modal operators at will and counts as well-formed a formula containing variables not within the scope of any modal operator.[1]

II

If we interpret the system M in this way, it soon becomes clear that neither it nor the related systems M′ and M″ are adequate to capture the logic of the 'can' of ability. This is most conveniently shown if we consider the systems in the axiomatic form in which they are presented in von Wright's appendix, in which it is shown that M′ is equivalent to the

Lewis system S4 and M″ to the Lewis system S5. (M is equivalent to the Feys system T.)

To illustrate the breakdown of these systems as applied to the logic of ability, I shall consider the system M′ or S4. I choose it in preference to the others as being initially the most plausible, and because as I shall indicate later I believe that it does come close to capturing two ordinary language senses of 'can', whether or not it captures the notions of logical possibility and necessity for the sake of which it was originally devised.

The system M′ contains three axioms: *CpMp*, *EAMpMqMApq* (which are also axioms of M) and *CMMpMp* (the characteristic S4 axiom). Let us begin by considering the last. An interpretation of this on the lines suggested would be 'If I can bring it about that I can bring it about that I am speaking German, then I can bring it about that I am speaking German' or, equivalently, 'If I can acquire the ability to speak German, I can speak German'. This is clearly false. If, when applying for a post in a German department, I am asked whether I can speak German, it would hardly be proper for me to reply 'yes', starting from the premise that I can acquire the ability to speak German (say by attending courses for three years) and reasoning with the aid of *CMMpMp* and *modus ponens*. Hence *CMMpMp* ought not to be a law of the logic of ability.

What of the two axioms which are common to M′ and M? *CpMp* at first looks unimpeachable. If I am speaking German, surely I can speak German. P. T. Geach, talking of concepts, has this to say in his book *Mental Acts*:

> To say that a man has a certain concept is to say that he *can* perform, because he sometimes *does* perform, mental exercises of a specifiable sort. This way of using the modal word 'can' is a minimal use, confined to a region where the logic of the word is as clear as possible. *Ab esse ad posse valet consequentia* – what is can be, what a man does he can do; that is clear if anything in modal logic is clear. (*Mental Acts*, London 1959, p. 15)

But is it so clear? Perhaps, we may imagine, it is inconceivable that someone should speak a language without being able to speak it. In fact, it is quite often done. The late Pope Pius XII used to give audiences to American servicemen at the Vatican. The gracious speech which he delivered on these occasions had been composed, I was told, by an Irish monsignore and learned by heart under the coaching of an elocutionist. At those audiences the Pope spoke English; but he was not, in the normal sense, able to speak English.

The example may be contested (is such parroting really 'speaking English'?). But others are beyond dispute. A hopeless darts player may, once in a lifetime, hit the bull, but be unable to repeat the performance because he does not have the ability to hit the bull. I cannot spell 'seize': I am never sure whether it is an exception to the rule about 'i' before 'e'; I just guess, and fifty times out of a hundred I get it right. On each such occasion we have a counter-example to *CpMp*: it is the case that I am spelling 'seize' correctly but it is not the case that I can spell 'seize' correctly.

Counterexamples similar to these will always be imaginable whenever it is possible to do something by luck rather than by skill. But the distinction between luck and skill is not a marginal matter in this context: it is precisely what we are interested in when our concern is ability, as opposed to logical possibility or opportunity. Of course it is on the basis of people's performances that we attribute skills and abilities to them; but a single performance, however successful, is not normally enough to establish the existence of ability. (I say 'not normally' because a single performance may suffice if the task is sufficiently difficult or complicated to rule out lucky success. Pushing one's wife in a wheelbarrow along a tightrope stretched across Niagara Falls would be a case in point.) But it would only be if a single performance always established an ability that we could offer *CpMp* as a law of the logic of ability.

Let us turn to the remaining axiom: *EAMpMqMApq*. One half of this is *CAMpMqMApq*. This, as mentioned earlier, looks at first like a genuine law of the logic of ability to the effect that if I either can bring it about that *p* or can bring it about that *q*, then I can bring it about that either *p* or *q*. For, one might argue, bringing it about that '*p*' is true is *eo ipso* bringing it about that '*p* or *q*' is true, and so is bringing it about that '*q*' is true; so any exercise of the ability to bring it about that *p*, and any exercise of the ability to bring it about that *q*, will *eo ipso* be an exercise of the ability to bring it about that *q*. But this is not correct, as is most easily seen by considering an example where '*q*' is 'not *p*'. The President of the United States has the power to destroy Moscow, i.e., to bring it about that Moscow is destroyed; but he does not have the power to bring it about that either Moscow is destroyed or Moscow is not destroyed. The power to bring it about that either *p* or not *p* is one which philosophers, with the exception of Descartes, have denied even to God.[2]

The other half of the biconditional equally is not a law. This may seem surprising. Surely, if I can either do *X* or do *Y*, then either I can do *X* or I can do *Y*. For instance, if I can either walk to the door or crawl to the door, then either I can walk to the door or I can crawl to the door. The claim ' I can take it or leave it' is surely a stronger claim than either 'I can take it' or 'I can leave it'. Surely each of the weaker claims severally – let alone their disjunction – can be inferred from the stronger claim.

This is correct, but it does not show that *CMApqAMpMq* is a logical law if 'M' is interpreted as suggested. In ordinary English 'I can do *X* or *Y*' is commonly equivalent to 'I can do *X and* I can do *Y*'.[3] 'I can take it *and* I can leave it'. But if we take '*p*' as 'I take it' and '*q*' as 'I leave it', then *CMApqAMpMq* must be read: 'If I can bring it about that either I take it or I leave it, then either I can bring it about that I take it, or I can bring it about that I leave it'. This may perhaps be true, but 'I can bring it about that either I take it or I leave it' is not what is normally meant by 'I can take it or leave it'. If my wife is worried about my smoking, and thinks I have become addicted, I may try to reassure her by saying 'Don't worry about the cigarettes: I can take them or leave them alone'. No doubt the reassurance will be unsuccessful; but it would be downright dishonest if my only grounds for making the statement were my knowledge that, complete addict as I am, I nevertheless make true 'Either I am taking a cigarette or I am leaving cigarettes alone' every time I compulsively reach for the pack.

If we are careful in interpreting *CMApqAMpMq* we see that it does not express a logical law. Given a pack of cards, I have the ability to pick out on request a card which is either black or red; but I don't have the ability to pick out a red card on request nor the ability to pick out a black card on request. That is to say, the following (*MApq*) is true:

> I can bring it about that either I am picking a red or I am picking a black

but the following (*AMpMq*) is false:

> Either I can bring it about that I am picking a red or I can bring it about that I am picking a black.

Similar counterexamples can be constructed in connection with any other discriminatory skill (e.g., one may have sufficient skill at darts to be quite sure of hitting the board, and yet not be at all sure of obeying either the command 'Hit the top half of the dartboard' or the command 'Hit the bottom half of the dartboard').

The failure of ability to distribute over disjunction is a particularly serious matter. There are alethic modal systems which, like the logic of ability, lack the law *CMMpMp*. And like the logic of ability deontic logics lack the law *CpMp* – in an imperfect world it is unsafe to assume that whatever is the case is permitted to be the case. But there are no modal systems in which the weak operator fails to distribute over disjunction so that *EAMpMqMApq* fails to be a law.

III

It is time to make more precise the nature of the 'can' that we have been trying to fit to formal systems. In recent years philosophers and linguists have offered a number of distinctions between senses and uses of 'can' and between corresponding different types of possibility. Drawing on their work one can offer an incomplete list of ten distinguishable 'can's, which can be set out in Table I.[4]

Most if not all of these classifications reveal different senses of 'can': uses of 'can' in which different syntactical and semantical rules apply. Even within the ten different classes there are significantly different sub-classes of instances, as the examples illustrate. Fortunately it is not a matter of present concern to investigate these differences in detail: the point of the table is to illustrate by contrast the particular type of 'can' under discussion, the 'can' which is used to report those human abilities, exercisable at will, of which the voluntary movements of the body and the speaking and thinking of units of language are the standard examples.

Some of the types of possibility which I have listed are capable of being confused with ability, and some of them are deliberately identified with ability by some philosophers. The possibilities most frequently thus confused or identified are circumstantial and epistemic possibility, corresponding to the 'can' of opportunity and the 'can' of consistency with known data. There are, I think, good philosophical reasons for refusing the identification; but one does not need to be grinding a philosophical axe in order to draw the distinctions, and indeed in English they are clearly marked linguistically. The epistemic can – where 'it can be that *p*'

TABLE I

Example	Type of possibility, etc.	Type of modality
(1) (a) *Modus ponens* cannot lead from true premisses to a false conclusion (b) Nine can be divided by three (c) Equals can be substituted for equals	Logical or formal possibility	Alethic
(2) (a) Men cannot survive without oxygen (b) Smoking can cause cancer	Physical possibility; dispositions, natural powers	Dynamic
(3) (a) She can speak Russian (b) I can't touch my toes (c) Anyone can learn to drive a car	Ability, mental and physical powers; personal powers; human possibility	Dynamic
(4) (a) I couldn't cross the road (b) We can't expand the economy indefinitely	Circumstantial possibility; opportunity	
(5) (a) I could have sunk the putt (b) I was able to overtake the car	Particular possibility; 'all-in' can; natural possibility	
(6) (a) He can be very stubborn (b) You can't take a joke	Volitional possibility; character	
(7) (a) Can you pass the salt? (b) I could have slapped her face	Willingness, particular inclination	
(8) (a) I can hear a strange noise (b) I can't feel any pain	Perception and sensation	
(9) (a) Stonehenge could be a primitive computer	Epistemic possibility, consistency with known data	Epistemic
(10) (a) You can import one fifth duty free (b) You can get down now (c) I cannot condone perjury	Legal, moral possibility	Deontic

is equivalent to 'For all we know to the contrary, *p*' – unlike the 'can' of opportunity or ability, is replaceable by 'may', and in British English usually is so replaced. The 'can' of ability and the 'can' of opportunity differ from each other in the way they form the future tense. 'I can speak Russian', in the present, according to context, may express either an ability or an opportunity. Not so in the future.

> I can speak Russian tomorrow, we have guests coming from Moscow

is correct; but not

> *I can speak Russian next spring; I'm taking a beginner's course this fall.

The future of the 'can' of opportunity may be either 'I can' or 'I will be able'; the future of the 'can' of ability must be 'I will be able'. Similarly with conditionals. If an ability is attributed conditionally, it must be expressed by 'will be able' or the like; an opportunity can be attributed conditionally by the plain 'can'. Compare:

> If you give me a hammer, I can mend this chair

with;

> *If you teach me carpentry, I can mend this chair.

It is not difficult to see philosophical reasons for this and connected linguistic differences. A skill or ability is a positive explanatory factor in accounting for the performance of an agent; an opportunity is rather a negative factor, the absence of circumstances that would prevent or interfere with the performance. Many abilities are states that are acquired with effort; opportunities are there for the taking until they pass. Whereas I have to possess an ability before I can exercise it, I may have an opportunity to do something which passes away before the time for taking it arrives: that is to say, it may be that now nothing prevents me from øing at *t*, but before *t* arrives something will have transpired to prevent me.

An ability is something internal to an agent, and an opportunity is something external. It is difficult to make this intuitive truth precise. The boundary between external and internal here is not to be drawn simply by reference to the agent's body: illness, no less than imprisonment, may take away the possibility of my exercising some of my abilities without necessarily taking away the abilities themselves. One thing that seems clear is that the presence of absence or an opportunity must be something external to an agent considered as a locus of current volition of wanting: of current decision, intention, choice and desire.[5] The mere lack of a desire to do something, the mere presence of a desire to do the opposite, does not by itself remove the opportunity to do it. I am away from home for three weeks and I fail to write to my wife: when I return home I can hardly avoid her reproaches by saying "I had no opportunity to write: every time I had a spare moment I was prevented by a strong desire for a Martini".

Abilities and opportunities are, of course, interconnected. Abilities can be exercised only when opportunities for their exercise present themselves, and opportunities can be taken only by those who have the appropriate abilities. The greater one's ability the less one needs in the way of opportunity: a cliff which would be impossible of ascent for the normal person presents the skilled mountaineer with an opportunity for a good climb. Conversely, some opportunities may be so good that one needs no great ability to make use of them: if the ball is only 1 mm from the edge of the hole it will not take a very skillful golfer to sink the putt. In the limiting case, omnipotence needs no opportunities; or, to put it another way, omnipotence can make an opportunity out of anything. On the other hand, it does not seem that we can say that if an opportunity is good enough no ability at all will be needed to exploit it. The opposite pole from omnipotent ability seems rather to be the necessary exercise of natural powers, where what we have is not so much an opportunity for action as a sufficient condition for a reaction. Perhaps we should say that the realm of application of the two concepts of opportunity and ability coincides, with omnipotence and necessitation marking the extremes on either side.

The fifth of the types of 'can' listed above was isolated in a famous controversy between J. L. Austin and P. Nowell-Smith: the 'all-in can' defined as ability plus opportunity. Von Wright, in his more recent work, has made a suggestion which is tantamount to the proposal that the logic of ability should be approached via the logic of an 'all-in "can"'. In *An Essay on Deontic Logic and the General Theory of Action* he introduces a modal operator '*M*', to be read as 'it is possible that', such that an expression consisting of '*M*' followed by an action-description or a biography[6] says that a certain action or life is possible. He continues:

The possibility of a certain action or life may be said to depend on two factors. It depends first of all upon the agent's *ability*, upon what he *can do* in the various acting-situations. But it also depends upon which acting-situations are possible in nature, upon the opportunities for action which nature (inclusive of other agents) will 'allow' (p. 49).

The aspect of possibility relative to ability von Wright calls 'human possibility'; that relative to opportunity he calls 'physical possibility'; the 'all-in' possibility he calls 'natural possibility'. As a formalisation of natural possibility he proposes the system M. But he adds:

The question may be raised, whether the system M is the modal logic which best serves the purposes of [the] logic of action. On this question I shall only say that the modal logic we need must, in my opinion, be *at least as strong* as the system M. Perhaps it could be some stronger modal logic such as Lewis's S4 or S5.[7]

However, if what we have said so far is correct, the system M is *far too strong* for the logic of natural possibility and the 'all-in "can"'. For something is only naturally possible for an agent if the agent has both the ability and the opportunity to bring it about. Therefore the logic of natural possibility must be no stronger than the logic of ability. But, as we have seen, the logic of ability must be weaker than M in that it must not contain as theses either *CpMp* or *EMApqAMpMq*.

The system M, however, is far more promising as a candidate to formalise the logic of the 'can' of opportunity. I have an opportunity to ø iff nothing prevents me from øing, i.e., if nothing compels me not to ø. If something compels me to ø, then I cannot help øing. These (approximations to) conceptual truths resemble the interdefinability of '*L*' and '*M*' in the von Wright-Lewis systems. Remembering that compulsion is determination by external factors, we can interpret '*L*' as 'External factors make it unavoidable that...' and '*M*' as '*NLN*', i.e., as 'It is not the case that external factors make it unavoidable that not...'. Mention of external factors shows that, as we should expect, the notion of opportunity is a relativised one for whose formalisation we shall need some such device as subscripts. Using subscripts we can read '$M_a p$' as 'A can (opportunity-wise) bring it about that *p*' or 'There is an opportunity for A to bring it about that *p*' where this is to be taken as meaning '$NL_a Np$', i.e., 'No factors external to *A* have made it unavoidable that not *p*'. 'External to *A*' here means, as indicated earlier, 'external to *A* considered as a locus of current volition'.

It will be obvious that the upshot of these suggestions is that '*M*' represents a notion considerably broader than the ordinary notion of opportunity. For instance, wherever 'p' is a necessary truth, it will follow from these definitions that $M_a p$ is true, whereas no agent A can bring it about that a necessary truth is true. Given the definition of 'M' in terms of 'L' and the translation of 'L' into ordinary language, there is nothing paradoxical about the result; but it shows that the English reading suggested for 'M' is only a very rough approximation. Again, if an agent is about to be necessitated to ø by a determining antecedent condition, he will *a fortiori* not be compelled by external factors not to ø, and therefore in this sense of 'can' he can ø. This conflicts with the point made above that necessitation by a sufficient condition does not seem to count as the exploitation of an opportunity. What this shows, however, is not that M is the wrong logic to apply to the formalisation of opportunity, but that it captures a form of possibility which lets in as well as opportunity in the normal sense the kind of possibility which von Wright calls physical possibility. But this broadening, if I am not mistaken, is harmless and no more objectionable than the broadening of the use of 'if ... then' which takes place when these words are used as a reading of the sign for material implication.

If the modal operators are interpreted in this manner, then the axioms of M recover the plausibility which they lost when interpreted in terms of the logic of ability. Let us, as before, for simplicity's sake drop the subscripts and read 'L' and 'M' as first person operators. Then $CpMp$ seems an obvious truth; if p is the case, the nothing made it unavoidable that it was not the case, and a fortiori nothing external to me made it unavoidable that it was not the case. We can no longer provide counterexamples to the distribution of 'M' over disjunction, and so we can accept the second axiom $EAMpMqMApq$.

Perhaps, following von Wright's suggestion in an analogous context, we should raise the question whether a logic stronger than M would be the appropriate one to formalize opportunity. What of the S4 axiom $CMMpMp$? Interpreted on the present lines, it would state that if no factors external to me have made it unavoidable that factors external to me have made it unavoidable that not p, then no factors external to me have made it unavoidable that not p. Is this correct? Suppose that factors external to me have made it unavoidable that p, and yet it is not the case that factors external to me have it unavoidable that factors external to me have made it unavoidable that *p*. Then either the factors which have brought about the constraining factors are internal to me; or there are no such factors, and the constraining factors were not unavoidably brought about by anything. The first possibility can be ruled out: while a present state of compulsion may have been brought about by factors which *were* internal to me, these factors must now have ceased to be internal or the present state would not be one of compulsion. The second possibility is less easily dealt with: to rule it out would need, *inter alia*, an account of the notion of 'bringing about' and of the appropriate notion of 'unavoidability'.[8] On these topics von Wright's work has thrown much light; but they are beyond the scope of the present paper. I shall not try to decide whether the S4 axiom *CMMpMp* should be accepted as a law of the logic of opportunity. If it is, there would not here be any clash with the ordinary use of 'opportunity': an opportunity for an opportunity is itself a kind of opportunity; this at least is a feature of the concept of opportunity we use when we talk of equality of opportunity.

Though the S4 axiom may be acceptable in the logic of opportunity, the characteristic S5 axiom is not. *CMLpLp* if accepted as a law of the logic of opportunity would mean that every opportunity for a constraint was itself a constraint; it would mean that wherever it is possible for me to be forced to do something I am forced to do it. A world in which the S5 axiom held in this logic would be a nightmare world of unremitting compulsion.

The suggestion that S4 may be the appropriate system for the logic of opportunity is given some support by the parallel between the logic of opportunity and the logic of knowledge; between the 'can' of opportunity and the epistemic 'can'. The weak operator '*M*' in epistemic logic is sometimes read as 'It is credible that...' or 'It may, for all we know, be that...' or simply 'For all we know to the contrary...' It can be defined in terms of a strong operator '*L*' understood as 'It is known that': '*Mp*' thus comes out as 'It is not known that not *p*'. If we make the artificial assumptions that we know the logical consequences of what we know, and that whatever we know we also know that we know, then the appropriate system for epistemic logic appears to be S4. Both the opportunity 'can' and the epistemic 'can' seem to be essentially negative notions: the absence of knowledge, or constraint, in the opposite direction. Like the logic of opportunity an adequate epistemic logic has to be relativised, as it is by Hintikka in his *Knowledge and Belief.*

IV

Let us return to the problem of formalising the 'can' of ability. This 'can', like the two operators we have just been considering, is essentially a relativised operator, unlike the corresponding operator for logical possibility. R. Hilpinen has devised a system of relativised modality, but the rightly does not offer it as a formalisation of dynamic modality: considered as such it would be open to a number of the objections made earlier to von Wright's system *M*.[9]

In one respect the formalization of the 'can' of ability should be a simpler matter than that of the other [two] 'can's we have considered. Opportunities are things which come and pass away; they are not like logical truths which remain for ever the same. Clearly, a full formalisation of the logic of opportunity would need to be combined with a tense-logic, or a time-logic, to allow for an indication of the time at which an opportunity occurred. Similarly, abilities come and go; what we are now able to do we may not always have been able to do and we may not always continue to be able to do. Once Falstaff could slip through a ring; some day we may be able to cure cancer. But the temporal modifications necessary in a logic of ability are simpler than those in a logic of opportunity.

In a logic of opportunity it is not only the opportunity-operator which needs to allow for temporal qualification. Consider the following examples:

(1) Now I can see you; a few moments ago I was busy, and couldn't.
(2) I can dine with you tomorrow, but not on Tuesday.
(3) Yesterday I could lecture on 5 May, today I can't (my engagement book has got filled up in the meantime).

In the first example the modality is temporally qualified but not the action; in the second the action is dated but not the modality; in the third, the action and the modality are both qualified but the temporal qualification of each is different. Clearly, an adequate formalisation of opportunity-sentences will have to allow for independent dating of the sentence modalised and of the modalisation. So too with epistemic logic, where we want to be able to formalise such sentences as 'It is known today that there will be a meeting tomorrow, but it was not known yesterday that there will be a meeting tomorrow'. With the 'can' of ability no such double dating is necessary. The ability-operator needs temporal specification, but the description of the exercise of the ability should not be temporally specified. For abilities are inherently general; there are no genuine abilities which are abilities to do things only on one particular occasion. This is true even of abilities, such as the ability to kill oneself, which of their nature can be exercised only once.[10]

In fact, despite the superficial expectation of similarity, the logic of ability is fundamentally different from that of the other 'can's. The inapplicability of the axioms of the Lewis-type systems which we saw in earlier sections is a symptom of a deeper feature of ability. To see this we must turn briefly from the syntax to the semantics of modal systems.

Kripke and Hintikka and their followers have shown how the semantics of a modal system may be formalized with the aid of the notion of a set of possible worlds and of an alternativeness relation between members of the set. In this type of account the proposition 'Lp' is true in a given possible world if the proposition 'p' is true in every possible world alternative to that possible world; the proposition 'Mp' is true in a given possible world if the proposition 'p' is true in some possible world alternative to that possible world.

The philosophical interest of possible-world semantics is that it enables us to systematize our intuitions about the truth-conditions of propositions containing various modal operators. Formal semantics does not enable us to dispense with intuition: we still have to use our intuitions as rational users of language to decide whether or not a given formal semantics captures the informal meaning of an ordinary language modal word. But we can apply our intuitions not just piecemeal to particular formulae – which may well result in contradictory upshots – but to systems as a whole. In the light of this one is then able to make a rational decision between conflicting intuitions in particular cases.

In effect it is the alternativeness relation on which we have to focus the beam of philosophical intuition. The Kripke-Hintikka approach permits the alternativeness relation to have a wide variety of properties: the two-placed relation of alternativeness may or may not be, for instance, transitive, symmetrical or reflexive. There are necessary relationships between the properties of the alternativeness relation in the semantics and the different syntactic systems: thus a semantic system in which the alternativeness relation is also transitive and symmetrical will make true under every interpretation all and only the theses of the Lewis system S5.

In relating the formal modal operators with the modal words of ordinary language, consequently, it is important to direct one's attention to the interpretation of the alternativeness relation. For epistemic logic, for instance, a world *W*2 will be alternative to a world *W*1 if it is a world in which whatever is known in *W*1 is true. A world *W*2 will be deontically alternative to a world *W*1 if in world *W*2 all obligations holding in *W*1 are fulfilled. A world *W*2 will be alternative to *W*1 in the logic of opportunity if in *W*2 all constraining forces operative in *W*1 have achieved their effect.

Now what would be the corresponding intuitive account of the alternativeness relation for a logic of ability? One suggestion which comes to mind is that in the logic of ability *W*2 is alternative to *W*1 if in *W*2 all the abilities present in *W*1 have been exercised. Analogy with the other cases would suggest that if this were the appropriate relation then the 'can' of ability should be represented as a strong modal operator ('*L*') not, as we have so far supposed, a weak one like the other 'can's. At first sight this seems reasonable enough. But reflection shows that there is something wrong with the idea of a world in which all *A*'s abilities are exercised. For suppose that for some ø *A* is able to ø and is able not to ø: John, say, can be a smoker and can also be a non-smoker, i.e., not be a smoker. Then in a world in which all John's abilities are exercised, it will be true both that John is a smoker and that he is not a smoker. And that is not a possible but an impossible world.

It is true that people may have inconsistent beliefs and may be under incompatible obligations: so that a world in which all a person's beliefs were true, and a world in which all his obligations were fulfilled, may be as impossible as a world in which all his abilities are exercised. That is why it has been found convenient in doxastic and deontic logic to adopt the assumption that one is dealing with rational belief and reasonable obligation. This is a justifiable simplification, because it is a defect in beliefs and obligations to be inconsistent: a defect which calls in question *pro tanto* their genuineness as beliefs and obligations. But with ability it is not so. That I have the ability to ø in no way weakens the claim that

I have the ability not to ø: it is a merit, not a defect, in an ability that it is accompanied with an ability of a contrary kind and is therefore an ability which can be exercised at will: indeed it is a mark of full-blooded ability, as I have been using the term, as opposed to natural power, that it should be a two-way ability of this kind.

The difficulty in applying possible world semantics to the logic of ability goes further than the problem of finding the appropriate alternativeness relation, however. In the different modal logics some principles follow from special assumptions about the nature of the alternativeness relation while others follow from the basic framework of possible world semantics. But one of the principles which we earlier gave reason for rejection – the distribution law, *EAMpMqMApq* – is a principle of this kind. Given the customary semantic analysis, this says that if a disjunction is true in some possible world, then one of the disjuncts must be true in some possible world. This principle will hold no matter how we choose our possible worlds or specify our alternativeness relation.[11] Hence, if we regard possible world semantics as making explicit what is involved in being a possibility, we must say that ability is not any kind of possibility; or, as I put it at the beginning, that dynamic modality is not a modality.

V

In this final section I shall consider briefly two possible lines of escape from the impasse we have reached in exploring the formal properties of the 'can' of ability. One way leads through the logic of action, the other through the logic of volition. I shall consider the two in turn.

It is not surprising, it may be suggested, that the 'can' of ability should prove recalcitrant when considered as a modality: for it represents a complex concept where the theories of modality and of activity intersect. The way out of the difficulties, therefore, may be to separate out, in formalisation, the motions of possibility and action. Suppose, for instance, that instead of representing 'I can bring it about that...' by the operator '*M*' alone, one introduced an operator '*D*' corresponding to 'bring it about that...' so that 'I can bring it about that *p*' was symbolised by '*MDp*'. This would give us the requisite symbolic multiplicity to cope with the apparent failure of ability to distribute over disjunction. The counterexamples given earlier would show that *CMDApqAMDpMDq* wasn't logically true; but this is not a counterexample to *CMApqAMpMq*. The important question would be whether *CMADpDqAMDpMDq* was a logical truth, and this of course would depend on the theory of action incorporated in the rules for '*D*'. Similarly the counterexamples to the rule that *ab esse ad posse valet consequentia* could be explained as counter-instances not to *CpMp*, but to *CpMDp*; and the interesting question would be whether *CDpMDp* held.[12] The problems thus become more tractable; but rather than disappearing they return as problems in the logic of action.

This can be seen if we consider a recent interesting paper of von Wright's, 'Deontic Logic Revisited' (*Rechtstheorie* 1973, 37ff). In that paper von Wright modifies his earlier position on deontic logic and introduces a system in which the connectives which bind descriptions of actions are not the same as those which connect propositions – verb phrases, or descriptions of action being, as in von Wright's earliest deontic work and as in the interpretation of M1 proposed above, the intended substitutions for the basic variables of the system. Von Wright's new system can be presented in the style of the present paper in the following manner. Let *p*, *q*, etc. represent verb-phrases for actions, and let '*L*' '*A*' '*N*' symbolize conjuncion, disjunction, and negation as applied to such phrases to make molecular verb-phrases out of them. Let '*D*' followed by an atomic or molecular verb-phrase represent the proposition that I perform the action described by the verb-phrase. '*K*' '*A*' '*C*' '*N*' outside the scope of the operator 'D' are to be truth-functional connectives in the normal way. The system has the following special axioms:

(A1) *CDNpNDp*;
(A2) *EDNNpDp*;
(A3) *EDKpqKDpDq*;
(A4) *EDNKpqAADKNpNqDKpNqDKNpq*.[13]

In his paper von Wright does not explicitly link his system for the description of action with a system of modality: but in the course of distinguishing between the non-performance and the omission of an action (between *NDp* and *DNp*) he has this to say:

> Omission is here understood as something stronger than the mere fact of not-doing. It is not-doing in a situation when that which is not done *could* have been done by the agent in question.

This suggests that a system combining modality with action theory should contain, on von Wright's view, the equivalence

EDNpKNDpMDp.

Interestingly enough, this equivalence plus the characteristic S4 theses (in particular *CLCpqCMpMq* and *CMMpMp*) will yield

EMDpMDNp

which has a reasonable claim to be a formalisation of the Aristotelian principle about two-way powers, *eadem est potentia oppositorum*.[14]

Another interesting feature of von Wright's system in the present context is that it lacks the law *CDpDApq*. This is exploited by von Wright in his treatment of Ross's paradox, to avoid the conclusion yielded by some deontic logics that an obligation to mail a letter is an obligation to mail a letter or burn a letter. In the present context the failure of this law enables us to solve the difficulty mentioned earlier that such things as the impossibility of bringing about a logical truth seemed to provide counterexamples to *CAMpMqMApq*. The counterexamples, we can now say, are really counterexamples to *CAMDpMDqMDApq*; and this is a principle which, in the absence of *CDpDApq*, we have no reason to accept.

But the other half of the troublesome biconditional *EAMpMqMApq* is not so easily dealt with. For when we attempt to apply von Wright's system to the solution of this problem we discover that it has a defect as a logic of action which is precisely analogous to the defect of M as a logic of ability. The difficulty arises over disjunctive actions. Von Wright's fourth axiom yields (given that iterations of '*N*' within the scope of '*D*' can be cancelled in the light of axiom (2)) the following equivalence:

EDApqAADKpqDKNpqDKpNq.

Von Wright uses this to explicate the notion of disjunctive action.

Consider, when we can truly attribute to a man the *action* of, say, reading or writing. If we do not know for sure what *a* is doing this afternoon, but think he is engaged in studies, we may say we think he is reading or writing. But this, usually, means no more than affirming the disjunction of the two propositions that he is reading or that he is writing. But suppose *a* is my student and that I order him to spend the afternoon reading or writing. Then he will have to consider what to do, paying attention to both alternatives. Assume he decides not to comply with the order. Then he omits both actions. Assume however that he complies. Then he omits the conjunctive omission of both the actions. This he can do in one of three ways. Either he reads but omits to write, or he writes and omits reading, or he does a little of both. Whichever course he chooses, decides to do, his action is intentional under the description '*a reads or writes*'. He performs the disjunctive action of *reading or writing*.

If, as this passage suggests, von Wright's system is meant to be a logic of action as intentional under a certain description, then the equivalence stated above does not hold. Our earlier counterexamples will show this. If I am given the order 'Either pick a red card or pick a black card', then I can obey it by picking out a card at random from the downward facing pack. My action is then intentional under description 'Either picking a red card or picking a black card'; but it is not intentional under the description 'Both picking a red card and picking a black card' (because that is not a true description of the action at all) nor is it intentional under the description 'Picking a red card and omitting to pick a black card' nor the description 'Picking a black card and omitting to pick a red card'. My action is not intentional under either of those descriptions, for each description, severally, can fail to hold without my intention being thwarted; indeed at the time of performing the action it is not even conscious under either of those descriptions. I conclude then that as it stands von Wright's *Rechtstheorie* system will not enable us to solve the difficulties which led to our earlier impasse.

I turn to the other way out. Many philosophers have maintained that, in the words of J. L. Austin, all 'can's are constitutionally 'iffy': that is to say, that any statement of an ability to ø on the part of an agent is to be analysed as a conditional to the effect that if certain conditions are fulfilled the agent will ø. If such a suggestion is to be worth considering, clearly the conditions of øing will have to be interpreted broadly enough to include the agent's wants. We cannot say that 'I can ø' is equivalent to 'If I have the opportunity to ø, and I do my best to ø I will ø': I agree with Austin against Nowell-Smith that even all-in power plus will does not necessarily suffice for successful action, because abilities are inherently liable to fail on occasion for no reason. However, 'I can ø' surely does entail 'If I have the opportunity to ø and I do my best to ø, I most likely will ø'. Abilities which were never, or rarely, exercised at will in the appropriate circumstances would be suspect.

Some of the arguments of this paper could be looked on as givings upport to a conditional analysis of 'can'. On such an analysis, it might be claimed, one might expect there to be the breakdown of the distribution of ability over disjunction on which we have so often remarked.[15] Certainly there can be no doubt that the logic of human abilities is connected with that of trying, wanting, intending and choosing. But on one point the argument whether 'can's are constitutionally 'iffy' has been based on a confusion. Both parties to the debate have often assumed that in the expression 'I can if I choose' the 'if' clause expresses a condition on the ability. They have then gone on to argue whether it was an ordinary 'if' or a special 'if' analogous to the 'if' in 'there are biscuits on the table if you want them'. Davidson has recently argued that in such cases no special 'if' is called for.[16] Be that as it may, it is a mistake to read 'I can if I choose' as equivalent to 'If I choose, I can'. 'I can if I choose' is elliptical for 'I can ø if I choose', where the appropriate substitution for 'ø' will be given by the context; and in 'I can ø if I choose' the 'if' clause is to be taken with the ø, as qualifying the exercise, not the ability. (Consider the capacity to weep: a child does not have this capacity at all when newborn; I do; but I don't have the ability which an actress would have, to weep-when-I-choose or weep-if-I-choose.)[17]

This confusion, of course, is no essential part of the conditional analysis of ability. Still, linking the logic of ability with that of choice will not enable one to circumvent the problems raised in the present paper. The logic of volition – of practical inference – is a region where almost total darkness reigns. I will mention just two of the problems which the notion of volition involves over and above the obvious problems about ability. First of all, the notion of wanting like that of belief is an intentional notion and therefore raises problems of referential opacity. Thus, if the notion of ability is to be analysed in terms of wanting or trying it too will be affected with referential opacity. Secondly, the direction of fit between language and what language is about is different in the case of the expression of wants and in the case of the expression of beliefs. The onus is on a belief to conform to the facts; an expression of a want casts an onus on something non-linguistic to conform to something linguistic. When we match beliefs against the world we appraise the beliefs as true or false; when we match the world against our wants we appraise the world, or bits of it, as satisfactory or unsatisfactory. It is a matter of dispute whether this different direction of fit should make a difference between the logic of theory and the logic of practice. To the extent that it should the analysis of abilities in terms of volition will lead to complications in the ability which are not – like the complications of referential opacity – paralleled in epistemic and doxastic logic.

I conclude that despite the spectacular progress of modal logic in the last two decades we are still as far from a satisfactory formalisation of the 'can' of ability as we were when von Wright wrote his pioneering *Essay*.

Balliol College, Oxford

NOTES

* The present paper was read, in various stages of development, to seminars at Stanford University, Cornell University and the Universities of Massachusetts, Lancaster and Durham. I am indebted to those who took part in those discussions, and to my fellow symposiasts at Helsinki, for much valuable criticism and stimulation. In particular I am indebted to Professors Stalnaker, Aune and Kanger and to Dr. P. J. FitzPatrick.

1 The line of thought explored in this first section was suggested by corresponding explorations of von Wright in the area of deontic logic. Over the years von Wright has wavered between reading a formula of deontic logic such as '*Op*' as 'one ought to do *p*' and reading it as 'it ought to be the case that *p*'. (See, for instance, 'Deontic Logic Revisited', *Rechtstheorie* 1973, p. 37). A way of reading '*Op*' which he sometimes adopts is 'it is obligatory to see to it that *p*' (*An Essay in Deontic Logic and the General Theory of Action*, Amsterdam 1968, p. 37). My suggestion here is in the same spirit; but I prefer the expression 'bring it about that *p*' because 'bring it about that' does not carry the suggestion of intentionality, of purposively bringing it about that *p*, which to my ear 'see to it that *p*' does.

2 Here I am indebted to Dr. P. J. FitzPatrick.

3 "He can speak Spanish or Portuguese" as Prof. A. Brod has pointed out to me, may mean "Either he can speak Spanish, or he can speak Portuguese; I don't know which".

4 For the distinctions used in the table, see F. R. Palmer, *A Linguistic Study of the English Verb* (London 1965); B. Aune, article 'Can' in *The Encyclopedia of Philosophy*; J. L. Austin, 'Ifs and Cans' in *Philosophical Papers* (Oxford 1961); P. Nowell-Smith, 'Ifs and Cans', *Theoria* **26** (1960) 85–101; M. R. Ayers, *The Refutation of Determinism* (London 1968); R. Gibbs, 'Real Possibility', *American Philosophical Quarterly* **7** (1970) 340–348; A. M. Honoré, 'Can and Can't', *Mind* **73** (1964) 463–479; J. P. Snyder, *Modal Logic and Its Applications* (New York 1971) and G. H. von Wright, *An Essay on Modal Logic* (Amsterdam 1951).

5 The qualification 'current' is important: clearly one can be constrained by past desires whose effects may now be unalterable.

6 These are both technical terms of von Wright's theory, defined in such a way as to approximate to an idealisation of the intuitive meaning of the terms.

7 Von Wright maintains basically the same position in *Explanation and Understanding* (London 1971).

8 For instance, it might be that a satisfactory analysis of the notion of bringing about would show that if *A* brings it about that *p* then *A* brings it about that *A* brings it about

that *p*. If so, then the second possibility would be ruled out (for an agent who brings about a state of affairs *a fortiori* renders it unavoidable).

9 'An Analysis of Relativised Modalities' in *Philosophical Logic*, ed. by J. W. Davis, D. J. Hockney, and W. K. Wilson, (Dordrecht 1969), p. 181.

10 For some ingenious but inconclusive arguments to the contrary, see Honoré *art. cit.*

11 For this important point I am indebted to Prof. Stalnaker.

12 To deal with these matters adequately would need a whole logic of action; but I think it can be seen that no combination of the system M with an action-operator will serve the purpose. For if the notion of agency represented by '*D*' is taken to be mere brute agency, with no suggestion of intentionality or voluntariness, then *CMADpDq-AMDpMDq* will be a logical truth, but it will be possible to find counterexamples to *CDpMDp* and thus to *CpMp*; whereas if the notion of agency represented by '*D*' includes a type of intentionality strong enough to safeguards *CDpMDp* from falsification, then it will be possible to find counterexamples to *CMADpDqAMDpMDq* and thus to *CMpqAMpMq*.

13 The fourth axiom appears to be misprinted in the *Rechtstheorie* article. As stated above it conforms to von Wright's explanation in the accompanying English text, rather than to his formulation (in Russellian style) of the axioms.

14 At least to some versions of this principle; Aristotle does not appear consistent in its enunciation.

15 If we put '*T*' for 'try to bring about that' then a first attempt at the conditional analysis of 'can' replaces '*Mp*' by '*CT*pp'. Then the law to which we have found counterexamples is *CCTApqApqCAppCTqq* which is not a distribution law. '*T*' does not in fact distribute over disjunction (*ETApqATpTq* is not a law); but in a logic of rational attempts, *ETKpqKTpTq* would no doubt be a law so that '*T*' would distribute over conjunction. The trouble with '*CTpp*' for '*Mp*' is that one becomes omnipotent by never trying.

16 In *Essays on Freedom of Action*, ed. by T. Honderich (London 1973).

17 It is a merit of von Wright's *Rechtstheorie* system that it enables one to distinguish between *CpMDq* and *MDCpq*.

LARS HERTZBERG

ON DECIDING

I

How are decisions made? A long-standing tradition in philosophy has it that we arrive at decisions through a process of reasoning called deliberation. If we deliberate correctly, we will reach the wisest decision possible. In *Explanation and Understanding*, G. H. von Wright puts forward the view that human actions can be explained by appealing to so-called practical inferences. In an indirect way, his view is a continuation of that philosophical tradition.[1]

My concern in this paper is with certain difficulties that seem to be associated with a view like von Wright's. They center around the concept of indecision: the conditions in which a person can be said to be undecided, and the way in which thinking may help resolve a state of indecision. My feeling is that the way one solves, or fails to solve, the problem of indecision has consequences for one's view of how actions may be understood.

What I shall have to say may appear not to have a direct bearing on the kinds of issues that von Wright discusses in his book. He is not concerned with the question of how we make decisions. His concern is rather with the understanding of actions, and so, ultimately, is my own. However, in taking over the traditional concept of practical inference, I feel von Wright has also taken over certain limitations to which the traditional view is subject. This is what I shall try to show. These limitations, I feel, have a bearing on the use of practical inferences for explaining actions. (It seems likely to me that Aristotle, who first spoke of practical inferences, was aware of these limitations. And so, it seems, is G. E. M. Anscombe, in her book *Intention*.)[2]

Generally speaking, I want to suggest that indecision is possible only because normally we do not hesitate about our actions. When we cannot make up our minds, that is because of a conflict between the ways in which we would otherwise act as a matter of course. And similarly, in-

Manninen and Tuomela (eds.), Essays on Explanation and Understanding, 233–247.

decision can only be overcome by our being able to relate the problematic situation to one in which we would act without hesitation.

However, it seems to me that, if we were to apply a view like von Wright's to the nature of decisions, it should be possible to conceive of a person for whom indecision was the rule rather than the exception. Such a person could still make decisions as long as he had desires and beliefs: all he would need to do is derive an action from them – make a practical inference. This is what deliberation consists in.

The principles of deliberation, on this view, form the basis of human action – and also the basic clue for understanding actions. This, I believe, is a mistake. We begin by learning to *act* – deliberation can only come later. And similarly, our understanding of conduct begins with our learning to recognize the ways of acting that are appropriate in various circumstances. It is only against this background that we may get to understand the way people deliberate.

II

Let us begin by trying to get clear about the nature of indecision. We may be struck by the fact that if a decision to be made is important enough and difficult enough, we sometimes feel agony in the face of it. But what is 'difficult enough'? Tasks to be done, it seems, may be difficult for one of two reasons: because they demand a great effort, or because, even with our best efforts, the risk of failure remains high. Which of these do we have in mind in calling a decision difficult? It seems obvious that it could not be the latter. For the limiting case of a difficulty of this kind would be one in which I could do nothing to improve the chance of success. This might be the case where I have to make a choice between two unmarked roads that are unfamiliar to me. But though it would be hard to do the right thing in such a case, this is hardly a paradigm of the way decisions may be difficult – precisely because every effort I could make counts for nothing. There is no room for deliberation, hence it would be strange for anyone to feel agony over this choice. (The agony of deciding must be distinguished from the fear of what may happen. The former can only be felt by those responsible, the latter by all those affected. Contrast, for instance, the predicament of the general who has to decide about an attack with that of his troops whose lives may be at stake.) Von Wright, too, wants to disregard such cases, referring to them as 'gratuitous choices'[3] as they cannot be seen as determined by the agent's wants and beliefs.

It seems, then, that we must opt for the other interpretation of what makes decisions difficult: the effort required to make the right choice. The difficulty of deliberation, then, is the difficulty of calculation: a decision is the more difficult, the more laborious the calculation that has to be carried out. (In von Wright's terms, the calculation is that involved in deducing the correct thing to do from one's wants and beliefs.)

Having to perform a laborious calculation might produce ennui, fatigue, or vexation – but agony? This reaction would seem out of place – unless its source be the fear of making an error. But what could be the basis for this fear, except a feeling that avoiding the error might demand an effort – or that little could be done about it? And why should either of these be agonizing in its turn?

It is beginning to look as if the concept of an agonizing decision cannot be explained. But I would contend that this is an illusion arising from a false dichotomy. For a decision may be difficult, not simply because it involves a risk, nor simply because it demands an effort (nor even by doing partly one and partly the other). It may be difficult by as it were doing both at the same time. For the dichotomy between decisions which are laborious but safe and those that are gratuitous but risky rests on a misleading picture of the facts. When we think of a decision as a putting into effect of what we believe, or consider right or valuable, it is easy to think that we do this by, as it were, holding up our beliefs and opinions against reality to see how it measures up. From this comparison we then derive our conclusion. The point, however, is that there is no unique way the comparison has to be made, and so everything depends on *how* we make it. For this reason, though an action *seems* to be the best thing we could do, we may remain in doubt as to whether it *is*.

The problem of indecision, I want to say, arises when a person is not certain whether the action he contemplates would be his best effort – not because he has not yet made the necessary computation, but because he is not clear about the way to think about the present case. (He may not be clear, for instance, what computation is to be made. Once he knows that, he is no longer properly undecided – to compute is to act.) There is no guarantee that thinking things over will make things clearer for him –

there may not even be a single 'wisest' thing to do. But then again, there is no guarantee that it will not make them clearer as there is when he makes a random choice. – What such a person fears is later coming to realize (or being made to realize) that he should have acted differently, and, more important, that *he might have realized this to begin with.* In cases where the wisest thing to do is obvious, as well as cases where there obviously is *no* 'wisest' thing, one can rest assured that he will have nothing to blame himself for. To be undecided is to lack this assurance; hence the agony.[4]

III

From the spectator's point of view, to understand a case of indecision is to detect a possible difficulty concerning the rules (beliefs, principles of action, etc.) in terms of which he would otherwise be able to understand what people do – a difficulty concerning the way those rules are to be applied in the case at hand. The situation must be seen as a *possible* borderline case. (Thus, indecision could not explain why a man allowed his hand to be burned, rather than pull it out of the fire.) And similarly, he can understand the deliberations of the undecided only in so far as they appear relevant to removing such a difficulty – as by letting the situation be viewed in such a light that the thing to do will become obvious.

Consider, for instance, the way a U.S. Supreme Court justice might go about deciding whether a certain procedure violated a defendant's right to due process. (Say, police asking him to show them the scene of a crime before advising him of his right to remain silent.) Obviously, no simple inference could be used to solve the case; otherwise, it would not have reached the Supreme Court. In most cases, the question of due process is not even raised, or is solved in the lower courts. Unless there were such simple cases, no problem would exist; the situation would then be like that of a random choice. The difficulty arises because there are obvious cases on either side of the matter, all of them bearing some striking similarity to the present case. The problem lies in seeing which similarity is most important. One might do this, perhaps, by turning the case over on all sides, maybe imagining it slightly different, in the hope of finding an angle that will permit the solution to stand out as obvious. (Consider the way we sometimes decode a handwriting by comparing an illegible character to one that is clearer.)

What will appear as obvious, of course, will not be the same for all persons, but unless agreement were fairly common, no legal system could exist. It is important, however, to see that this agreement could not rest on a rule, since the problem only arises when the rules give way. (A justice might of course attempt to extract a general rule from his decision. But the rule is then based on his decision, not the other way round.)

At other times, indecision might concern the way a situation was to be viewed in relation to a rule of conduct, a moral standard, or some value one was trying to achieve. – Can I trust this man? Is that an ashtray? Is this action compatible with my promise? Would that be a generous thing to do? – Such questions can be asked, and answered (if at all) only because we have learnt to apply these rules and concepts in situations where no question arose.

IV

Could we imagine a person who was immune to the agony of indecision? While he took no fewer pains than others to do the right things, he would shrug his shoulders at the idea that a decision, once made, might not be the best one. If his decisions had bad consequences, he would invariably claim that this could not have been avoided. He would put off decisions only for the time it took him to gather the relevant facts or perform the necessary calculations. He would find the notion of being paralyzed by a difficult choice unintelligible. It would never occur to such a person that a situation could be seen in different ways. If he had overlooked some aspect of a situation, he would later claim that it was impossible to notice it.[5]

This man, on the average, might be right as often as anyone. Perhaps he would even be more efficient than most of us are at making decisions. Yet he might strike us as inhuman, machinelike. We might perhaps try to justify this feeling by saying that such a person could not be fully aware of what it means to be responsible for one's actions.[6]

A related point is made by Peter Winch in 'The Universalizability of Moral Judgments'[7], where he suggests that the *way* a decision is reached is sometimes more important for our view of a man's character than the decision itself. His example is from Herman Melville's story *Billy Budd*, in which Captain Vere is faced with the decision whether to sentence to death a man he knows to be morally innocent though guilty in the sense of the law. After a hard inner struggle he decides in favor of the punishment. The reader of the story may disagree with the captain's decision, yet he cannot fail to recognize, from the way it was reached, that the captain was a man of conscience, as surely as if he had decided the opposite way. Yet if someone else had decided as the captain did, but without feeling any anguish, we might have seen in that a sign of callousness or cruelty. (Similarly, some of Melville's readers might feel that, had the captain reached the opposite decision *without qualms*, that would have been a sign of sentimentality, not of high moral standards.) So in a case like this, our opinion of a person's character may depend more on the way he relates to his decision than the decision itself. (The existence of such cases may be part of the reason why the inability to agonize over a decision strikes us as inhuman.)

How could a view like von Wright's allow for this observation? If the only difficulty in deciding is that of the calculation required, what is to keep us from saying: "The faster the decision is made, the better – as long as it is the right one"? A view like that seems to leave no room for genuine dilemmas – moral or otherwise. For if one's beliefs and attitudes, and only they, uniquely determine the correct decision to be made, only two possibilities exist. Either they dictate a solution in the present case – and then there is no dilemma, or they leave the solution open – but then no dilemma exists either; for if everything depends on them, it cannot matter what we do unless they tell us what to do.

Generally speaking a difficulty I seem to detect in a view like von Wright's – and this is a feature it shares with many philosophical accounts of what it is like to make a decision – is that it does not leave any room for what one would like to describe as the personal contribution that is often involved in making a decision. (On this account, we might as well hand over our decisions to be made by someone else – say, someone more skilled in mathematics.) The solution lies in seeing that, while a person's beliefs and attitudes do have a bearing on the way he acts, they may be expressed in a variety of behavior. Consequently, something beside those beliefs and attitudes comes into play; for in the end, as we have seen, everything depends on what the agent does about them. The fact that I am committed to some goal, or to some belief about the way to achieve that goal, does not by itself show me how I am to act; rather, I have to apply my commitments myself. The way I do that shows what kind of person I am, and how I have learnt to act.

V

On von Wright's view, reaching a decision must be similar to making a discovery: the correct answer is already determined before we find it. What determines it is the fact that we have certain intentions, beliefs, etc. – "Given these intentions and beliefs, the one thing for a rational man to do is –." Our task is to complete that sentence, and the method is practical inference. However, we could do this only if our beliefs and attitudes could be taken to determine how they were to be applied independently of us. This view of the matter becomes plausible because we think of all the cases in which the way to act on some principle presents no problem – and so, it appears, there is no room for a personal contribution. Hence we conclude that there must be a definite way of acting on the principle in the borderline case too, although it is harder to detect. On this conception, the question of what is the rational, consistent thing for a man to do turns out to be impersonal, ahistorical, and culture-independent. In selecting the correct action, the agent's intentions and beliefs are *factors to be taken into account*; the correct way to do this is determined once and for all.

The approach I want to recommend takes a different view of intentions and beliefs and their connection with action: what the agent gives expression to in announcing his intention is not a premise from which a conclusion is yet to be drawn, but is *itself a commitment to acting a certain way*. What we take account of in acting is not the fact that *we have* certain intentions or beliefs, but rather the facts of the case as we perceive them. To have an intention is to be prepared to take facts into account in a certain way – if I never did, I simply would not have that intention, however much I protested I do. For in making decisions, I am not concerned with doing whatever it takes to be rational; I am simply concerned with doing whatever it is that I happen to intend, desire, consider right, etc. Yet my rationality – or lack of it – may show itself in the decisions I make. (Just as the behavior of a sane man is not to be thought of as a successful attempt at sanity. If I have to try, I am not really sane.)[8]

But we cannot infer what is involved in a commitment from the words used to express it; this in the end is something that can only be seen from the way a commitment expressed that way is commonly carried out. So in a sense the correct way of acting on a given belief or intention depends on the way people actually do things. This is the context of use that shows what interpretation to put on words like, "I intend to ---". Obviously, the context, and so the interpretation, may vary among different societies.

To see this point more clearly, let us consider von Wright's concept of *practical necessity*.[9] We commonly explain our plans by saying that such and such an action has to be performed for the sake of some goal. The picture involved here seems to be this: for a person who has a certain intention, discovering that a given action must be performed for the sake of fulfilling it will give his behavior a specific focus. In this way, the belief that an action is required mediates between intention and behavior. This point is thought to be of peculiar importance for the explanation of actions. For only when you know why an action was the *sole* thing a person could consistently do, von Wright claims, have you succeeded in explaining it. This requirement is analogous to the deductive-nomological model of explanation which von Wright accepts for natural science, but rejects for social science. But in either case, he says, to explain is to show what took place to have been inevitable. And so a perspective which makes it seem as though the agent had several courses of action open to him does not yet enable us to understand why he id what he did.

My feeling is that this account may involve a misunderstanding of the role of practical necessity in explaining actions. If practical necessity really were what it is made out to be on this view, it would rarely be useful for trying to understand behavior. How often does it occur that we have some goal in mind, and could think of only *one* way of trying to achieve it? If someone told me that the only way I could get a pound of caviar would be by going to a store and paying one hundred dollars, I could perhaps retort that I could also do it by snatching the caviar from the counter and running off with it, holding up a restaurant, or taking up sturgeon fishing on the shores of the Volga. The same result might be 'technically possible' in all of these and any number of other ways. The same is usually true in most other cases where a claim of practical necessity is made: our imagination is the only limit to the number of possible methods. – The point of this observation is not facetious; what I am trying to suggest is that the commonly accepted notion of practical necessity is misguided, and does not answer to the way we ordinarily speak. To say that an action is practically necessary is usually not to claim that the result could not possibly be achieved any other way. That is why the statement that I could get caviar by fishing for sturgeon would not even count against the claim of practical necessity. Rather it is simply irrelevant: "That is not the way things are done". Knowing what is relevant and what is not is part of understanding what it means for a person in our culture to say, in a certain setting and tone of voice, "I would like some caviar". For to understand such expressions we must know their place in the lives of those who use them: we must know what kinds of actions a person would commit himself to in using those words, and what actions would show that he had changed his mind.[10]

Judgments of practical necessity, then, involve more than a knowledge of empirical fact. We learn to make such judgments by learning the correct way of acting in various contexts – that is, by growing up into a social framework. A discussion of practical necessity has sense only against the background of a shared way of life.

Practical necessity, we might say, is not so much to be thought of as the basis for our actions, as something which shows itself *in* the ways we act and find it appropriate to act. Generally speaking, expressions of intention, desire, belief, and the like, do not automatically make clear the sense of the actions that express them; for *their* sense, in turn, depends on what kinds of actions are commonly taken to be connected with them. But on von Wright's account, to grasp the sense of an action *is* simply to know what beliefs and intentions it derives from.

VI

At this stage, someone might object that we have overlooked the following point: in their everyday use of language, people do not aim at logical completeness of expression, but rather tacitly assume that certain facts are understood. When someone says, "I want some caviar", we can normally take it for granted that he does not intend to commit a crime, or go to every expense or effort whatsoever to get it. So what rules out robbing a restaurant to get the caviar is not the meaning of such an expression itself, but simply another, unmentioned intention. Though sometimes we mention only a single intention in explaining an action, the action will in most cases have been guided by other intentions as well. Many of those, however, are commonly shared, which is why we need not always mention them. But they too are involved in our understanding of the action.

Thus far the objection. Arguments of this form are quite common in philosophy; yet I believe they are often misconceived. What makes this type of argument suspect is the impression it gives of adjusting the facts to fit the theory. The way people actually speak is given a rather high-handed treatment: "In everyday discourse we use forms of expressions that are logically incomplete because they are more convenient and rarely cause confusion." Hence we assume that the conditions for logical completeness may somehow be found by looking beyond the actual use of language, as if the *logic* of a way of talking did not show itself in – and only in – that way of talking itself. But it is precisely this view of the matter that often enables philosophical difficulties to arise.[11]

Yet it might seem as if this approach had something to recommend itself in this case. The picture invoked is this: though we could not deduce from the goal mentioned in explanation that the agent had to do precisely what he did, we understand that he chose the route he did in order to avoid stepping on the toes of all his other, unmentioned intentions. But those background intentions too would have to be stated in a logically complete explanation. And this, it will be added, is important, because it is just the fact that the agent has them that enables us to understand his actions at all.

It is true that, in most of our everyday affairs, we do not pursue a single goal at a time. The ways we act can usually be seen as compromises between a variety of demands imposed on our behavior. In writing a paper in philosophy, one's concern is with making points that are valid, but also original; with finding a form of expression that is clear without being tedious; and (maybe hardest of all) with completing the work under severe limitations of space and time. And we are all familiar with this necessity to accommodate competing demands in our lives – sometimes to the point of frustration. Some find this easy to do, others – zealots, puritans, bohemians – refuse even to try (at least to the same extent as the rest). Society looks askance at those who refuse: living without compromise is not – nor could it become – a generally accepted way of life. Such an attitude, we want to say, represents a shortcoming, an escape from responsibility. This, of course, is a weakness we all share at times, as when our present concerns absorb us and make us lose our sense of proportion. The important point, though, is that it *is* considered a weakness.

However, I believe these observations help refute the view we are discussing rather than support it. Looking more closely at these compromise situations, we can see again that what is primary is the activities themselves – the ways of doing things – not the intentions. For usually we cannot simply derive the correct compromise from the goals that we have. There is no simple formula to be used, for instance, by a surgeon who is torn between the necessity for operating soon and the need to get a reliable diagnosis before he does. If there were, his work would be easier than it is. Of course different surgeons would not decide such matters in precisely the same ways – thus displaying differences in their personalities or their education. But there are some decisions that, say, a medical review board would unanimously consider wrong; and hopefully also some it would consider right. But this being so is in the end simply a matter of its members having a shared way of judging – a shared practice. And so it appears that the question about what would be a good compromise in a case like this is one of those that can only be discussed against the background of a common way of doing things.

As a matter of fact, the ways competing demands are accommodated are among the most striking differences between cultures. In some cultures, the ability to 'keep one's cool' is highly valued, while other cultures put emphasis on 'letting go'.[12] Westerners are often astonished by what is reported to be a nearly complete disregard for time among peoples in the East. Such a difference cannot be explained simply by saying that making good use of time is one of the goals that a Westerner has, and, say, a Bedouin lacks. It is much more a question of what *is considered* a 'good' use of time in the two cultures. Nor is this something that members of a culture have decided on (whether consciously or otherwise): it is simply the way they do things. Indeed, only through being brought into contact with other ways of doing things is the Westerner made aware of the importance he gives to time. Besides, not even he always gives it the same importance: cutting down funeral services to five minutes, say, is not an acceptable way of saving time. (And not because of a 'weighing' of the factors involved.) The point could be expressed as follows: in a manner of speaking, the Westerner does not hurry because he values time, but rather he values time because he hurries. *Why* this life-style has developed is a different story; one that would perhaps involve a mention of industrialism and the new social roles it created.

VII

So far, this paper has concerned itself with areas where I disagree with von Wright's position, even though I feel that, if it is erroneous, the error is *deep*. However, I also agree with many of the points made in *Explanation and Understanding*, among them one I consider of special importance. This concerns a basic difference between two kinds of explanation. In explaining events in nature, we usually rely on our *recognition* of patterns of events as having occurred before. But in a social context, what we need is rather an ability ot look at behavior in the light of social *requirements* – to distinguish between a right and a wrong way of doing things. We can see more clearly what this difference is like if we think of a context where we are trying to predict an event. If, say, a prediction in physics misses the mark, we only have ourselves to blame (say, for our ignorance; for having overlooked something, or having relied on too crude measurements). But when we fail to predict behavior in the social realm, we can sometimes blame it on the *agent*.[13] – "The reason we expected the car to stop was that the light turned to red." It was the driver's fault, not ours. (Or maybe the brakes' – but then perhaps the mechanic's, or the manufacturer's.) – Such is often the knowledge that we have of human behavior, on which we base our own actions. One could even imagine a philosopher – someone more eager for definitions and essences than I am – making this his definition of society: "a collective in which expectations are primarily based on mutual demands, rather than empirical laws of behavior."

The idea of a practical inference helps bring out this aspect in two ways. First, where necessity is concerned, a practical inference invokes a different concept of necessity than that appealed to, say, in physical explanations. The action is not shown to have been necessary *because* of what *was* the case, but *for* something to *be* the case. It was demanded by the goal the agent had, not caused by his having the goal.

Secondly, the concept of a practical inference emphasizes the analogy between deliberating on actions and explaining them; since it makes out the latter as involving a retracing of the steps of the former. In this manner, it brings out the close connection that holds between learning to act in society, and learning to understand the actions of other members. Both of these are usually taught side by side, and both are equally important for a person to be said to participate in an institution. (A striking example: language.)

But as I have tried to show, the details of this approach raise some questions. And my feeling is that these questions cannot be brushed off as trivial. Rather, they urge us to reverse the underlying picture. Let me try to outline what I take the picture to be, and why it is misleading.

The picture concerns the way we learn to act. An engineer and a shopkeeper, for instance, have learnt to do the various things required by their respective trades. Each has a special skill, but they also have some skills in common: both, say, have learnt some mathematics, though they apply it to different things. Now by the picture of rationality I have in mind, rationality is similar to such a common skill, although it is even more common than mathematics, for it is shared by all who have learnt to act rationally. To learn to be rational is to have one's natural behavior replaced by (or channeled into) rational ways of acting: one's life comes to be guided by means-end relationships. (This might either be taught or the outcome of a natural development.) In acting rationally, we all use the same 'method' for different ends. Once you have learnt what it means to be rational in one context, you can apply it in whatever you do.

What I have tried to show in this paper is that matters are being turned upside down in this view. Different ways of life are not derived from some common conception of rationality, for there is no conception of rationality that exists independently of the ways people actually live. Rather, the criteria by which we appraise behavior grow out of (and are part of) the various activities that make up human society. By this token, to describe human behavior in general as rational (contrasting it, say, with animal behavior) is not to point to some peculiar characteristic of the way people do things. Rather, it is to draw attention to the importance of such criteria for human actions. This is connected with the view we have of human conduct as something that has to be learnt, which entails that we may ask how well a person has learnt it. People being rational says something about the kinds of questions that we may ask; but nothing about the answers that will be given.

In criticizing those who fail to see the difference between the ways we understand human actions and events in nature, a view like von Wright's seems to go too far towards the other extreme. It is true that we often understand conduct in the light of social requirements. But the requirements imposed on behavior must be tempered by an understanding of what can be expected of a person. And hence, not everything that would make an action seem correct can be used to explain it. (Obviously, the fact that the lottery ticket you picked carried the biggest prize does not explain why you picked just that one. For that was not something you could have been expected to take into account.) But to know what can be meaningfully demanded of a person is to know something about the kind of person he is, the kind of life he lives and the kind of society he belongs to. What we appeal to in attempting to understand conduct in this way, then, is not rationality in the abstract, but a view of things which presupposes knowledge about the lives of flesh and blood people.[14]

University of Helsinki

NOTES

1 Among the best-known representatives of this tradition are Aristotle, Max Weber, and R. G. Collingwood.

2 Oxford: Basil Blackwell, 1963.

3 *Explanation and Understanding*, p. 99; p. 165.

4 We might note in this connection that it is common after one has made an important decision to try to convince himself that he *had* to choose as he did, as a method of fending off self-recrimination. The agony that follows a decision may sometimes be no easier to bear than that which preceded it.

5 He might, perhaps, be taken to be a first cousin of the aspect-blind man discussed by Ludwig Wittgenstein, *Philosophical Investigations*, pp. 213–14. This man was unable o take note of the fact that some pictures can be seen in different ways.

6 This point, of course, is strongly emphasized in the existentialist tradition. This is one of the main factors that mark it off from the mainstream of Western philosophy.

7 In Peter Winch, *Ethics and Action* (Routledge and Kegan Paul, London, 1972).

8 For a discussion of moral conduct which runs parallel to this, see Winch, 'Moral Integrity' (in Peter Winch, *op. cit.*).

9 Von Wright does not use this term in *Explanation and Understanding*, but he discusses it in a later paper, 'On So-Called Practical Inference', *Acta Sociologica* **15**, pp. 42ff.

10 Similarly, we understand that someone who asks for caviar would not be satisfied if told that he could have some – ten years from now. Here lies a difference between the use of 'want', say, in "I want some caviar", and "I want to be a doctor". Ignoring distinctions like these is part of the humorist's stock-in-trade.

11 To point this out is not to advocate the supremacy of colloquial speech. The same point holds whatever the context in which words are used, even technical terms in

science: in order to talk about a use of language, we must indeed talk about that use of language, and not some other thing.

[12] For a classical treatment of such differences between cultures, see Ruth Benedict, *Patterns of Culture* (Houghton Mifflin, Boston, 1934).

[13] Similar points have been made several times in the literature. Cp. Anscombe, *op. cit.*, pp. 56–57; Peter Winch, *The Idea of a Social Science and its Relation to Philosophy* (Routledge and Kegan Paul, London, 1958), pp. 91–94.

[14] My thanks are due to Mitchel Axler and Risto Hilpinen who commented on an earlier draft of this paper, and helped me say what I was trying to say. They are not responsible for any errors or obscurities that remain.

JAEGWON KIM

INTENTION AND PRACTICAL INFERENCE

I

According to Professor von Wright, a sharp methodological division exists between the science of natural phenomena and the study of human action. Natural phenomena are explained, and predicted, on the causal model: to explain a natural event is to specify its cause, and such an explanation is thought to yield predictions for similar events. One modern form of this causal approach, according to von Wright, is the so-called deductive-nomological model of explanation: to explain an event is to subsume it under a law. Human actions, on the other hand, are held to be subject to a wholly different methodology of investigation. Von Wright argues, vigorously and often persuasively, that the correct model of understanding actions is not the causal-nomic model, but rather the following fundamental schema of practical inference:[1]

> *A* intends to bring about *p*.
> *A* considers that he cannot bring about *p* unless he does *a*.
> Therefore, *A* sets himself to do *a*.

It will be convenient for our discussion to think of actions as cases of *bringing about a state or event p*.[2] Although it is not clear that von Wright would accept this approach, it appears compatible – in fact, congenial – with his general framework. So we reword von Wright's schema as follows:

(PI) *S* intends to bring about *p*.
S believes that he cannot bring about *p* unless he brings about *q*.
Therefore, *S* sets himself to bring about *q*.

What does 'setting oneself to...' mean? Von Wright explains: "*Setting* oneself to do something I thus understand in a way which implies that behavior has been initiated" (p. 96).

Throughout Chapter 3 of *Explanation and Understanding*, von Wright is much concerned with the question of the logical validity of (PI) as a

Manninen and Tuomela (eds.), Essays on Explanation and Understanding, 249–269.

schema of inference; all of the revisions, refinements, and qualifications he suggests for (PI) are obviously intended to give us a deductively conclusive schema of inference. That of course is no surprise: it is the heart of von Wright's teleologist viewpoint that there is a relation of logical dependence between the premises and the conclusion of an intentional explanation of human action, and that the presence of such a logical connection excludes a causal relation between the events or states represented by them. Now there is the problem of reconciling this aim of getting as tight a logical connection as possible between the premises and conclusion of (PI) and the implicit but indispensable assumption that the events or conditions referred to in the premises are distinct from the action specified in the conclusion. It is a debatable point whether such a reconciliation is possible here; a close enough 'logical connection' between two events may lead to the collapse of their distinct identities as events. Whether there really is a problem of this sort and how, if there is, it is to be resolved will of course depend on the theory of events and actions presupposed, or preferred, as a general framework. But it *is* a problem to worry about.

We shall, however, move on to another question. Why is von Wright so concerned with the logical validity of (PI)? Apart from his concerns with the 'logical connection argument', he appears to think that the schema would not be adequate as an explanatory model unless it were logically valid – at least, valid '*ex post actu*', as he puts it (p. 117). But why should logical validity be a requirement of any explanatory schema? Why does von Wright think – at least he appears to think so – that it is necessary, or highly desirable, to have a version of (PI) that is logically conclusive? One possible line of thought is this: If (PI), or some version of it, is logically valid, the volition-cognition structure represented in the premises will *conclusively determine* the action issuing from it; in fact, to show that the action issues from it amounts to showing that the inference in question is logically valid; and the existence of a determinative relation like this is necessary to sustain an explanatory relation from the volition-cognition complex to the resultant action.

Those who would interpret the relation of volition and cognition to action as causal, namely those whom von Wright calls 'causalists', would agree that volitions and cognitions *determine* actions. However, they will say that the determinative relation involved here is none other than the causal relation; volitions and cognitions cause actions. So the controversy between the teleologist and the causalist, or between the methodological dualist and the monist, boils down to a dispute about the nature of this determinative relation. At this stage of the dispute, the causal theory appears to have more definite contents: for it asserts that the volition-cognition complex explains an action because it is the cause of the action. How the notion of 'cause' is to be explained is another matter, but there is a great deal of initial plausibility in thinking that causal relations explain, that to know the cause of an event is to understand, in one possible way at least, why or how the event came about.[3]

Contrasted with the causal theory, it's not quite clear how the teleological theory accounts for the explanatory power of volitions and cognitions with regard to actions. Clearly, a logical connection alone does not suffice to generate an explanatory relation; for example, the event of Herbert's making a jumpshot logically entails the event of Herbert's jumping,[4] but we would hardly think the former explains the latter. If, therefore, the existence of a logical connection does not by itself generate an explanatory connection, then the nature of the things connected – that is, the fact that the logical connection connects a volition-belief system and an action – must be crucial to the explanatory efficacy of (PI) and similar explanatory models. But what explains this? What explains the alleged fact that a logical connection of the sort exemplified in (PI) between a volition-belief system and an action is explanatory? I do not believe we get a clear answer from the teleologists here. Perhaps, they will say that this must be accepted as a primitive, fundamental fact about the way our concepts of explanation, action, and the like work, just as we must accept as unexplainable the fact that causal relations explain.

But this response is not completely satisfactory. The causalists would indeed be in the same situation if it were the case that not all causal relations generate explanatory relations, and that whether or not they do depended essentially on the nature of the items that are causally connected. But that is not the case. It is the burden of the teleologists, but not the causalists, to go beyond the claimed relation between the explanans and the explanandum to give an account of why the explanans indeed explains its explanandum. But we leave aside these issues in this paper.

II

The starting point of our subsequent discussion is an interesting instance of (PI), due to Nicholas Sturgeon, which appears to be a prima facie counterexample to (PI) as an explanatory schema. Suppose I am going to Cincinnati to read a philosophy paper at University of Cincinnati, and since I will be there, I plan to look up my in-laws. There is no problem about the following instance of (PI):

(1) I intend to read a philosophy paper at University of Cincinnati.
I believe that I cannot read a philosophy paper at University of Cincinnati unless I go to Cincinnati.
Therefore, I set myself to go to Cincinnati.

It is plausible to say that (1) is an intentional explanation of the action of my going to (or setting myself to go to) Cincinnati. But further, given what I have said about my plan for visiting my in-laws, all the statements comprising the following instance of (PI) are also true:

(2) I intend to visit my in-laws in Cincinnati.
I believe that I cannot visit my in-laws in Cincinnati unless I go to Cincinnati.
Therefore, I set myself to go to Cincinnati.

This is Sturgeon's counterexample. If von Wright is right about the logical validity of (PI), (2), too, must be a valid argument – at least, on the point of validity it should be as good as (1). And, as noted, all of the statements in this inference are true. But we would hardly think that its premises explain my going to Cincinnati. What explains it is the intention-belief structure of the earlier practical inference (1).

One sign that something is amiss with (2) is that both of the following counterfactuals are false:

If I had not intended to visit my in-laws in Cincinnati, I would not have gone to Cincinnati.
If I had not believed that I could not visit my in-laws unless I went to Cincinnati, I would not have gone to Cincinnati.

The falsity of these counterfactuals is indicative of the absence of an appropriate determinative relation between the premises and the conclusion of (2) that might sustain an explanatory relation between them. In contrast it is obvious that the explanatorily efficaceous practical inference (1) passes the counterfactual test: the two counterfactuals corresponding to the above are true – e.g., it is true that had I not intended to read a philosophy paper at University of Cincinnati, I would not have gone to Cincinnati.[5]

I want to point out that the absence of a determinative relation in (2) is compatible with the logical validity of (2). For in general the fact that P logically implies Q is consistent with the falsity of the counterfactual 'If P were false, Q would be false'. 'George is a bachelor' implies 'George is a male'; however, 'If George were not a bachelor he would not be a male' obviously isn't true. We earlier observed that the presence of a logical relation between two events is not sufficient in itself to yield an explanatory relation between them. What all this suggests is that, in spite of von Wright's concern with the 'logical conclusiveness' of (PI), the logical validity of (PI) or its brethren may not be crucial to the question of its explanatory adequacy. If we agree with von Wright that (PI), or some schema like it, is logically valid, then both (1) and (2) are logically valid inferences. The first appears to be explanatory; the second is not. Further, the explanatory (1) exhibits a determinative relationship, as evidenced by the truth of certain counterfactuals, whereas the nonexplanatory (2) does not. What does this suggest if not that the 'logical conclusiveness' of (PI) may be wholly beside the point as far as the explanatory efficacy of (PI) is concerned?

One might say in reply that logical validity is at least a necessary condition of the explanatory adequacy of (PI). But this seems false – at least the point must be demonstrated in view of the fact that causal explanations in general do not satisfy it. Under the deductive-nomological view of causation, the statement of the cause event *taken together with a law* is said to imply the statement of the effect event. That's clearly another matter, however.

Now the causalist has available to him a plausible initial move to explain why (2) isn't explanatory, while (1) is. He would say that there is no proper causal connection between the intention-belief system and the action in (2), but such a connection does exist for (1). Whether or not the causalist wants to concede that (1) is an explanation is another question – especially if he is prepared to accept (PI) as a logically valid schema. One possibility is to reject the assumption, shared by von Wright and other teleologists, that logical connections preclude causal connections. Indeed, under the counterfactual analysis of causation,[6] the intention-belief complex of (1) is shown straightforwardly to be a cause of the action specified in the conclusion, in virtue of the truth of the counterfactual 'If I had not intended to read a paper at University of Cincinnati or had not believed..., then I would not have gone to Cincinnati'; and the intention-belief complex of (2) is shown, equally straightforwardly, not to be a cause of the action. Most teleologists will be unimpressed with this demonstration of the causal nature of (1); for the point of their contention is not that the intention-belief complex is not causally related to the action in the sense of 'cause' explained by the counterfactual theory of causation, but rather that this relation is not *nomological*, that no empirical, or contingent, laws connect the two.

Another possibility, therefore, is for the causalist to deny the explanatory efficacy of (1), and in general that of (PI), altogether. This isn't to imply that he would reject the explanation of actions by volitions and cognitions; his point would be that volitions that explain are not *intentions*, but rather *wants*, *desires*, *valences*, and the like. He will argue that there is no deductively valid inference from a person's want to bring about *p* and his belief that *q* is necessary for *p* to the conclusion that he sets himself to bring about *q*. Wants and desires admit of degrees of strength; they can be in conflict with one another; they can be overridden by other wants and desires. But intentions have none of these properties.
I shall not further pursue here the controversy between the causal and the noncausal interpretations of (PI), or more broadly, between the causal *vs.* the noncausal theory of explanation of actions. What I propose to do in the remainder of this paper is to explore ways of dealing with the Sturgeon-style counterexamples without worrying about how they affect, or are affected by, this controversy. We shall, however, be assuming what appears to be eminently intuitive, namely that (1) is explanatory and (2) is not.

III

I propose, first of all, to consider von Wright's (PI) as consisting of two stages: (1) inference from the two premises of (PI) as given to an intention to bring about the state or event deemed necessary to bring about the state initially intended, and (2) inference from this newly formed intention to the intended action. Thus, the modified (PI) will look like this:

Modified (PI)

Phase 1:

(1) *S* intends to bring about *p*.
(2) *S* believes that unless he brings about *q*, he cannot bring about *p*.
(3) Therefore, *S* intends to bring about *q*.

Phase 2:

(1) *S* intends to bring about *q*.
(2) Therefore, *S* sets himself to bring about *q*.

The first phase of the modified (PI) represents one important way in which intentions give rise to other intentions. The principle of intention formation expressed here is one that many philosophers appear to have recognized implicitly or explicitly.[7] Similar principles have been formulated with regard to wants and desires. Goldman suggests the following schema for wants:[8]

S wants it to be the case that *p*.
S believes that if *S* bring about *q*, then *p* will be the case.
Therefore, *S* wants it to be the case that *q*.

Goldman does not claim that this is a schema of logical inference: in fact, he suggest that the connection is causal – the want-belief system of the first two statements causes the want expressed in the third. Whether Goldman is right or not, it's evident that the schema is not a logically valid inference pattern: cases in which the first two statements are true and in which the last is false are easily constructed. One typical case would be one in which although a person has a strong want for *p*, the only means *q* open to him is deemed by him too costly in terms of other consequences. In such a case, even though he believes that bringing about *q* is a means – necessary and sufficient under the circumstances– for bringing about *p*, he may not desire to being about *q*. It seems that this sort of case cannot be constructed for Phase 1 of the modified (PI): If I believe that I am not prepared to pay the price of bringing about *q*, which I deem to be a necessary condition for bringing about *p*, then I cannot be said to *intend* to bring about *p*. Intention to bring about an end represents a total commitment to do whatever one firmly believes to be necessary to achieve the end – or else one must give up that intention. In this way, Phase I of the modified (PI) appears a good deal tighter logically than Goldman's schema, or other possible schemas, for wants and desires. But this is not to say that there are no problems with it. I shall mention three below.

First, consider a case like this: I intend to drive to the market, and believe that unless I cause wear on the tires of my car I cannot drive to the market. But is it correct to say I intend to cause wear on the tires? The tire wear is, and is believed by me to be, a necessary condition for doing the intended action, but it seems odd to say that the tire wear is intended.[9] I want to make two remarks: one, whatever oddity there is here appears to be present also for von Wright's (PI): it is odd to say that under the circumstances as described, I 'set myself to cause wear on the tires'. Two, cases of this sort can be excluded by explicitly stating (2) as an assertion about the means-ends relationship, thus:

(2*) *S* believes that bringing about *q* is a necessary means for bringing about *p*.

Causing tire wear presumably isn't a *means* for bringing about the envisaged end, although it is an unavoidable consequence of it.

My second comment on Phase 1 is this: we would have a more useful form of practical inference if (2) were replaced by something like the following:

(2**) *S* believes that bringing about *q* is the best (most economical, optimific, etc. – or one of such) means of bringing about *p*.

The reason for such a change is simple: Phase 1 of the modified (PI), or von Wright's (PI), applies only to cases where bringing about *q* is seen by the agent to be a *necessary* condition for the intended end; however, we often do things, and do them intentionally, when we fully believe that they are not necessary, but only sufficient, for bringing it about. ('Sufficient', too, would be too strong, however; we should say something like 'sufficient under the circumstances'.) For example, I took an Allegheny flight to come to Cincinnati, but it is false that I believed that unless I took that flight I could not reach Cincinnati. In fact, I believed that I could take a United flight with more or less similar results. It may be replied[10] that the only thing that can be explained and needs to be explained in such a case is why I intended, or set myself, to take one flight or the other, but not why I intended, or set myself, to take the Allegheny flight. The rejoinder to this would be that there clearly was an intention to take the Allegheny flight, which would go unexplained, and that a suitable modification of (PI) would easily accommodate it without difficulty.

My third and last comment again concerns the second premise of Phase 1, and also to the same in von Wright's (PI). Suppose I am in a city with which I am relatively unfamiliar, and want to reach the nearest hospital for an emergency. The only route I know of getting there is to go down State Street, make a left turn onto Ann Street, and then go up the hill to the hospital. And suppose that is what I in fact do. It seems to me that this is a typical sort of case in which an intentional action takes place, and in which an intentional explanation of action is called for. However, it is obvious that the second premise of Phase 1 is not satisfied. Not only did I not believe that I couldn't get to the hospital unless I took that particular route – but also I didn't believe it was the best way – the fastest – way of getting there. I took that route because it was the only way *I thought I knew* of getting to the hospital. Moreover, it could be that it was the only way that I believed had *any chance at all* of accomplishing the intended result.

Reformulation of the second premise of (PI), or of Phase 1 of the modified (PI), by an explicit use of counterfactuals in the belief clause may help meet the foregoing point, perhaps thus:

(2***) *S* believes that should *S* fail to bring about *q*, *S* would not (or might not) be able to bring about *p*.

It is perhaps possible to understand (2) in the sense of (2***), in which case the reformulation can be taken as picking out the most plausible reading of it.

von Wright has the following to say about Phase 2 of the modified (PI), although he himself does not distinguish these two stages and does not discuss the question of intention formation:

For the case when the action itself is identical with the object of intention, and not a means to the attainment of this object, one cannot construct an explanation of the form of a practical inference. The second premise is missing. There is only the first premise and the conclusion (explanandum). The first premise is: *A* intended to press the button. The conclusion, depending upon the peculiarities of the case, is either: *A* set himself to press the button, or: *A* pressed the button, or: *A* would have set himself to press (or would have pressed) the button, had he not been prevented. Assume that it is the second. We can then form a 'mutilated' inference:

> *A* intended to press the button.
> Therefore *A* pressed the button.

This sounds pretty trivial. Can it be the 'explanation' of anything? It would not be quite correct to say that it is the explanation of an *action*. The action of pressing the button is not explained by saying that it was intentional, willed. For that it was this is already contained in calling it an action (pp. 122–123).

von Wright grants that we might call the above inference a 'rudimentary teleological explanation', but insists that it is better viewed as 'understanding' a piece of behavior as an *act*, namely classifying it as an intentional action. I am not concerned to defend the explanatory power of this 'mutilated inference'. The only point I note is that von Wright never questions the validity of the inference; it appears that he grants to it at least the same degree of inferential validity that he argues for his (PI).

IV

Having noted these possible revisions and improvements, we will continue to use the modified (PI) as stated, since they would not make a material difference to the discussion to follow. As we noted, Phase 1 of the modified (PI) gives us a principle of intention formation – it shows how a given intention, in conjunction with a suitable belief, gives rise to another. In fact, we can think of a tree-like structure of generated intentions. For example:

Intention to ventilate the room

↙	↘
Intention to open the window	Intention to open the door
↓	↓
Intention to turn the knob	Intention to push the door with the knob
↓	↓
Intention to grasp the knob	Intention to walk to the door and grasp the knob
↓	
Intention to move my right hand counter-clockwise	

We have omitted the associated series of beliefs needed to generate the tree.

We will say that an intention generated in this fashion by another is *subordinate* to it. An intention-tree like the above will presumably terminate in intentions to perform 'basic actions' in the now current sense of Danto, Goldman, and others. Under the usual characterization of this notion, moving my hand counter-clockwise would count as a basic action, namely an action which the agent can perform at will and which the agent performs not *by* performing another. The generation pattern of intentions we have just sketched is, as we noted earlier, one that is familiar in philosophy, especially writings in moral philosophy. We will call this type of generation 'Pattern A', and speak of an intention 'A-generating' another intention.

In the Sturgeon-type counterexample, an intention appears to be generated by another, but the generation does not conform to Pattern A. My intention to visit my in-laws is not formed because I believe that doing so stands in the means-to-end relation with respect to my giving a philosophy lecture in Cincinnati. Let us speak here of 'primary' and 'secondary' intentions: my intention to give a philosophy lecture in Cincinnati is *primary* with respect to my intention to see my in-laws, and the latter is *secondary* with respect to the former. A primary intention in this sense could be secondary to another, and a secondary intention could be primary to yet another. For example, given my intention to visit my in-laws, I form the intention to take along with me a book I had borrowed from them. And a secondary intention can have explanatory power: my intention to see my in-laws can explain why I am hailing a taxi in front of my hotel. What that intention cannot explain is why I am setting about to go to Cincinnati; nor can it explain why I am buying my ticket; nor why I am driving to the Detroit Metropolitan Airport.

Let us look more closely at the details of this mode of intention formation. First of all, I have the intention to go to Cincinnati, which is subordinate to my intention to read a philosophy paper at the university there. Now, that intention gives rise to a *belief* in me that I will be in Cincinnati. I am inclined to accept the following rule as generally true: *If S intends to bring about p, then S believes that p will be the case*. This rule appears to hold in most cases. If I have a serious doubt as to whether *p* will in fact be realized, it would be incorrect to say that I *intend* to bring it about. This seems to be another respect in which intentions differ from wants, wishes, and hopes. There is, however, the following sort of case which may appear to be a prima facie counterexample to the general rule:[11]

A passenger on an aeroplane is discovered to be a foreign spy and a murderer, and he realises that if he is carried to the plane's destination he will be denounced, arrested and convicted. He therefore produces a revolver, seizes a woman passenger as hostage, and, by threatening to shoot her, compels the captain to alter course towards a State that will give him asylum. The crew of the aircraft do not know whether they will be ordered by the criminal to bring down the plane on the foreign airfield, or whether the criminal will make a parachute jump; but the steward takes the precaution of secretly removing the rip-cord from a parachute. Shortly afterwards the criminal demands a parachute; he is given the damaged one, and jumps with it. He is killed in the fall.

It would be natural to say, I think, that the steward removed the rip-cord *with the intention* of causing the highjacker's death, and that he in fact *intended* to cause the death. But it may be false that the steward believed that the death would occur; he may have thought that there was only a very small chance of the highjacker's demanding a parachute – he removed the rip-cord with the merest hope that the highjacker would ask for one.

I am not anxious to defend here the principle under discussion, although I believe that the above putative counterexample can be explained away. What is important is that in the sort of case we are concerned with, the intention to bring about a state of affairs in fact involves the belief that the state of affairs will be realized. The presence of this belief is the first stage in the formation of a secondary intention. But how does this belief give rise to an intention? How does my belief that I will be in Cincinnati generate my intention to visit my in-laws?

One simple and elegant way of handling this matter is to use the concept of *conditional intention*.[12] A conditional intention is an intention to *bring about q should a certain condition p obtain*. We are all familiar with such intentions: for example, I have the intention of paying off my mortgage if I win the Michigan lottery next week. I have the intention of staying at Hotel Basilea if I go to Florence next summer. You may have the intention to order a steak if you go to one restaurant, and to order a lobster if you go to another, and so on. We can now say that I had a standing conditional intention of visiting my in-laws if I am in Cincinnati, and that this conditional intention and the belief that I will be in Cincinnati gave rise to my intention to visit my in-laws. In general, the following rule of 'detachment' seems valid:

> If S intends to do q if p, then if S believes that p will be the case, then S intends to do q.

According to this rule, a conditional intention when joined with the belief that the condition is – or will be – realized leads to an unconditional intention to do the action in question.

Now, in general, it is not necessary that the standing conditional intention be conditional on the condition p, where the primary intention is the intention to bring about p. It is possible that the primary intention is one to bring about p, and the conditional intention is one to bring about q if r, as long as the agent's belief that p leads to his belief that r. This is illustrated by the following example: as before I have the intention to visit my in-laws in Cincinnati, and this leads to my belief that I will in fact see them. Given my other beliefs about relevant matters, this belief leads to the belief that I will be walking past a drug store around the corner from my in-laws' house. I have the standing conditional intention to look up the owner of the drug store should I find myself in the neighborhood, and as a result I form the intention to look up the owner of the drug store. So the main point about this mode of intention generation is that the primary intention is in some way responsible for creating a belief that some condition r will be realized, where r is the condition of some standing conditional intention. In any case, schematically the second mode of intention generation will look like this:

> S intends to bring about p (primary intention).
> S believes that p will be the case.
> S's belief that p leads to the belief that r (where p could be the same as r).
> S intends to bring about q if r.
> Therefore, S intends to bring about q (secondary intention).

We will call this mode of intention formation 'B-generation'. Thus, B-generation represents the formation of secondary intentions from primary intentions. Compared with A-generation, a stronger case can be made for thinking that B-generation is a causal process – especially when the process whereby the belief that p leads to the belief that r is nontrivial; e.g., r is not a trivial logical consequence of p.

Let us now summarize our findings about intention formation:

(i) Any intention can generate secondary intentions in accordance with Pattern B, given appropriate beliefs and conditional intentions. This point applies to intentions that are secondary to other intentions, and also to intentions that are subordinate to others.

(ii) Any intention, regardless of its origin, can, and generally will, generate an A-tree of subordinate intentions, unless the intention is an intention to do a basic action. Thus, a secondary intention, like my intention to see my in-laws, will generate its own A-tree, i.e., a series of subordinate intentions to carry out actions which I believe will culminate in achieving the intended result.

(iii) An intention issues in an action of 'setting oneself', to use von Wright's term, to bring about the state or event intended. We assume whatever qualifications that are necessary (e.g., the proviso 'unless the agent is prevented') may be added to it.

(iv) Having an intention to bring about p is more than having a wish or even a strong desire that p be the case; it involves having an *action plan*[13] – there must be some cognitive element with regard to the means for bringing about p; at least the agent must believe that it is within his power to come up with a reasonably effective strategy. A-generation of intentions clearly presupposes that sort of cognitive component: for an intention to bring about p to A-generate an intention to bring about q, the agent must have the belief that bringing about q is a necessary means for bringing about p.

Now, if bringing about q is necessary for bringing about p and, further, bringing about r is necessary for bringing about q, then bringing about r is necessary for bringing about p. We are of course not logically omniscient; that is, we often fail to see, and hence fail to believe, logical consequences of our beliefs. But we will assume here that in the context of an action plan, the following is true: if an agent believes that r is necessary for q and that q is necessary for p, then he also believes that r is necessary for p. Given this assumption, we have the following:

(v) If an intention is subordinate to a second, which is subordinate to a third, then the first is subordinate to the third. That is to say, A-generation is a transitive relation.

These five points constitute our apparatus for dealing with some questions raised by von Wright's treatment of practical inference as an explanatory model, especially in light of Sturgeon's counterexample.

V

von Wright's PI – or the modified (PI) – makes use of what we have called 'A-generation' of intention. Essentially, his model of explanation of action comes to this:

(W) If an action issues directly from an intention *I* – that is, if *I* is the intention to do that action – then it is explainable, intentionally, by any intention *I** to which *I* is subordinate in conjunction with an appropriate associate belief.

In other words, if an intention *I** generates an intention *I* by one or more steps of A-generation – that is, *I* is reachable from *I** on an A-tree – then the action that issues directly from *I* is explainable by *I** and the associated belief to the effect that the action is necessary for bringing about the intended goal of *I**. Now, how does this principle (W) fare with regard to Sturgeon's counterexample? Let us state (2) in the form of the modified (PI):

(2*) *Phase* 1:
I intend to visit my in-laws in Cincinnati.
I believe that I cannot visit my in-laws in Cincinnati unless I go to Cincinnati.
Therefore, I intend to go to Cincinnati.
Phase 2:
I intend to go to Cincinnati.
Therefore, I set myself to go to Cincinnati.

The virtue of the modified (PI) is now clearly seen: we see that Phase 1 of (2*) should not count as a case of intention generation. Our intuitive sense of generation – what is dependent on, or determined by, what – clearly tells us that the direction of generation goes the other way. In fact, my intention to go to Cincinnati B-generates my intention to visit my in-laws. Thus, Phase 1 of the modified (PI) is not a generally correct schema of A-generation, if this notion of generation is to have any explanatory value – that is to say, if something like (W) is to remain a plausible explanatory principle.

Thus, we need not consider Sturgeon-style examples as exceptions to the principle (W) but rather as showing the inadequacy of our present characterization of A-generation of intentions. (W) is a prima facie plausible principle of action explanation – at least within the noncausalist framework – and it would be desirable to keep it simple and elegant. But how should we modify the characterization of A-generation? As a start we might try something like this:

(3) Phase 1 of the modified (PI) is a case of A-generation only if the intention to bring about *q* does not B-generate the intention to bring about *p*.

This is fine as far as it goes, but it's not enough. For consider: my intention to go to Cincinnati A-generates an intention to make flight reservations. Now, all of the statements in the following instance of Phase 1 are true:

(4) I intend to visit my in-laws in Cincinnati.
I believe that unless I make flight reservations to Cincinnati I will not be able to visit my in-laws in Cincinnati.
Therefore, I intend to make flight reservations to Cincinnati.

However, this is not a case of intention generation, though it is not excluded by (3). The intention to make flight reservations was there independently of the intention to visit my in-laws, and this is reflected in the falsity of the counterfactual 'If I had not intended to visit my in-laws in Cincinnati, I would not have intended to make flight reservations to Cincinnati', or the truth of 'Even if I had not intended to visit my in-laws in Cincinnati, I still would have made flight reservations to Cincinnati'.

One might at this point suggest the following counterfactual test: an instance of Phase 1 is a case of A-generation only if the counterfactual 'If *S* did not intend to bring about *p*, *S* would not intend to bring about *q*'. The main problem with this suggestion is that it would not suffice to distinguish the problematic sort of case under discussion from cases of overdetermination. For example, it could have been the case that I had both the intention to visit my in-laws and the intention to give a philosophy talk in Cincinnati, and either would have been sufficient to motivate me to go to Cincinnati. In such a case, two intentions *severally* A-generates the intention to go to Cincinnati, and both the following counterfactuals would be false: 'If I had not intended to visit my in-laws in Cincinnati, I would not have intended to go to Cincinnati' and 'If I had not intended to give a philosophy talk in Cincinnati, I would not have intended to go to Cincinnati'. Cases like this are obviously different from those that we want to eliminate, like (2*) and (4).

Going back to (4), let us suppose the situation looks like this:

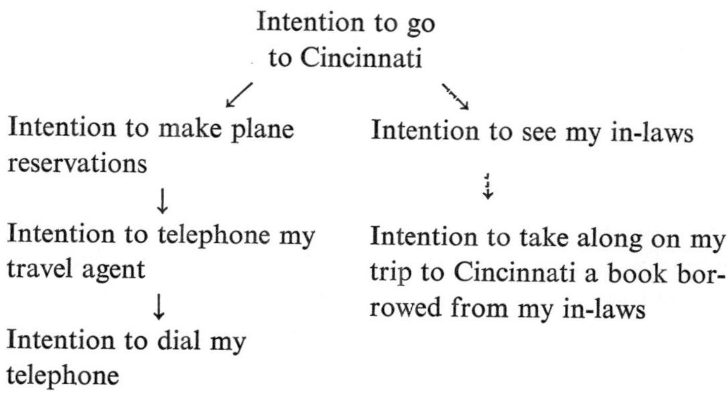

The unbroken lines indicate A-generation, and the broken lines B-generation. A brief reflection on this diagram shows that the intention to see my in-laws, which is B-generated by the intention to go to Cincinnati, cannot A-generate any of the intentions A-generated by the intention to go to Cincinnati, although in each case, an instance conforming to Phase 1 of the modified (PI) can be constructed. The same is true of the intention to take along on my flight to Cincinnati a book borrowed from my in-laws: this intention cannot A-generate any of the intentions on the left-hand branch of the tree, although, again, instances of Phase 1 can be constructed.

Further, it is not difficult to imagine situations fitting the following sort of schematic diagram:

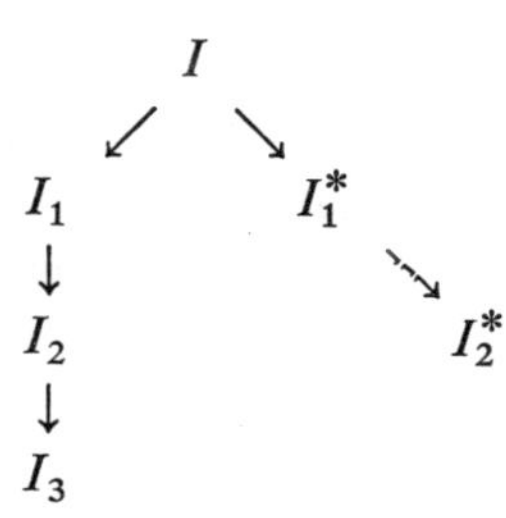

where I_2^*, which is B-generated from I_1^*, which in turn is A-generated by I, can be seen not to generate any of the intentions $I_1, I_2, I_3, \ldots$, even if instances of Phase 1 are constructible.

Let us say that a generational path of intention is *B-mixed* just in case the path contains at least one instance of B-generation; let us also say that an intention is *on a B-mixed path from* another just in case the path reaching from the second to the first is a B-mixed path. We can then state a general restriction on Phase 1, as follows:

(5) No intention that is on a B-mixed path from an intention I can A-generate any intention that is A-generated by I.

Here we are supposing that there is at most one path from any given intention to any other; this is a simplifying assumption that isn't likely to hold in all cases. However, we will not consider here more general cases in which generational paths may merge as well as divide.

VI

In this last section I want to introduce yet another sort of counterexample to Phase 1 of the modified (PI). Imagine the following story: I am going to Rome and my plane will have a stop-over in London overnight. My going to Rome is intended – that's what I am doing. However, it isn't my intention to stop over in London. But since I'll be in London in any case, I reason, I will look up my friends, the Reillys, who live just outside the city. And I now begin to take certain steps to bring that about, for example, consulting a train schedule and writing to the Reillys about my plans. Consider the following instance of Phase 1:

(6) I intend to visit the Reillys.
I believe that unless I go to London I cannot visit the Reillys.
Therefore, I intend to go to London.

Here, the third statement is false. *Ex hypothesi* I do *not* intend to go to London. It is only that I know – or believe – that I will be in London on my way to Rome. So Phase 1 is in fact fallacious as a schema of logical inference. And if we replace the third statement of (6) with 'I set myself to go to London', we have a counterexample to von Wright's (PI) which is somewhat different from Sturgeon-style counterexamples.

Here, too, the source of the difficulty is the insufficient amount of attention given to the origin or formation of intentions. As may be recalled, the upshot of the discussion in the preceding section was that the genealogy of an intention can affect its explanatory role. The obvious trouble with (6) is similar: my belief that I will be in London was a generative factor (whether causal, logical, or some other kind we will not worry about here) of my intention to visit the Reillys. It may be recalled that in the problematic Phase 1 of (2*), my belief that I will be in Cincinnati played a similar role in the formation of my intention to visit my in-laws. The only difference is that in the latter, but not in the former, the belief in question was induced or accompanied by an intention.

Generally, we can think of an intention – or having the intention to bring about a state or event – as having two components: an *action plan* involving the agent's beliefs about the steps that must be taken if the intended goal is to be realized, and a readiness to carry out that plan (some will suggest something even stronger: the actual initiation of the plan).[14] A *conditional intention*, on this view, would contain the first element, namely an action plan, and a *conditional* readiness to execute the plan, *i.e.*, readiness to execute it when a certain condition is realized. And there may be freefloating action plans, too, that are not elements of any intention, whether conditional or categorical.

We can identify the forming of an intention with the adoption of an action plan for execution. Now there may be various factors that determine why we adopt a certain action plan at a given time, but our beliefs about the world and ourselves, including beliefs about our abilities, capacities, etc., are surely important determinants of the adoption of an action plan. My belief that I will be in London was the crucial factor in my forming the intention to visit the Reillys; and similarly my belief that I will be in Cincinnati gave rise to my intention to see my in-laws. Generally we can regard our intentions as being determined by a complex of beliefs and desires – and perhaps also aversions if they are not a species of desires. How our intentions are in fact determined by them is likely to be a complex story, which is clearly beyond the scope of the present paper.

In any event, I would suggest something like the following to handle the present case:

(7) If the belief that p is part of the belief-want complex that deter-

mines the adoption of an action plan corresponding to the intention *I*, then bringing about *p* is not part of the action plan, and the intention *I* does not A-generate an intention to bring about *p*. In fact, this latter intention may not exist.

The rationale of (7), I believe, is obvious. If the antecedent of (7) is satisfied, the belief that *p* is a *presupposition* of the adoption of the action plan, and in consequence bringing about *p* cannot in turn show up as a step in the plan. Of course, this does not mean that it cannot be part of another action plan – in fact, this was the case with Sturgeon-type counterexamples. (5) and (7) together are unlikely to cover all anomalous cases, but I hope they are correct as far as they go.

My treatment of the anomalous cases has been more or less on a trial-and-error basis, and it would be highly desirable to have a simple theoretical account that would give them a unified treatment. I hope that my discussion in this paper gives at least some hint as to how one might begin to construct such an account.

The University of Michigan

NOTES

1 G. H. von Wright, *Explanation and Understanding* (Cornell University Press, Ithaca, N.Y., 1971), p. 96. Hereafter we will refer to this work by citing page numbers in the body of the text.

2 See, e.g., Roderick M. Chisholm, 'Freedom and Action', in Keith Lehrer (ed.), *Freedom and Determinism* (Random House, New York, 1966).

3 See Donald Davidson, 'Actions, Reasons, and Causes', *Journal of Philosophy* **60** (1963), 685–700.

4 I am assuming here that the relation of entailment or implication can be explained in an adequate way for *events, actions,* and the like – or, at least, that such talk is an oblique way of talking about their descriptions.

5 For a somewhat more detailed discussion of counterfactuals and determinative relationships, see my 'Noncausal Connections', *Nous* **8** (1974), 41–52. Also see David Lewis, 'Causation', *Journal of Philosophy* **70** (1973), 217–36.

6 See Lewis, *ibid.*

7 For example, see Aristotle, *Nichomachean Ethics*, Bk. III, ch. 3, and of course Kant, who says, "Whoever wills the end wills also (so far as reason decides his conduct) the means in his power which are indispensably necessary thereto. This proposition is, as regards the volition, analytic...," *Fundamental Principles of the Metaphysics of Morals*, sec. 2 (tr. by Thomas K. Abbott).

8 Alvin I. Goldman, *A Theory of Human Action* (Prentice Hall, Englewood Cliffs, N.J., 1970), p. 102.

[9] For a more detailed discussion of cases like this, see Jack W. Meiland, *The Nature of Intentions* (Methuen, London, 1970), ch. 1.

[10] This was Professor von Wright's reply in a conversation with me.

[11] Glanville L. Williams, *The Mental Element in Crime* (Hebrew University Press, Jerusalem, 1965), p. 51.

[12] For a useful discussion of conditional intentions, see Meiland, *op. cit.*, ch. 2.

[13] For possible notions of 'action plan', see G. E. Miller, E. Galanter, and K. Pribram, *Plans and the Structure of Behavior* (Holt, Rinehart & Winston, New York, 1960), and Goldman, *op. cit.*, esp. pp. 56–62.

[14] Cf. Miller, Galanter and Pribram, *op. cit.*, p. 61:

> What does it mean when an ordinary man has an ordinary intention? It means that he has begun the execution of a Plan and that this intended action is a part of it. 'I intend to see Jones when I get there' means that I am already committed to the execution of a Plan for traveling and that a part of this Plan involves seeing Jones.

FREDERICK STOUTLAND

THE CAUSAL THEORY OF ACTION[1]

The great revival of interest among analytic philosophers in questions about action stems largely from the work of Wittgenstein, who set forth a number of powerful considerations against a causal theory of action. Because of this that theory has been on the defensive, and the sheer bulk of papers criticizing it made it seem in full retreat. The tide has turned recently, and the causal theory is no longer on the defensive. Some of the criticisms have been shown to be mistaken, and others have been turned aside by more subtle developments of the theory. Another reason the theory has regained its status is that its opponents, though they made cogent criticisms, failed to develop a cogent alternative, and the causal theory was given the benefit of the doubt, not so much because it could answer the criticisms but because it was believed that it *had* to be right, there being no reasonable alternative.

It is one of the achievements of von Wright to have presented an alternative, which speaks cogently to the issues and enables us to evaluate the causal theory without having to think that it is the only plausible theory on the horizon. The result is an interesting and healthy philosophic situation, with no theory of action holding a specially favored position among so-called analytic philosophers. There is no analytic position on these issues, only a characteristic way of stating the questions and a characteristic concern for certain distinctions. And though this may continue to make discussions with philosophers from other traditions difficult, the situation is now such that these discussions will be fruitful and mutually advantageous, and that insures that they will in fact take place, however difficult they may be.

My primary intent in this paper is to criticize the causal theory of action. But I shall do this with the consciousness that there is an alternative theory, and I shall therefore show points of agreement and difference between the causal theory and a theory which I understand to be von Wright's. My special concern is to give the causal theory its due by dealing with it in its more subtle forms, and for that reason I shall

Manninen and Tuomela (eds.), Essays on Explanation and Understanding, 271–304. *All Rights Reserved.*

focus on the formulation given by Donald Davidson. Davidson has been perhaps the major figure in restoring respectability to the causal theory; in a series of papers he has given both a subtle articulation of the theory and cogent arguments for it.[2] These papers have been widely influential and have stimulated a number of competent defenses of the theory.[3] I shall not discuss criticisms current in the literature which Davidson's formulation turns aside, but shall concentrate on criticism the theory has difficulty meeting.

I. THREE ISSUES IN THEORY OF ACTION

A theory of action does not seek to resolve a single set of issues, and a philosopher might resolve one set of issues in terms of the causal concept, for example, and another set in a different way. There are, I believe, three kinds of issues to which theories of action seek to speak in a reasonable and systematic way.

The first concerns the question of how we should *understand* intentional action. This may be called the question of the correct analysis of the concept of intentional action – the attempt to state the conditions necessary and sufficient for an agent to have acted intentionally – provided we are careful to avoid prejudging the question of the form an analysis must take, for this is one matter on which the theories divide. Fundamental to any theory of action is its thesis about how we should understand intentional action.

A second issue concerns the *explanation* of intentional action. What conditions does a propositional scheme have to meet in order to constitute an adequate explanation of an act? This question is logically subsidiary to the first but not easily separable, and many discussions approach the first question by way of the second. The second, of course, bears most directly on how we should conceive the human sciences and stands to gain most from reflection on the methods, logic, and results of these sciences.

A third issue concerns how it is possible for agents such as ourselves to perform an intentional act. I understand this issue as separable from the first, for we might know the necessary and sufficient conditions for an intentional act but not understand how we are able to meet them. It raises some of the most fundamental, as well as most difficult, questions in the theory of action; my concern in this paper, however, is not to discuss these questions but to set them aside in order not to confound them with questions about understanding and explanation. The causal theory often fails to make this distinction, a main motivation for giving a causal analysis of intentional action being a conviction that what makes it possible for agents such as ourselves to act are certain causal powers we possess. This conviction is normally not argued for, however, though it is not self-evident. I shall not argue for or against it; apart from a few remarks at the end, I shall set it aside in order to consider on their own merits questions about the understanding and explanation of intentional action.[4]

II. PRELIMINARY DISTINCTIONS

In this section I intend to develop briefly some distinctions and lay down some vocabulary in terms of which to discuss them.

My first point is that the concept of intentional action is more basic than the concept of action and that it is futile to attempt to characterize action except by reference to intentionality. This means that whenever an agent acts he also acts intentionally, and that it is only by reference to intention that we can draw a non-arbitrary line between acts an agent performs and changes that occur in him or that he causes to occur without acting. This does not mean that every act is intentional; persons often act nonintentionally, for example, when by mistake they turn off a light. My argument is that this will be an act only if the agent did it in doing something intentional, for example, turning a switch in a way he thought would turn on a light. And if that too was unintentional, then it will be an act rather than a reflex movement only if it was done in doing some intentional act, for example, the agent's moving his hand in some intentional way.

Davidson makes this same point by saying that "unintentional acts are intentional acts under other descriptions"[5]. Davidson is here using a distinctive locution and affirming an ontological thesis, both of which entered current discussion through Anscombe's *Intention*. The locution is 'under a description'. The ontological thesis concerns the individuation of acts and says that acts are concrete particulars, which can have diverse descriptions. An agent, for example, who calls for the floor by raising his hand and in so doing bumps his neighbor performs not three acts, but one act with three descriptions, which is intentional under the first two but not under the third.

I wish to avoid the ontological issue; it is important but will not affect my argument. I shall avoid it in the following way. To the question, "what did *S* do?" there will always be, if *S* acted at all, more than one true answer: "he raised his hand", "he called for the floor", "he bumped his neighbor", "he disturbed the air". With respect to each of these answers we can ask whether the act described was intentional or not. But that does not commit us either to the position that these answers give different true descriptions of the same act or to the position that each description describes a different act. On either alternative the question "did *S* act intentionally?" will be, if taken out of a specific concrete context, useless, for the answer to it will always be equally "yes", given that acting presupposes acting intentionally, and "no", given that every act has consequences which are nonintentional. We should ask, not whether *S* acted intentionally, but ask of some designated act whether he did *it* intentionally. Did he raise his hand intentionally? Did he bump his neighbor intentionally? The issue I propose to avoid is whether these are questions about a single act described differently or about different acts related in some significant way.[6]

It is difficult to deal with the substance of Davidson's position without the locution 'under a description'; I find it confusing, however, and shall shy away from it in favor of von Wright's distinction between the result and the consequence of an act, which does much of the work of this locution and more besides. Von Wright understands acts to include events logically, in the sense that the performance of an act entails the production (or prevention) of a change: that an agent raised his hand entails that his hand rose. The event which is in this sense intrinsic to an act, part of its logical structure, is the *result* of the act. *Consequences*, on the other hand, are events extrinsic to an act's logical structure, whose occurrence is not entailed by the performance of an act; a consequence of my raising my hand, for example, is that molecules of air are disturbed. Since a result is logically intrinsic to an act, an act cannot cause its result; it may cause its consequences. Raising my hand does not cause my hand to rise; it does cause the air to be disturbed.[7]

This distinction is relative: a state of affairs which is a consequence of one act may be the result of another. The air being disturbed is a consequence of my raising my hand, a result of my disturbing the air. I shall understand this distinction as relative to the agent's intention: the result of an act is what is intended in it, its consequences are what occur because of the act but are not intended in it. If I intentionally raise my hand, then that my hand rise is the result of the act. If I did not intentionally disturb the air, then the air's being disturbed was a consequence of but not the result of my act. This means that the result of an act will be (a) that (generic) event whose occurrence is logically required for the performance of the act provided; (b) the agent intended (aimed at) that event.[8]

A term that is unavoidable in discussing this topic is 'behavior'. It is a treacherous term, easily equivocated on, but it fulfills as well as any the role of referring in a non-committal way both to intentional acts and to those bodily movements of the agent which may or may not constitute acts. A man's head is nodding; he may be giving a signal or he may be falling asleep. If we say that his behavior is that his head is nodding, we leave it open whether he is acting – to give a signal – or whether there is no action here and his behavior is *mere* behavior. Behavior then may be intentional or it may merely be bodily movement, and a fundamental task of a theory of action is to state the difference.

I understand 'behavior' as a *process* term rather than an *event* term – as giving what goes on rather than what takes place. Since behavior is process like, even intentional behavior does not have a result; there is no event logically intrinsic to it (it is what von Wright calls an 'activity' as opposed to an 'act').[9] But when an agent performs an act his behavior *eventuates* in a result – a term which is broader than 'cause'. If I close a door, then my behavior causes the door to close. If I put my hand on the shelf, the behavior of moving my hand does not cause my hand to be on the shelf, but the behavior *eventuates* in that. I shall use the latter term to cover both these relations. The behavior which eventuates in the result of the act I shall call 'the behavior in the act'.

III. TWO CONTRASTING THEORIES OF ACTION

I want now to give a very brief statement of the causal theory's analysis of intentional action and then a longer statement of von Wright's approach, so we may have on hand an alternative when developing criticisms of the causal theory.

The causal theory's central thesis is that the necessary and sufficient conditions for an agent's performing an intentional act are: (1) that the agent's behavior eventuate in the result of the act; (2) that there be some end the agent wants and which he believes his behavior will bring about; and (3) that this want and belief cause his behavior. Thus an agent will have intentionally opened a door (1) if his behavior eventuates in the door's being open; (2) if he wants to let in some fresh air and believes his behavior will result in that; and (3) if this want and belief cause this behavior.

Results will be distinguished from consequences in that a result is what is wanted and the wanting of which is a necessary cause of the behavior in the act. The causal theorist sees the result of an act as the 'object' of a want which is a cause of the behavior in the act. A consequence of an act, on the other hand, is not an 'object' of the want which causes the behavior, though it may in fact be wanted (or at least not unwanted). In opening the door I may let out a fly, but that will be only a consequence of my act, even if I wanted the fly out, if the want was not a cause of the behavior.

Davidson's analysis, which I shall discuss below, is somewhat more complex than this, but it shares with it the aim of decomposing the concept of intentional action into its basic logical elements, with neither the concept of intentional action nor the concept of intentionality being taken as basic. The basic concepts rather are the concepts of (mere) behavior, of beliefs and desires, and of causality. The analysis partitions the description of an intentional act into three parts: a physical description of mere behavior as eventuating in a certain event; a mental description of the agent's beliefs and desires; and a statement that the object of the first description is caused by the object of the second. It might be said that its aim is to understand the concept of intentional action by eliminating it in favor or other concepts.

Von Wright states his central claim as follows:

> To establish that the agent's causing *A* to come about is a case of his doing *A* is not to establish, in addition to the happening of *A*, a different event which so to speak occurs 'inside the agent'. It is to *understand* (the meaning of) the agent's conduct, *i.e.*, to see that by certain changes in his body or changes causally connected with changes in his body the agent is *aiming* at this result. If he aims at it without achieving it, we shall have to say that the agent tried but failed....[10]

The thesis is that what is required for an intentional act is (1) that the agent's behavior eventuate in the result of the act; and (2) that *by this behavior* the agent have intended (or aimed at) the result. Thus an agent will have intentionally opened a door if by the various movements of his body which did cause the door to open he was aiming at that result – if that was what he intended (aimed at, meant) by his behavior. On the other hand, if whatever behavior the agent was engaged in at the time did not result in the door's opening, but if, nevertheless, by his behavior he intended that result, then, though he did not open the door, he was *trying* to.

The *result* of an act will be that event brought about by the behavior which the agent was intending *by* that behavior. *Consequences*, on the other hand, will be those events which occurred because of the agent's behavior but which he did not intend or aim at *by* his behavior. When an agent, in opening a door, accidentally lets in some flies, flies coming in is a consequence of his act, because that was not something he was aiming at *by* his behavior (even if he could foresee it would happen).

The causal theory's three conditions for intentional action are in this way reduced to two. That the behavior eventuate in the result of the act continues as a necessary condition. But instead of the two requirements that the agent want the result and that this cause the behavior, there is substituted the single requirement that *by his behavior* the agent intend (or aim at) the result. This reduction is possible because von Wright thinks that an agent's intending a result by his behavior is not a matter of the behavior's being caused. "Intentionality", he writes, "is not anything 'behind' or 'outside' the behavior. It is not a mental act or characteristic experience accompanying it."[11] Intentionality rather should be said to be *in* the behavior, for normally we see the behavior as action.

In the normal cases we say off-hand of the way we see people behave that they perform such and such actions.... Many of these actions we ourselves know how to perform; those, and others which *we* cannot do, have a familiar 'look' or 'physiognomy' which we recognize.[12]

In the normal cases what we see directly is not mere behavior but intentional action. That is, we see the behavior as aimed at an end. We see people unlocking doors or opening windows, and that seeing is not hypothesizing about or inferring the causes of their behavior.

We can be in doubt that what we observe is intentional action – it may be mere behavior – and we can be in doubt what the agent intends by his behavior. Now we must inquire whether we *do* observe an intentional act, whether the agent *is* aiming at anything by his behavior and if so what. But, argues von Wright, our inquiry here is not for the causes of the behavior, for the logic of this inquiry is represented by a 'practical inference' (P.I.), which is not causal in character.

We see an agent making some movements at a window; that is his behavior. Is he doing anything intentionally? (Perhaps he is asleep or drugged). If so, what is he doing? At this point we cannot tell by observation, for we do not see what he is doing, that is, we do not see what he is intending by this behavior. Then we notice he has a glass cutter. A practical inference emerges:

> *S* intends to get into the house.
> *S* believes that he cannot get into the house unless he now cuts a hole in the window.
> ———
> *S* cuts a hole in the window.

What *S* is doing (or trying to do) is cutting a hole in the window. We have determined that he is doing something intentionally, and now we can *see* what it is.

This is not a causal inquiry, von Wright argues, for these premisses (with suitable qualifications) *entail* the conclusion.[13] If the premisses described causes of what is described in the conclusion, then the connection between them would be effected by a causal law, and such a law must be contingent. This inference, on the contrary, is not contingent.

It is crucial to notice that the conclusion of the P.I. describes an intentional act, not *mere* behavior, and that the premisses describe intentions and beliefs which have as their objects not mere behavior but intentional action. The second premiss, for example, involves a belief with regard to one *act* being a (necessary) means to another; if I perform an act because I believe it to be necessary, then my act *must* be intentional. This means that if our inquiry into the intentionality of the agent is accurately represented by this P.I., it can be said not to be an inquiry into the causes of behavior because not an inquiry into mere behavior at all, but action. In asking whether the agent intended anything by his behavior, we consider what sort of *act* he might be performing in this setting, given that he had this tool, was (possibly) looking in a certain direction, etc. This is the act he might be performing; let us consider his behavior in that light and see if what results from it constitutes acts done *by* doing the first act. Our inquiry into the agent's intentionality concerns the wider context of action, not the causes of his behavior, and the latter therefore is not a mark of intentional action.

Von Wright's mode of analysis here is different from that of the causal theorist. His aim is not to decompose the concept of intentional action into its more basic elements, for he regards the concept of intentional action and of intentionality as in a crucial sense irreducible. One does not understand the concept of intentional action by first understanding a concept like (mere) behavior and then adding to it other concepts like causality or desire. To understand the concept is not to eliminate it in favor of other concepts but to see its place in a larger conceptual structure, by possessing which we are agents who can act and see others acting, and who can explain our action and the action of others.[14]

Von Wright's analysis of the nature of the P.I. and the role it plays in understanding the intentionality of an agent's behavior also entails a thesis about the nature of *explanation* of action. For if the P.I. yields as a conclusion a true statement of what the agent's intentional act was, its premisses will also constitute an explanation of his act. This kind of explanation von Wright calls 'teleological', for like traditional teleological explanation it subsumes diverse behavior not under laws which connect it nomically with antecedent conditions but under an end toward which the behavior is directed. Cutting a hole in a window may involve a wide diversity of behavior: using the left or right hand, swinging the elbow in or out, holding the fingers in various ways, pointing the cutter up or down. To know what caused the right hand rather than the left hand to move, what caused the elbow to swing in rather than out, provides no illumination here, for it is indifferent to the act whose explanation we seek that it was the right hand rather than the left that moved, or that it was an inward rather than an outward swing of the elbow. We are interested rather in the end under which these various movements, whatever they may be, are subsumed, the intention which makes them to be the act of cutting (or trying to cut) a hole in a window.

Unlike traditional accounts of teleological explanation, however, von Wright restricts the *explanandum* of this kind of explanation to intentional action, for what subsumes diverse behavior under an end is simply the agent's intending that end, and his behavior being understood in terms of it. But to intend an end by one's behavior is to act intentionally, and teleological explanations, therefore, *presuppose* intentional action, they presuppose, that is to say, that the agent's behavior be understood in terms of a result which it intends.[15]

There is, therefore, a necessary reciprocity between inquiry into the intentionality of an agent's behavior and explanation of his act, and this is the key to von Wright's conception of the P.I. For the premisses of the P.I. imply (a statement about) an intentional act without the addition of any causal or law-like statement, and therefore explain that act in a teleological rather than a causal way, just because those premisses lay down conditions in terms of which to *understand* (the intentionality of) the agent's behavior. The P.I. is valid because its premisses constitute, "a set of conditions under which the conduct of an agent has to be interpreted or understood in a certain way, *viz.*, as the doing of *A* or as aiming at the result".[16] "The formal validity of the practical inference requires that the item of behavior mentioned in its conclusion is described (understood, interpreted) as action... . In order to become *teleologically explicable*... behavior must first be *intentionalistically understood*."[17] The premisses imply (a statement about) the act the behavior is understood to be, regardless of what the (mere) behavior is. If it is true that an agent intends to get into the house, and believes that cutting a hole in the window now is necessary for that, then whatever his (mere) behavior may be, his intentional act is cutting a hole in the window (or trying to). That is what he intends (means) by his behavior, whatever it is, and that is how we must understand it. Whatever behavior is occurring, he is doing something intentionally. It may not be cutting a hole, for he may be failing at that, hence only trying. But to try to cut a hole is itself to act, to do something with the intention that it result in a hole's being cut.

Von Wright then offers a non-causal account both of the intentionality of behavior and of the explanation of action, arguing that the understanding and explanation of action differ from the understanding and explanation of physical events and behavior. Causality is essential to the latter, intentionality to the former. To subsume behavior under causal laws is to understand it as mere behavior; to subsume behavior under an agent's intention is to understand it as action. The human sciences, therefore, take a form different from the natural sciences. Causal judgments play a role, but not a fundamental role, for causal explanations of an intentional act will presuppose that the agent's behavior has been set in an intentionalist perspective, and that is not a perspective in which causal concepts play the fundamental role.

IV. DAVIDSON'S CAUSAL THEORY

Davidson's version of the causal theory is more complex and subtle than the one briefly sketched out in the last section. One of the problems with that version is its appeal to wants (or desires) as a necessary condition of intentional action, which is to confuse the concept of an intentional act with a voluntary act. Not all intentional acts are voluntary; I may intentionally do *A* as a means to doing *B*, even though I do not want to do either *A* or *B*, but do so because I feel obliged or am compelled to do so.[18]

Davidson repairs this defect by replacing 'want' with 'pro attitude'. An explanation which explains an action "by giving the agent's reason for doing what he did" he calls a *rationalization*. A reason "rationalizes an action", he writes, "only if it leads us to see something the agent saw, or thought he saw, in his action – some feature, consequence, or aspect of the action the agent wanted, desired, prized, held dear, thought dutiful, beneficial, obligatory, or agreeable."[19] 'Pro attitude', is intended to cover not only desires but this whole range of attitudes that an agent may take toward actions of a certain kind (as having a certain result), insofar as those attitudes may be reasons for his action. Davidson in effect defines an intentional act as an act done for a reason, where 'an act done for a reason' also covers acts which, as we say, were done for no reason, with 'for no reason' meaning "not that there is no reason but that there is no *further* reason... besides wanting to do it."[20]

This last point means that Davidson, like von Wright, sees the question of the analysis of the intentionality of behavior as inseparable from the question of the explanation of action. Indeed he claims that a single scheme will serve "to give an analysis of what it is to act with an intention; to illuminate how we explain an action by giving the reasons the agent had in acting; and to provide the beginning of an account of practical reasoning... ."[21] We will need this scheme for everything that follows, so let us lay it out here.

In 'Actions, Reasons, and Causes' (pp. 181, 188) Davidson argues for two theses:

(1) *R* is a primary reason why an agent performed the action *A* under the description *d* only if *R* consists of a pro attitude of the agent toward actions with a certain property, and a belief of the agent that *A*, under the description *d*, has that property.

(2) A primary reason for an action is its cause.

Given these theses we might construct a Davidsonian scheme for intentional action as follows:

(1) *S* has a pro attitude towards actions of type *B*.
(2) *S* believes that *A* under description *d* is of type *B*.
(3) This belief and pro attitude cause *A*.

---

(4) *S* does *A* intentionally under description *d*.

Let me reformulate this, however, in order to facilitate comparison with von Wright's scheme (his P.I.),[22] eliminating the locution 'under a description':

(1) *S* has a pro attitude toward act *B*.
(2) *S* believes that behavior which eventuates in result *A* is necessary for *B*.
(3) This pro attitude and belief cause *S*'s behavior (the behavior in the act).

---

(4) *S* does *A* intentionally (intentionally performs the behavior which eventuates in result *A*).

The scheme could be instantiated as follows:

(1) *S* has a pro attitude towards getting into the house.
(2) *S* believes that behavior which eventuates in a hole being cut in the window is necessary for getting into the house.
(3) This pro attitude and belief cause *S*'s behavior (the behavior in the act).

---

(4) *S* intentionally cuts a hole in the window.

The claim is that each conclusion is entailed by its premisses, and that the latter constitute both a causal explanation of the act described in the conclusion and an analysis of the concept of intentional action in terms of a set of conditions necessary and sufficient for the agent's having acted intentionally. Given the truth of the premisses, it follows that *S* performed *A* intentionally and that he performed it for the reasons described in the first two premisses.

Discussions of the *explanatory* adequacy of the causal theory have usually focussed on the question of whether explanations of action exhibit the deductive-nomological pattern of explanation so carefully articulated by Hempel, Popper, Nagel and other philosophers. Causal theorists have said that they do exhibit that pattern and that adequate explanations of action, therefore, presuppose general laws. Explanations in history and everyday life do not normally mention any general law, but the claim is that each assumes one implicitly, whose truth is required for the adequacy of the explanation. Explanations in psychology or sociology, on the other hand, it is argued, seek to make explicit the laws which are presupposed, and the aim of these sciences is to state and systematize these causal laws. Action explanations are, therefore, of the deductive-nomological form, having as premisses statements of initial conditions plus a contingent general law statement, which, entailing as conclusion that a certain act has been done, explain that act.

This so-called 'covering law theory' of explanation has been attacked at a number of points. The most telling attacks have focussed on the fact that historians and the like seldom, if ever, mention any general laws in giving explanations. The covering law theorists reply that historians use them implicitly and do not state them, either because, given the present state of our knowledge, they are too complex (Hempel), or because they are so trivial as not to bear mentioning (Popper). But when attempts are made to state the laws an historian might be using, the complex ones turn out to be so detailed and specific as to do little more than reiterate that the historical factors in terms of which the act is explained *do* explain it, and the trivial ones turn out not to be relevant to significant historical explanation. When relevant laws are stated, they turn out to be suspiciously like necessary truths.[23]

It is a distinctive contribution of Davidson to have argued that explanations of action can be causal even though general laws do not figure in the explanation. The fact, therefore, that historians do not appeal to general laws, or that no relevant and significant general laws to which they might appeal have been formulated, is not a valid objection to the causal theory's claim that such explanations have the same form as causal explanations generally. Davidson's argument is that the third premiss in his scheme does not formulate a general law; it should be taken as it stands, as a singular statement affirming that a certain pro attitude and belief caused specific behavior, and not as a general law connecting attitudes and beliefs with actions of a certain kind. In this sense Davidson's theory is not a covering law theory of explanation.

Davidson's comments on causality are notably interesting and subtle. He does not simply abandon the notion that causality is nomological, that singular causal statements entail general (causal) laws. But he argues that the latter is ambiguous.

> It may mean that '*A* caused *B*' entails some particular law involving the predicates used in the descriptions '*A*' and '*B*', or it may mean that '*A* caused *B*' entails that there exists a causal law instantiated by some true descriptions of *A* and *B*.[24]

The first option is the one assumed in the controversy over the covering law theory, where the discussion is over attempts to formulate a law in the very terms which the agent or an observer would use in giving the agent's reasons for his action.

The second option is different. It allows Davidson to argue that a reason causally explains an action even if no general law can be formulated connecting the reason with the act. All that is required for a pro attitude to causally explain an act is that there be in principle a general law connecting events of a kind to which that specific pro attitude belongs with events of a kind to which the act belongs, though we may not know what law that is or in what terms it can be formulated. As Davidson puts it:

> The principle of the nomological character of causality... says that when events are related as cause and effect, they have descriptions which instantiate a law. It does not say that every true singular statement of causality instantiates a law.[25]

With regard to actions he argues:

> The laws whose existence is required if reasons are causes of actions do not, we may be sure, deal in the concepts in which rationalizations must deal. If the causes of a class of events (actions) fall in a certain class (reasons) and there is a law to back each singular statement, it does not follow that there is any law connecting events classified as reasons with events classified as actions – the classifications may even be neurological, chemical, or physical.[26]

Explanations of action, therefore, will exemplify the causal scheme, by virtue of a singular causal premiss that the agent's behavior has been caused by a given attitude and belief. That causal premiss in turn will entail that there is a nomological explanation why the given attitude and belief did, in these circumstances, cause the agent's behavior; but the latter explanation is not an explanation of *action*. It will, Davidson thinks, very likely be an explanation that belongs to the physical sciences, and will either state the physical (neurological) circumstances under which the given attitude-belief-behavior complex arose, or else will state the physical events to which this particular attitude-belief description happens to refer (the latter, as a version of the identity theory of materialism, being, in Davidson's view, more likely). Such an explanation is obviously not an explanation of action; is is the causal scheme taken as a whole that gives the explanation of action. Action explanations *presuppose* that the third premiss in the scheme is true and that in turn presupposes that there exists in the physical sciences the sort of nomological explanation sketched out above.

In order to understand and evaluate Davidson's theory it is crucial to see the role the third premiss plays in his scheme. We have seen that Davidson defines an intentional act as an act done for a reason. The first two premisses in the scheme specify the agent's reason in terms of his pro attitude and belief; the definition amounts to the claim that an agent's performing an act for that reason is a sufficient condition of its being intentional, while its being performed for some reason or other is a necessary condition of its being intentional.

Now to say an agent performed an act for a reason is to say he performed it *because* of that reason – that the reason *explained* his act. One of Davidson's most telling arguments has been directed against accounts of intentional action which simply ignore the question of what is required for an agent to perform an act *because* he had a reason. Such accounts fail to provide, he argues, "an analysis of the 'because' in 'He did it because...' where we go on to name a reason."[27] This is an important point: there is a decisive difference between, on the one hand, having (say) a desire *and* acting, and, on the other hand, having a desire and acting *because of* it, or, as I prefer to put it, acting *on* it. One may *justify* an act by citing a desire one has even if one did not in acting act *on* it, but one cannot *explain* an act by citing a desire unless one acted *because* of it or *on* it.

This point about explanation also yields a point about the analysis of intentional action. It does not follow from the fact that an agent regarded some act as, say, beneficial *and* performed that act that the act was intentional. For he may have done some act which he in fact regarded as beneficial, but whose beneficial character had nothing to do with his doing it; such an act could be accidental or in some other way non-intentional. Given that an agent regards an act as beneficial, it will follow that the act is intentional only if the agent performed it *because* it was beneficial, if he acts *on* his desire for some beneficial result.

The primary role of Davidson's third premiss is to give a *causal* account of this distinction between an agent's acting *and* having an attitude and his acting *on* his attitude. The thesis is that an agent acts on a pro attitude when and only when his attitude, together with his belief, cause the behavior in the act; it is this which the third premiss asserts.[28] When that premiss is true, an agent acts on the attitude and belief expressed in the first two premisses of the scheme, and his act will be intentional. If, on a given occasion, the first two premisses are true and the third not, so that the agent's attitude and belief are not on this occasion sufficient to cause his behavior, then the agent does not act *on* that attitude and belief, and though it may still be true that he acted intentionally, the falsity of the third premiss means that the intentionality of his behavior stems from other sources.

V. CRITICISMS OF DAVIDSON'S THEORY

Although Davidson's version of the causal theory is subtle and powerful, it faces some fundamental difficulties of its own, and I do not think it is adequate either as an analysis of intentional action or as an account of the explanation of action. These difficulties center on the third premiss in Davidson's scheme. That premiss takes the form it does in order to avoid the difficulties we have noted in the covering law version of the theory: it is not a general law linking attitudes of a certain kind with actions of a certain kind, but a singular causal statement linking the agent's attitudes on a specific occasion with his behavior on that occasion.

While this meets many of the objections made to the covering law theory, it raises new difficulties of its own. The most serious ones concern the claim that the third premiss in the scheme is adequate to provide an account of the distinction between merely having an attitude (or belief) and acting on it. That claim I believe to be mistaken, and I shall in what follows develop three criticisms of it:

(1) that it is not a necessary condition of an agent's acting on an attitude that the attitude cause the behavior in the act; (2) that it is not a sufficient condition of an agent's acting on an attitude, and therefore acting intentionally, that the attitude cause the behavior in the act; (3) that the premiss is, in any case, not plausible.

(1) It cannot be a *necessary* condition of an agent's acting on a pro attitude (or belief) that the pro attitude (or belief) cause the behavior in the act because if it were, then an agent will act on an attitude (or belief) only if the attitude (or belief) is the causally strongest at the time of his acting. This, however, is false; the attitude an agent acts on at a certain time may, causally speaking, be rather weak compared to an agent's other attitudes at the time.

Talk about the strongest desire is a familiar motif in contexts such as this, and controversies over determinism have often centered on whether an agent can ever act against his strongest desire. If we take the claim that an agent always acts on the attitude which is, at the moment, causally strongest to be *contingent*, it seems false. I may feel a very strong obligation to finish writing a paper this evening, know that I ought to stick with it, have very little desire to go ice skating, but decide that I will, just to please my children, even though they've had plenty of attention. In the middle of all this I may stop skating for a moment and smoke a cigarette. It is surely implausible to suggest that a graph might be plotted of the strength of my attitudes, showing them waxing and waning, so that my sense of obligation to finish the paper recedes as I decide to go skating, and that the latter then becomes subordinate to my urge to smoke a cigarette, which in turn recedes as I resume skating. In fact the urge to finish the paper almost obsesses me, causing me to have a headache and to be very distracted – surely causal consequences of my feeling that obligation and a pretty accurate measure of its causal strength – and yet I go skating. My wanting to smoke a cigarette is a purely trivial want – since I'm not a habitual cigarette smoker – and it doesn't causally override anything. The claim that whether I act on a desire is a matter of its causal strength is, considered as a contingent claim, surely false; similar considerations would show that the claim is also false for pro attitudes other than desires.

It may be replied to this objection that it is a *necessary* truth that agents always act in terms of their strongest desire (or other pro attitude). "Strongest desire", it might be argued, should be understood as "overriding desire" so that it becomes a *criterion* of a desire's being strongest that it be that desire (no matter how strong it in fact may be) on which the agent acts, so long as he acts intentionally. Understood in this sense the claim that an agent always acts on his strongest desire is analytic, for "strongest desire" just *means* "desire on which an agent acts".

The difficulty with this move is that on the causal theory this *cannot* be a reasonable criterion for determing an agent's strongest desire. Given the criterion, determining that *D* is *S*'s strongest desire requires determining that *S* acted on (because of) *D* rather than that *S* acted *and* had *D*. But on the causal theory, determining that *S* acted on *D* at *t* requires determining that *D* was *S*'s strongest desire at *t*; but to determine that *D* was *S*'s strongest desire requires, given the criterion, determining that *S* acted on *D* at *t*. On the causal theory, therefore, the criterion is circular.

It does not seem to me that it is in any case a reasonable general criterion of a desire's being strongest at *t* that an agent have acted on it at *t*. What is a reasonable general criterion is a question to which I do not have an answer, but it will surely be more complex than that, making reference to such things as an agent's tending to act on a number of occasions or to the degree of satisfaction he conceives its fulfillment would yield him. We might grant, nevertheless, that the fact that an agent acted on a pro attitude is in some sense logically adequate for determining that that was his strongest attitude at that time. But it does not follow that it was his *causally* strongest attitude. There is no reason to think that agents always act on that attitude which is causally strongest, whether we construe that as contingent or necessary. Indeed it seems untrue, and hence is a reason for thinking that the statement which entails it – namely that a necessary condition of an agent's acting on an attitude is that the attitude cause the behavior in the act – is also untrue.

(2) The difficulty in claiming that it is a *sufficient* condition of an agent's acting on an attitude (and therefore acting intentionally) that the attitude cause the behavior in the act is one Davidson himself discusses – the problem of 'wayward causal chains'. Consider an example of Chisholm's, who appears to have first raised this problem:[29] (1) a certain man desires to inherit a fortune; (2) he believes that only if he kills his uncle

will he inherit a fortune; and (3) this desire and this belief agitate him so severely that he accidentally runs over and kills a pedestrian who, unknown to the nephew, is none other than the uncle. This man does not intentionally kill his uncle, though he has a pro attitude toward an act of a certain type, he believes that certain behavior is required for that act, and this pro attitude and belief cause that behavior. Here the attitude (together with the belief) caused the agent's behavior and yet he did not act on it.

Chisholm's example involves what Davidson calls *external* wayward chains, and a number of defenders of the causal theory have proposed adding conditions to the scheme to meet this form of the objection. Shaffer, for example, writes: "The fortunate nephew intends the end but does not intend the means [running over the pedestrian].... We must add the further condition that where it is necessary to employ some means in achieving an end, to bring about the end intentionally one must also bring about the means intentionally."[30] Davidson suggests a similar reply in terms of requiring that if the act is to be intentional, "the wanted effect must be produced by a causal chain that answers, at least roughly, to the pattern of practical reasoning".[31] Here the nephew did not kill the uncle intentionally because the act which he did perform intentionally (driving his car) did not result in the intended outcome in the way envisaged in the agent's plan of action.

Whether or not these additional conditions resolve the problem of external causal chains, a slight modification in Chisholm's example will generate the problem of *internal* wayward chains. (1) a certain man desires to inherit a fortune; (2) he believes that only if he surreptitiously kills his uncle, now driving in the car with him, will he inherit a fortune, and he believes, moreover, that the only way to do that is to crash into one of the light poles on this very highway; and (3) this desire and this belief agitate him so severely that, without thinking, he swerves and hits a light post killing his uncle. Neither the swerving nor the killing here is intentional, and yet the case meets even the augmented set of conditions, for the killing did occur in the way envisaged in the agent's plan of action.[32]

Although this case meets all the conditions in the causal scheme, it is not a case of intentional action, for the nephew did not act on his desire to inherit a fortune, even though the desire was so strong and the circumstances so fortunate that the behavior in that act was caused to occur. Davidson suggests that in order to counter this objection additional conditions have to be laid down about *how* the attitude and belief cause the behavior – so they cause it in the right way – but not knowing how to do this he regards the objection as serious (if not sufficient for abandoning the causal theory). I believe there are reasons to think these conditions cannot be supplied, for what the counter-example shows is that if acting on an attitude is to be a matter of the attitude causing the behavior in the act, then the causation will have to be *direct* in some sense, and no causal concept of 'direct' can be forthcoming.

The nephew did not intentionally kill his uncle, the causal theory will have to argue, because his desire and belief did not *directly* cause his behavior but caused it by way of agitating him. This kind of case cannot be ruled out, however, by ruling out *all* intermediate causal links, because any causal theory is going to lay down a series of neural and physiological (e.g., muscular) links between the want and the behavior. 'Directly causing' cannot, therefore, involve a causal link which has no intermediate links. One might try ruling out 'abnormal' links, arguing, for example, that it is abnormal that agitation stand between believing you have to kill your uncle in order to inherit his fortune and killing him. Agitation, however, is a perfectly normal causal consequence of such a belief and desire. All such moves will, I believe, fail, for this kind of directness is not causal at all.

The case is analogous to a certain issue in theory of perception. One of the reasons philosophers have given for saying that we never *directly* perceive the material world is that if we do perceive the material world at all, then a complex causal chain has transpired between the object and the perceiver, and, it is argued, we cannot directly perceive what lies removed from our mind at the other end of a complex causal chain. One way to counter this argument is to say that it confuses an analysis of what it is to perceive directly with causal facts that must obtain if, being the kind of creatures we are, we are going to perceive directly. Central to perception are certain beliefs about the objects presently affecting one's senses, and if those beliefs are about objects in the material world, directly about them (as they surely are, even if as philosophers we may by indirect means have to show that they *are* directly about them), then we *directly* perceive objects in the material world.

This analogy has to be used with care but the point is to show that laying down special conditions for direct causation cannot yield a notion of *direct* perception (even if *perception* requires a certain kind of causation). Perception is direct by virtue of the object of the perception, a matter to which it is irrelevant whether or not the perception was caused directly. The case is similar for action. Acting on an attitude or belief cannot be defined by laying down special conditions for the direct causation of the act by the attitude or belief. What is required is not special conditions on the causation of the behavior but that the attitude or belief be directly related to the behavior by the behavior's being that *by which* the agent intends what is given by the attitude and belief.

This is von Wright's way of putting the matter, and his account has no difficulty with this kind of case. For the nephew's behavior will be intentionally killing his uncle only if *by that behavior* he intends that his uncle die. The causation of the behavior is not relevant; what is relevant is that the agent intend a result by it, so that the behavior is understood in terms of the intention. In this way the intention relates to the behavior in that direct way required to rule out the aberrant case.

On von Wright's account the nephew's act will be a case of his acting *on* his desire to inherit a fortune if and only if there is a scheme (the P.I.) with true premisses from which it follows that the agent intends to kill his uncle now.[33] That will require: (1) that the agent believe (know, think, assume) that killing his uncle now is necessary for satisfying his desire; and (2) that his desire to inherit a fortune have 'become' an intention to satisfy that desire – that is to say, that he now not only desire the fortune but intend to do what (he believes) is required to satisfy the desire. This belief and intention will imply further – in the way and for the reasons discussed in Section III – that the nephew kill his uncle (or try to), and if these premisses are true, then that the nephew kill his uncle (unless he is prevented or is unable to do the act) will also be true, and the premisses will constitute both a set of conditions in terms of which to understand his behavior and a teleological explanation of the intentional act his behavior is understood to be.

This discussion has centered on the question "when does an agent act on a pro attitude?" I have criticized the causal theory's view that it is necessary and sufficient that the pro attitude be causally strong enough to make the third premiss in the causal scheme true. I have now suggested an alternative answer – namely, when the pro attitude 'becomes' an intention, that is, when the agent acquires an intention to do what (he believes) is required to satisfy it. I want to say a little more about this matter.

Consider the nephew who desires to inherit a fortune. My point is that he will act *on* that desire if and only if he acquires an intention to satisfy it, and, therefore, an intention to do what he believes necessary to satisfy it. If he should see that having that intention will require him to do things he does not want to do (or cannot do or ought not to do), he will not cease to desire to inherit the fortune (being a normal person), but he will cease to intend to satisfy it. He changes his mind, as we say, and does not act on his desire. The desires (and pro attitudes generally) that an agent does not act on are ones he does not acquire (or ceases to have) an intention to satisfy.

Some things that Davidson says suggest that this is in fact the view he is assuming. Thus he writes, commenting on the problem of wayward causal chains, that "beliefs and desires that would rationalize an action if they caused it in the *right* way... may cause it in other ways. If so, the action was not performed with the intention that we could have read off from the attitude that caused it."[34] What Davidson is after is the right causal chain between attitude and act that will entail that the agent acted on that attitude; when we have that, then we can 'read off' the intention from the attitude that caused it.

'Read off' here surely means that the intention is a logical consequence of the attitude and belief. But if the intention with which the act was done is a logical consequence of the attitude and belief, then the attitude has 'become' an intention, for only intentions are 'transmitted' in this way. From the fact that the nephew wants the uncle's fortune and believes killing him now is necessary for getting it, we may not infer that he intends to kill his uncle, or that he acts with the intention of killing him, or even that he wants to kill his uncle (for he may find killing his uncle reprehensible, but continue to want his fortune very badly). What is required for the conclusion that the nephew intends to kill his uncle or acts with that intention is that he *intends* to do what is necessary to inherit a fortune and believes the murder *is* necessary.

Another way of putting this point is that there is a difference between (1) acting because of your desire; and (2) acting because you intend to satisfy your desire. It may seem that there is no difference here and that an agent's intention can be read off in either case. These phrases are often, perhaps normally, used in such a way that there is no difference, but whereas (2) is always synonymous with acting *on* your desire, (1) need not be.

Someone says, "the nephew is drinking a glass of water because of his desire to inherit his uncle's fortune." Can we read off from that statement the intention with which he is acting?[35] We can, if we construe it in sense (2): "the nephew is drinking a glass of water because he intends to satisfy his desire to inherit his uncle's fortune", which is puzzling because it's not obvious what beliefs he might have about how that act is required, but perhaps he thinks the water will give him the strength to do the deed. But we cannot, if we construe it in the way sense (1) differs from sense (2), that is, that he is not drinking with the intention of satisfying his desire to inherit a fortune. We do not now know the intention with which he's drinking; perhaps he's drinking to quench the thirst caused by the agitation caused by his desire for the fortune. We can read off from his desire the intention with which he is acting only if he also has the intention to satisfy the desire.

If the scheme for intentional action requires 'intention' rather than 'pro attitude', then von Wright's arguments about intentions not being causes become applicable, and those arguments seem to me cogent. Indeed Davidson appears not to disagree with them, for the intention with which an agent acts, he writes, is not a cause, "for it is no thing at all, neither event, attitude, disposition, nor object."[36] None of choosing, willing, intending, and trying, he writes, "is plausibly the cause of an action, because normally these are ways of characterizing the action chosen, willed, intended, or tried, not descriptions of further actions, events, or states."[37] Presumably it is this idea that intentions are not causes which leads him to speak instead in terms of 'pro attitudes'; what I have been arguing is that insofar as 'pro attitudes' is used to characterize what we act *on*, it is in effect doing the job of 'intention', for acting on a pro attitude requires that it 'become' an intention, that is, that the agent acquire an intention to attain the object of the attitude.[38]

If, therefore, the scheme for intentional action is set up with 'intention' in the first premiss rather than 'pro attitude', it appears that the first two premisses will imply the conclusion without any contingent principle like the causal scheme's third premiss. I want now to argue that this third premiss is in any case not plausible.

(3) In order to consider the question of the plausibility of the third premiss, we must look more carefully at what the premiss claims to be caused, rather than at what it claims to be causes. As we have formulated it, it speaks of behavior, but 'behavior' is a term which, as noted above, may be used in a variety of ways and which invites equivocation. There are three possibilities in this context: it may be used in the sense of 'intentional action', in the sense of 'nonintentional action' or in the sense of 'mere behavior'. In which of these three senses could Davidson be using the term, and is it plausible under any of them?

The third premiss cannot use 'behavior' in the sense of 'intentional action' because there can be no mention of intentional action in the scheme until the conclusion. An essential purpose of the scheme is to give conditions necessary and sufficient for an intentional act; it would be trivial to make it a condition of an agent's performing an intentional act that his intentional act be caused by his attitudes and beliefs. To put the point in another way, Davidson's argument is that attitudes and beliefs causally explain intentional acts. The third premiss is a necessary condition for this explanatory relation; but it does not (as the covering law theory holds) formulate a general law which has the attitude and belief as antecedent and the act as consequent, for no such law has been (and, Davidson argues, can be) formulated. The premiss rather states that the attitude and belief cause the behavior in the act. If 'behavior' here meant 'intentional act', then this significant thesis would be reduced to the triviality that it is a necessary condition of an attitude and belief causing an intentional act that they cause an intentional act.

Some philosophers have argued that this premiss must use 'behavior' in the sense of 'nonintentional action'. Richard Taylor, for example, writes that 'what, under this theory, is caused by certain beliefs and desires is not simply a bit of behavior or bodily motion, but an act of the agent",[39] where the context indicates that Taylor does not mean an intentional act but an act which may or may not be intentional, depending on (according to the causal theory) what its causes are. My raising my hand is an intentional act if and only if caused by my attitude and belief, otherwise a nonintentional act. But this cannot be correct, for there are no nonintentional acts except where there are intentional acts.[40] Acting intentionally is a necessary condition of acting nonintentionally; to hold that acting nonintentionally – whatever the cause – is a necessary condition of acting intentionally is, again, to reduce the scheme to triviality,

'Behavior' in the third premiss of the scheme must mean, therefore, 'mere behavior', that is, bodily movements of the agent. The premiss should read: "the agent's attitude and belief cause the bodily movements which eventuate in the result of his act." For example, "the agent's desire to get into the house and his belief that in order to do so he must now cut a hole in the window cause those bodily movements of his which eventuate in a hole being cut."

If we take this on its own merits, however, it is not plausible. For what it says is that specific bodily movements will occur just because an agent believes them necessary for some end he desires – that his hand will move in a certain way just because he believes that behavior necessary for getting into a house. Beliefs and attitudes certainly have effects – anger, agitation, nervousness – which in turn have as effects bodily movements. But it is effects of this kind which cause all the trouble about wayward causal chains, and they hardly provide evidence for the premiss the causal theory needs. There is no good reason to think that either beliefs or pro attitudes cause those bodily movements which occur when we act intentionally.

Consider the kind of analysis Alvin Goldman gives, while defending the causal theory in his book, *A Theory of Human Action* (p. 67), of what is required for someone to move his hand (intentionally) without doing anything as a means to that as an end. He says this (I paraphrase): *S* can move his hand as a basic act if and only if (1) If *S* were in standard conditions with respect to moving his hand (it wasn't tied down, etc.), then if *S* wanted to move his hand, his moving his hand would result from this want; and (2) this fact does not depend on any of *S*'s knowledge (or belief) about the causes of his hand's moving. The last clause is, of course, crucial to the act's being *basic*. If my hand were paralyzed, for example, I would have to move it with my other hand, and hence know that that is how I can cause it to move.

Now the ability to move one's hand as a basic act is both common and valuable, but what Goldman is saying here is really extraordinary. He is saying that persons who are able to move their hand 'at will' are persons whose desire to move their hand causally results in their hand's moving, so that when they desire to move their hand, it moves (if they desire it strongly enough, of course). That, however, may very well be an unfortunate power to have, for the mere desire to move one's hand will (given the right conditions) cause its moving. There are many circumstances in which I would rather my hand did not move; on Goldman's account I will have to be careful that the desire to move it does not develop, for as soon as I want to move it, it may move, and if conditions are (for all I know) right, it will move.

For most of us, fortunately, the mere desire to move our hand is not a sufficient cause of our hand's moving. Wanting to move my hand does not, in standard conditions, cause my hand's moving. That's not the sort of causal consequence a desire has. My desire to move my hand, if it is around a club and I'd like to do my uncle in, may cause me to tremble, but it won't cause my hand to move, and if it does, my act will be at best marginally intentional, for not under my control.

The only reason I can think of for Goldman's adopting the view that wants cause behavior in this way, even when it has this extraordinary consequence, is that he confuses the relation of a want to an intentional act with the relation of a want to mere behavior. We sometimes act intentionally just because we want to; our merely wanting to do something explains our doing it. Moreover, we can define a basic act as one involving wants but no knowledge (or belief) along Goldman's lines, but with a significant difference. Instead of saying, "If *S* wanted to move his hand, his moving his hand would result from this want", we must say, "If *S* wanted to move his hand, he *could* move it"[41] (his hand's moving would be the *result* of this act), adding Goldman's proviso about *S* needing no knowledge about the cause of his hand's moving. His wanting to move his hand is sufficient, not for its moving, but for his intentionally moving it, if he can move it as a basic act (whereas if he can't move it as a basic act, then he will need in addition knowledge about the cause of his hand's moving). It moves when he wants it to move, fortunately, only if he acts on that desire and moves it.

The same must be said about belief. Our performing an act because we believe it necessary for some end is common, and as a result, all sorts of events, bodily movements and others as well, occur. But to say we perform an act because we believe it necessary for some end is quite different from saying bodily movements occur because we believe them necessary for some end. For when we cite an agent's belief and attitude as his reason for acting and thus give an explanation in terms of what he acted on, what we explain is necessarily intentional. Goldman writes that, "We certainly know that wants and beliefs result in acts. Indeed... this is a part of our conception of wanting and believing."[42] This is true, but it is equally part of our conception of wanting and believing that what they explain is intentional acts; those are the *explananda* in our everyday appeal to an agent's attitudes and beliefs. Knowing an agent's wants and beliefs we can make reasonable predictions about his intentional acts, but not about his bodily movements, for the range of bodily movements that may occur when an intentional act is performed is wide and diverse. Knowing I am intentionally cutting a hole in the window does not tell you what bodily movements I am making.

It may be replied that the causal theory also holds that what we explain in terms of an agent's attitudes and beliefs are *necessarily* intentional acts. Thus Davidson writes that, "If rationalizing attitudes do cause an action of mine, then not only does the action occur, but it is, under the rationalized description, intentional."[43] The act, moreover, will *necessarily* be intentional, for the central thesis of the causal theory is that causation by rationalizing attitudes is both necessary and sufficient for an intentional act. Behavior caused (in the right way) by rationalizing attitudes will thereby meet the criteria for being an intentional act, and hence will necessarily be intentional.

But this way of construing the claim that the *explananda* of an agent's attitudes and beliefs are necessarily intentional acts is different from the claim I have been making. My claim is (a) that an act can be explained in terms of an agent's attitudes and beliefs *because* it is intentional. An agent's act is explained in terms of his attitudes and beliefs when he acts on them, and to act on an attitude *presupposes* that the act is intentional.[44] The causal theory's claim is (b) that an act is intentional *because* the behavior in the act can be (causally) explained by the agent's attitudes and beliefs. The intentionality of the behavior presupposes that it is explained by attitudes and beliefs (rather than the other way around as in (a)). Which of these two construals of the claim is correct?

I believe that (a) is correct, for (a) entails, whereas (b) does not, that in order to know which attitude and belief an agent is acting on, we must look to his intentional acts, reasoning that an agent who acts intentionally in *this* way is very likely acting on *this* attitude and belief. This means that behavior will not come into the picture except insofar as it is understood in the categories of intentionality, understood in terms of what an agent intends by it. The bodily movements an agent is making will not as such be any clue to his attitudes and beliefs, insofar as our concern is for the attitudes and beliefs that explain his act. It is, of course, true that a person's attitudes and beliefs do not manifest themselves only in the acts they explain. An agent may, for example, betray his attitude by some unintentional slip, some involuntary movement; here his bodily movements are the clue to his attitudes. But attitudes which play this role are playing no role in the explanation of his intentional action, so that this case is not relevant to our question.

It seems to me clear that knowing the attitudes and beliefs on which an agent is acting does require looking to his intentional acts in this way. But if (b) is correct, then in order to know an agent's attitudes and beliefs we must look to his bodily movements. For on the causal theory looking to an agent's intentional acts *means* looking to the bodily movements caused by his attitudes and beliefs, and if we subtract the attitudes and beliefs we are seeking to know, what is left is his movements. We will come to know which attitude and belief an agent is acting on by coming to know which attitude and belief caused his bodily movements, reasoning that if this is his behavior, then these are very likely the attitude and belief causing it.

But this procedure cannot even begin unless we can identify which behavior we are seeking the causes of. For whenever an agent acts on an attitude and belief, he engages in much behavior which is irrelevant to them. An agent who is cutting a hole in a window may be swinging his leg, sniffling, scratching his ear; these items of behavior are not relevant to his act, and we are not seeking their causes. The only reasonable characterization of which behavior is relevant is that it is the agent's intentional behavior; to identify which behavior we are seeking the causes of requires that we identify the agent's intentional behavior. But on the causal theory that just is behavior caused by the attitude and belief on which the agent acted. If claim (b) is correct, therefore, the inquiry into an agent's attitudes and beliefs cannot begin until we have identified what they are. I conclude, therefore, that (a) is correct: that explaining an act in terms of an agent's attitudes and beliefs *presupposes* that it is intentional. What follows from the true claim that it is part of our conception of wanting and believing that wants and beliefs result in acts is not that wants and beliefs cause behavior but that they explain intentional acts.

There is one other move I can think of that might be attempted at this point, namely to appeal to the fact that attitudes and beliefs explain intentional acts as evidence that they cause the behavior in the act. The move invokes the ontological thesis that in the context of a specific scheme 'bodily movements' and 'intentional act' both refer to the same state of affairs 'under different descriptions'. The argument is that, given this identity, if attitudes and beliefs *causally* explain an agent's intentional act, they also cause his bodily movements.

In this form the argument is invalid, for 'explain' is not extensional – *A* may explain *B*, and *B* may be identical with *C*, and *A* may not explain *C* – and 'causally explain' is, therefore, also non-extensional. But 'cause' is extensional – if *A* causes *B*, it causes anything identical with *B* – so if it could be shown (a) that a certain attitude and belief cause an intentional act and (b) that the intentional act is identical with certain bodily movements, it will have been shown that the attitude and belief caused bodily movements, which is that third premiss whose plausibility we have been examining.

I have my doubts about statement (b), partly because I'm not clear what it is asserting, partly because I'm not certain that it is contingent; if it is not contingent, then the argument fails for a number of reasons. But I have not been able to think through adequately the issue of whether or not the identity is contingent (or whether or not it is known *a priori*, which is a different question), so I shall not pursue the matter.

Granting (b) leaves (a) but that clearly involves a *petitio*. The reason for our inquiry into the plausibility of the third premiss in the causal scheme – that attitudes and beliefs cause bodily movements – was that the scheme was not acceptable unless its third premiss was plausible. But what we have here is an argument that the third premiss is plausible based on the claim that an agent's attitude and belief cause his intentional acts, which is precisely the central claim of the causal theory. The argument that the premiss is plausible assumes that the theory is acceptable, when our task was to show that the theory is acceptable because the third premiss is plausible. This has at least the advantages Russell noted that theft has over honest toil.

VI. CONCLUSION

I want to conclude by coming back briefly to the distinction I made in Section I between the kinds of issues which theories of action address. It is a striking thing that many of the arguments made by proponents of the causal theory of action are concerned not with showing that the theory *is* true, but with showing that it *can* be true.[45] A typical reply to my third argument would be that it does not show that the causal scheme's third premiss cannot be true. This, of course, is correct; indeed if the final argument I discussed is a *petitio*, then it is a *valid* argument for the truth of the third premiss (though it is not *sound* unless the causal theory is true, and does not, therefore, accomplish its purpose). My objection does not show that the causal theory cannot be true, but suggests that reasons for thinking it is are lacking.

Why should philosophers think that to defend the causal theory it is necessary only to show that it *could* be true? I suggested at the beginning that one reason is that alternatives are seldom developed very cogently. But I think a more important reason is that there is a hidden assumption made by proponents of the causal theory. That assumption concerns what I called the third issue in theory of action, namely the issue of how it is possible for agents such as ourselves to act intentionally. The causal theory assumes, I believe, that what makes it possible for agents to act causally are certain causal powers we possess – the power to cause events to occur, above all to cause events to occur in our bodies. The assumption is that unless wants and beliefs caused our bodily movements we could not act intentionally.

About this assumption I want to make two comments. One concerns what follows from it. It does not follow from it that a causal *analysis* of intentional action is correct, any more than it follows from the fact that perceiving the world requires that we be causally effected in certain ways that an analysis of perceiving must make the causal concept primary. Even if an agent cannot act intentionally unless certain causal facts obtain, an analysis of acting intentionally need not involve them nor need we know what they are in order to understand intentional action. A causal analysis of intentional action, therefore, has to stand on its own feet, and not rest on this metaphysical thesis about the conditions of the possibility of action.

The other comment is that this assumption *is* hidden, in the sense that it is seldom discussed and less seldom argued for. That it is true does not follow from a causal analysis of action any more than a causal analysis follows from its truth. Nor does its denial follow from the failure of a causal analysis. That it is not discussed rests likely on a failure to disentangle it from the other issues in theory of action; that it is not argued for rests likely on a sense that it *must* be true.

But why must it be true? Davidson has, on at least one reasonable interpretation of his 'Freedom to Act', seen this distinction between the issues. He concludes that paper with the candid confession that, "We must count our search for a causal analysis of '*A* is free to do *x*' a failure." But, he goes on, "Although we cannot hope to define or analyse freedom to act in terms of concepts that fully identify the causal conditions of intentional action, there is no obstacle to the view that freedom to act is a causal power of the agent."[46] One way of construing this is an expression of skepticism about a causal analysis of action and a reaffirmation of a causal theory about the conditions of the possibility of action. I have argued that this is a perfectly consistent position. But it is disappointing that no arguments are given for the causal theory of the possibility of action.

Theories about the conditions of the possibility of action, however, need explicit attention. Alternatives should be formulated and arguments about them developed. The causal approach to *this* issue has deep roots in the modern consciousness, due perhaps to Descartes' influence as much as anything. But it is not an approach universally shared; Kant, for example, did not share it (or at least tried to shake loose from it), which is why we find his discussion of action so difficult to understand. Nor did Wittgenstein, nor does von Wright. Let von Wright have the last word:

> Acting does not cause events in the world.
> To think that it does would be "animism."[47]

St. Olaf College

NOTES

[1] Members of the Northfield Noumenal Society read this paper and discussed it with me. I am grateful to them, as well as to Tom Carson, O.R. Jones, Raimo Tuomela and G. H. von Wright for helpful suggestions and for pointing out some mistakes; I'll take responsibility for the mistakes that remain. St. Olaf College deserves thanks for enabling me to get to the Colloquim in Helsinki.

[2] 'Actions, Reasons, and Causes' in Care and Landesman, eds., *Readings in the Theory of Action*, Bloomington, 1968; 'Agency' in Binkley *et al.*, eds., *Agent, Action, and Reason*, Toronto, 1971; 'How is Weakness of the Will Possible?' in J. Feinberg, ed., *Moral Concepts*, Oxford, 1969; 'Freedom to Act' in T. Honderich, ed., *Essays on Freedom of Action*, London, 1973.

[3] Especially noteworthy is Alvin Goldman, *A Theory of Human Action*, Englewood Cliffs, 1970.

[4] I have discussed the question of the conditions of the possibility of action at some length in 'Von Wright's Theory of Action', forthcoming in *The Philosophy of Georg Henrik von Wright*, ed. by Paul A. Schilpp.

[5] 'Freedom to Act', p. 145. Cf. 'Agency', p. 7: "A man is the agent of an act if what he does can be described under an aspect that makes it intentional."

[6] Goldman argues for the latter thesis in *A Theory of Human Action*, pp. 10ff.

[7] Cf. G. H. von Wright, *Norm and Action*, London, 1963, pp. 39ff.

[8] There are difficulties with this distinction, which cannot be resolved without raising the ontological issues I have set aside. It has, for example, the consequence of making every act intentional, which is no difficulty if one adopts a Davidsonian ontology so that every act is intentional under some description. Given this ontology we can make the same point by saying that every act has an intention in it whose object is the result of the act, so that when we specify what an agent did intentionally we specify the result of his act, and when we specify what he did non-intentionally we specify the consequences of his act. Where Davidson says that *S* acted intentionally under description *c* but not under description *d*, we can say that *c* was the result of *S*'s act but that *d* was a consequence.

[9] Cf. *Norm and Action*, p. 41f.

[10] 'On So-called Practical Inference', *Acta Sociologica* **15** (1971), p. 49.

[11] *Explanation and Understanding*, Ithaca, 1971, p. 115.

[12] 'On So-Called Practical Inference', p. 51.

[13] These qualifications are discussed in *Explanation and Understanding*, pp. 104–107, and in 'On So-Called Practical Inference', pp. 47–49. The latter paper is particularly helpful in the way it discusses the kinds of uses to which the scheme of P.I. may be put. It may be *used* for explanation, or for prediction, or for setting the conditions in terms of which to understand an agent's behavior as an intentional act. The qualifications required to make the scheme valid will vary with the use to which it is put.

It should be noted that von Wright has recently (in unpublished writings) changed his mind about the status of the P.I., no longer thinking it is logically valid, even with the qualifications referred to above. He continues to hold, however, that it is in no sense a causal inference. This means that while von Wright would no longer accept some of the arguments stated here, he would accept the basic conclusion, and that is what is crucial for my purposes.

[14] Cf. Thomas Reid's procedure in his analysis of the concept of conceiving. This, he says, is one of those "simple operations of the mind [which] cannot be logically defined"

His task, therefore, is not to discover its simpler elements but "to explain some of its properties, consider the theories about it; and take notice of some mistakes of philosophers concerning it". (*Essays on the Intellectual Powers of Man*, Essay IV, chap. I.)

[15] For von Wright's account of teleological explanation see *Explanation and Understanding*, pp. 84 ff.

[16] 'On So-Called Practical Inference', p. 50.

[17] *Explanation and Understanding*, p. 121.

[18] In one (important) sense of "could have done otherwise", therefore, I may do an act intentionally even if I could not have done otherwise. Cf. Luther's, "Here I stand; I cannot do otherwise".

[19] 'Actions, Reasons, and Causes', p. 179.

[20] *Ibid.*, p. 182.

[21] 'How is Weakness of the Will Possible?' p. 102.

[22] Cf. above, p. 278.

[23] Cf. von Wright's discussion in *Explanation and Understanding*, chap. I, esp. pp. 23ff.

[24] 'Actions, Reasons, and Causes', p. 194.

[25] 'Mental Events', in Foster and Swanson, eds., *Experience and Theory*, Amherst, 1970, p. 89.

[26] 'Actions, Reasons, and Causes', p. 195.

[27] *Ibid.*, p. 188. Davidson has Melden in mind particularly.

[28] Cf. *ibid*: "In order to turn the first 'and' to 'because' in 'He exercised *and* he wanted to reduce and thought exercise would do it' we must... [hold that a] reason for an action is its cause".

[29] Roderick Chisholm, 'Freedom and Action' in K. Lehrer, ed., *Freedom and Determinism*, New York, 1966, p. 30.

[30] Jerome Shaffer, *Philosophy of Mind*, Englewood Cliffs, 1968, p. 105.

[31] 'Freedom to Act', p. 153. Davidson is paraphrasing an unpublished paper of David Armstrong's.

[32] Shaffer's condition would rule out the killing's being intentional but it would not rule out the swerving's being intentional (as it surely is not) since it was not necessary to employ means to swerve the car.

[33] This way of putting it is misleading in the sense that I do not mean that acting on a desire requires that there exists some *linguistic* scheme; I mean that there exist the beliefs, intentions, etc., which such a scheme would formulate.

[34] 'Freedom to Act', p. 153.

[35] Omitting "with the intention of drinking a glass of water", which, if not trivial, may be false and hence cannot be read off. It is not true that whenever I act intentionally there is some intention with which I perform the act, unless one just makes 'acting with some intention' synonymous with 'acting intentionally'.

[36] 'Actions, Reasons, and Causes', p. 190.

[37] Freedom to Act', p. 147.

[38] That 'pro attitude' can be used to do the job of 'intention' is helped by the fact that the major pro attitude term, namely 'want', *can* be used synonymously with 'intend', as, for example, when Davidson writes that, "... When we know some action is intentional, it is empty to add that the agent wanted to do it". ('Actions, Reasons, and Causes', p. 182) That this statement will normally be taken to be true shows that 'wanted' here means 'intended' not 'desired'. For if it means 'desired', the statement is false, since agents may do an act intentionally even if they desire not to do it (e.g., dismiss a friend

from the company at the direction of one's superior). But from the fact that one wants to do an act in the sense of desires to do it, it does not follow that one wants to do it in the sense of intends to do it.

Under what conditions an agent who has a pro attitude *acquires* an intention to satisfy it is a large topic on which I shall say nothing. It is in that nexus where many of the problems of practical reasoning belong – reasoning which leads to decisions about whether and when to act on one's pro attitudes.

39 Richard Taylor, *Action and Purpose*, Englewood Cliffs, 1966, p. 252.

40 Cf. above p. 273f.

41 'Could' in this conditional does not refer to S's *ability* to move his hand, for his wanting to move it is hardly a sufficient condition for his having that ability. 'Could' here is rather being used in what von Wright calls the 'success' sense (cf. *Norm and Action*, p. 50f.); the conditional states that *S*'s wanting to move his hand is, given the way things are *on this occasion*, all that is required for his act's being successful (not for his doing it but for its being successful if he does it). What ascribes an *ability* to *S* is not the 'could' in the conditional, but the whole conditional.

42 *A Theory of Human Action*, p. 73; cf. also p. 114.

43 'Freedom to Act', p. 149.

44 Cf. above, p. 278f.

45 David Pears, 'Desires as Causes of Action' (in G. Vesey, ed., *The Human Agent*, London, 1968) is a typical example. In order to show that the theory could be true Pears postulates a "degree of feeling" which accompanies every act and which can be identified independently of the act. This may save the causal theory from incoherence but the price is implausibility.

46 'Freedom to Act', p. 155.

47 *Explanation and Understanding*, p. 199, n. 39.

REX MARTIN

EXPLANATION AND UNDERSTANDING IN HISTORY*

One of the main themes of von Wright's book *Explanation and Understanding* is the claim that "the practical syllogism provides the sciences of man with... an explanation model in its own right [and one] which is a definite alternative to the subsumption-theoretic covering law model". He continues, "Broadly speaking, what the subsumption-theoretic [or Hempelian D-N] model is to causal explanation and explanation in the natural sciences, the practical syllogism is to teleological explanation and explanation in history and the social sciences" (*EU*, 27).[1]

Now von Wright makes it clear that 'practical syllogism' is a term with a long history and, of course, a variety of meanings. So we might do well, accordingly, to turn to his own definitive formulation of this basic 'inference schema', as he sometimes calls it. We can put the schema, as von Wright himself does, in the form of an argument where there are several premises which together lead to a single conclusion or, to speak more pointedly, to the performance of an action.

(1) The agent intends to bring about a certain end *E*.
(2) The agent believes (considers, thinks) that unless he does action *A* in time, i.e., at time *t* and no later than *t'*, he cannot achieve *E*.
(3) At time *t*, so-called normal conditions prevail:
   (a) The agent is not prevented from acting.
   (b) He has not forgotten about the time.
   (c) His intention is still in effect (e.g., he has not forgotten it); he has not otherwise changed his mind, etc.
(4) Therefore, the agent undertakes the doing of *A*.[2]

The function of the so-called normal conditions here is to spell out the boundary conditions under which the two key premises have the deductive force indicated in the schema. Or, to put the matter differently, where the intention premise and the belief premise are *fulfilled* under normal conditions then, necessarily, the agent *does* the action specified. Here we see

Manninen and Tuomela (eds.), Essays on Explanation and Understanding, 305–334.

the force of von Wright's use of such terms as 'inference' and 'syllogism' with reference to his schema and the force of calling it 'practical' – because the conclusion is an action done or undertaken. But we could with equal justice describe this syllogism as von Wright's schema for the *explanation* of actions.

Von Wright's schema is a general account which tells us what sorts of things are involved, in his opinion, in giving a non-defective explanation of an action. So, where an investigator says 'the agent did *A* because *T*' (where *T* is an expanded statement of the 'thought' of the agent of the sort we are familiar with), the explanatory adequacy or force is provided by the fulfilling of the criterial conditions expressed in the schema, subject always to the proviso that the fulfilling statements are consistent with the body of available evidence. We could, accordingly, rewrite any such explanation, to exhibit the point more clearly: 'the agent did *A* because' and here we would have a conjunctive list of statements each one of which satisfied a premise of von Wright's schema.

What von Wright is contending, I take it, is that his schema in effect states, more or less accurately and completely, the necessary conditions of adequacy of a certain kind of explanation, the kind he called teleological. The formula 'The agent did *A* because (1)–(3)' provides an operative sketch of the principle of connection for explanations of actions. It does give us the set of necessary conditions which conjunctively are sufficient to warrant the adequacy of explanations of the form '*A* because *T*'. We might describe this formula, then, as representing the 'theoretical commitment' of teleological explanations.

Let me add, parenthetically, that on this analysis what von Wright calls the schema of practical inference is itself simply another way of stating those same conditions, this time as premises to which the action done stands as a (logical) conclusion. It would be pointless to regard either of these two forms (the action-explanation sketch or the argument from premises to conclusion) as *the* groundform; rather, the use of either of the two is a license to use the other (see here the argument of Ryle's paper "'If', 'So', 'Because'").

It would seem, then, that an adequate or non-defective teleological explanation is, in effect, an instantiation of this general action-explanation formula. At the risk of oversimplification, we can say that the logical structure of teleological explanations has these two levels: (a) the abstract general formula (or schema) and (b) particular explanations, each of which can be construed simply as a detailed application of the basic explanation sketch (see *EU*, 142).

Now, clearly, one problem raised by von Wright's account of teleological explanations is to determine the character of 'the *tie*' that holds "between premises and conclusions [in] a practical inference" (*EU*, 107). Or, alternatively, we can say that the problem here is to determine the nature of the connective that is implicated in explanations on the order of 'the agent does *A because* he thinks *T*' (i.e., under so-called normal conditions). Is the tie represented here by the 'because' one that holds in virtue of a general law and, hence, is it ultimately one of a causal and empirical sort? Or is it, in contrast, the tie of logical entailment and hence neither causal nor empirical but, rather, conceptual in nature? But it is important to notice that this is a question about the *logic* of the schema, in its form as a 'practical syllogism' *or* as an explanation sketch, and nothing in it points specifically to anything problematic in von Wright's schema when it functions as an *explanation* sketch in particular.

In this paper I want to focus specifically on the issue of explanation. Accordingly, I will not be concerned with this question of logical status or even, directly, with the question of whether the schema is a true one. I will simply *assume* that it is true and that, if true, it can be given a suitable analysis as to its logical status (empirical, conceptual, etc.). I will turn instead to the substantially different question, whether von Wright's account of teleological explanations is adequate as an account of the logic of explaining *actions*. Specifically I want to consider whether his account provides a suitable theory of what is involved, logically, in the giving and in the understanding of explanations of actions. Hence, I want to determine whether it does provide, as he claims it does, the eligible model for explanations "in history and the social sciences" (*EU*, 27).

Von Wright appears to believe that a given explanation is acceptable as a full-bodied action-explanation if the facts it cites can be taken, in effect, as substitution instances for the 'variables' of the basic schema. We should add, of course, that these substitutions are consistent (or 'univocal') throughout and are supported, as true, by the available evidence. (See *EU*, 120 and 142.) In contrast, I would argue that where a given statement of fact is said to 'fulfil' or 'satisfy' one of the necessary conditions of explanatory adequacy in the schema that claim itself requires further analysis.

I am suggesting that something problematic in the concept of action-explanations remains even after we have adduced their ground of explanatory force – the basic sketch or schema – and shown how individual explanations can be regarded as instantiations of this schema. Specifically the problem is to show exactly how the investigator can vindicate his selection of particular facts where these are said to operate, in effect, as the filling in of the corresponding 'place' in the schema for action-explanations.

The problem here is generated by the nature of the schema or sketch itself. For it is wholly formal; all it can do is exhibit how talk about actions relates to talk of certain beliefs, intentions, etc. The schema justifies our treating particular intentions, beliefs, scruples, and so on as explanatory of particular actions; it does not tell us, however, what *particular* facts to include but only what *kinds* of facts would satisfy the formula. The formula is inert with respect to the facts that are said to instantiate it.

In short, we are not yet clear on what is involved, logically, in *applying* this formula. The problem is how we can say, justifiably, that a particular insertion of facts does instantiate the schema.

It could be replied that there really is not a problem here. Let us suppose we wanted to explain a particular action, say, Brutus's joining the conspiracy of Cassius against Caesar. On what basis can we claim, justifiably, to see a connection between, for example, the purpose or end-in-view of the agent and the deed of his we are trying to explain? The answer, it might be maintained, is ready at hand.

We simply take the agent's *beliefs* about means to the end he intends (insofar, of course, as we have evidence for these), and this means/end relationship *in the context of the agent's belief* is what underwrites our judgment that certain particular facts (Brutus's joining the conspiracy, his resolve to save the republican constitution of Rome from the threat of Caesarism) can justifiably be inserted into the relevant 'places' under the formula. Once this point is made, what remains of the logic of action-explanations is a matter of simple consistency. So long as we consistently substitute the *same* facts throughout the explanation, in the places where they belong, then we have satisfied the formula and produced the explanation called for in the case at hand.

This answer to the problem I have posed is persuasive and has, in fact, been advanced on a number of sides. Von Wright has, in my opinion, put

DESCRIPTION

Volume 72. This is an ex-library book and may have the usual library/used-book markings inside.This book has hardback covers. In poor condition, suitable as a reading copy. No dust jacket. Library sticker on front cover.

the reply particularly cogently:

> The second [belief] premise can be said to 'mediate' between the primary intention of the first premise and ... the conclusion. One can also speak of a transfer or *transmission of intention*. The 'will' to attain an end is being transmitted to (use of) the means deemed necessary for its attainment [i.e., to an action]. This principle of 'transmission of intention from ends to means' is basically identical, it seems, with a principle which Kant thought analytically (logically) true and which he expressed in the following words: "Who wills the end, wills (so far as reason has decisive influence on his actions) also the means which are indispensably necessary and in his power" (*PI*, 45).[3]

The logic of von Wright's point is clear enough. Relying on the idea that the basic feature in a teleological explanation is the means (action)/end (intention) relationship, he argues that the agent's *belief* that action *A* is a means to his end-in-view *E* 'transmits' his intention to include the action. Hence the agent can be said to *intend A* as well. Thus the agent's means/end belief serves as a sort of glue to hold the agent's end-in-view and his action together. Furthermore, we also know it to be the case that where the agent does intend *A* then, under normal cases (as we have already specified them), he *does A*. In fine, the agent's means/end belief binds his action (i.e., the will to do *A*) to his intention (that he pursues *E* as an end) and, in the normal case, leads him to do *A*. (See *PI*, 47–48.)[4]

Now the implications of this for a theory of explanation are striking. For, as should be clear, in the case of attempting to *explain* an action *A* we already have the fact that *A* has happened; hence, we can dispense with so-called normal conditions (for *ex hypothesi*: these have held good in the case at hand; otherwise there would be no action to explain). Accordingly, we do not really need to mention them explicitly in the explanations we give. (See *EU*, 119 and *PI*, 50–51.) This leaves us then simply with the basic explanatory axis itself: *E* (intention or end-in-view) $\rightarrow$ *B* (means/end belief) $\rightarrow$ *A* (action). And since the action is 'already there', in von Wright's phrase (*EU*, 117), we need determine only the remaining two elements. It is at this point that von Wright introduces the contention, or thesis, that I am principally concerned to dispute in this essay. He says:

> It is ... taken for granted that the agent considers the behavior [*A*] which we are trying to explain causally relevant to the bringing about of [*E*] *and* that the bringing about of [*E*] is what he is aiming at or intending with his behavior. Maybe the agent is mistaken in thinking the action causally related to the end in view. His being mistaken, however, does not invalidate the suggested explanation. What the agent *thinks* is the only relevant question here (*EU*, 97).

Now, in a way, what von Wright is claiming here makes perfect sense.

For, if we do understand the notion of a means/end relationship, then we can understand how someone who intends *E* would do *A*, if in fact he *believed*, all things considered, that *A* was the means to *E*. The issue, though, is not whether von Wright has correctly described this important relationship or even whether this relationship figures in our understanding of actions (for, clearly, it often does). Rather, the question is whether the citing of that relationship, in any given case, has the explanatory power claimed for it, as, for example, in von Wright's remark that "what the agent *thinks* is the only [note this *only*] relevant question here."

It seems to me that part of the force of von Wright's point resides in the particular examples used. In many cases we do in fact already see, as we do in the Brutus case, an intelligible connection between intending *E* and doing *A*.[5] I mean that we can see intending a particular *E* as a clear and comprehensible reason for doing that particular action *A*; we can say that the doing of *A* is the sort of thing done when someone intends *E*. One wonders, accordingly, whether adding that the agent *believes A* to be a means to *E* really matters all that much in these cases. And, more to the point, one wonders whether *merely* showing that the agent *believes A* to be a means to *E* is sufficient, for explanatory purposes, in those cases where we *cannot* see the connection as an intelligible one in the first place.

If it is not enough in these latter cases, then it will not do, logically, *simply* to say that Brutus *believed* – even if in fact he did – that such-and-so action (joining Cassius's conspiracy) was the way of achieving his resolve to save the Roman republic. For it could be said, and rightly I think, that unless we can *understand* how someone with this particular purpose could believe that doing this exact thing would be a way of achieving his goal, then we lack the essential standpoint from which we could explain his action. And to attain this we would have to be able to show that such an action does 'make sense' in the light of the intention attributed to the agent.

I think that something in support of this point could be adduced by considering examples where, on the one hand, a perspicuous connection does *not* appear to hold between, say, the agent's intention and the deed he performs and, on the other, we do credit as fact that the agent *believed* a means/end relationship to hold between these things.

Consider the following case. A primitive man, a 'savage' or a 'peasant', has inflicted a knife-wound, accidentally, on his own leg. He attempts to 'cure' the wound by cleaning the *knife* but he leaves the wound itself unattended. We are now offered as an explanation this (true) statement: the man wants his wound to heal and he *believes* that cleaning his knife is the means to this end. If the von Wright thesis is correct, then we ought at this point to understand why the man cleans his knife and lets his wound be. Now, clearly, there is a sense in which we can 'see' this explanation and understand what is going on. But, also, there is a sense in which the connection between the stated intention and the deed performed (cleaning the *knife*) is not entirely clear; it is for this reason that we might say we don't understand the deed, despite our possession of the fact of a means/end belief on the part of the agent.

It should be clear that the problem is not in the *evidence* for the facts we cite but, rather, in the nature of the facts themselves. The problem is that the deed and the intent do not cohere; their connection, even though mediated by a means/end belief, is still not clear. To remedy this defect in our understanding we seek additional information. In the case we are considering, the following bit of anthropological lore might serve to render the connection perspicuous:

> The savage and the peasant who cure by cleaning the knife and leaving the wound unattended, have observed certain indisputable facts. They know that cleanness aids, dirt on the whole impedes recovery. They know the knife as the cause, the wound as the effect; and they grasp, too, the correct principle that treatment of the cause is in general more likely to be effective than treatment of the symptom.... [T]hey fall back on agencies more familiar to themselves, and use, as best they may, the process of magic intertwined with that of medicine. They carefully scrape the knife; they oil it; they keep it bright.

Hence, the injured man 'treats' the cause of his wound (the knife) in order to bring about the desired healing of the inflicted wound. (The example and quoted passage are from Kroeber, 'The Superorganic', 175.)

This little story, added to the bare account with which we started, does seem to provide a satisfactory explanation of the action performed. It does this by providing additional information. One could say that this insertion of the additional material 'fills in' on the agent's *belief* about the means/end relationship involved. Or one could say that it 'fills in' on our initial description of what he did and what he intended.

The important thing is that the information provided 'fills in' on our picture of the deed, a picture which our initial description had only sketched. This 'filling in' exhibits the original elements more fully by in effect redescribing them; and, most important, it brings these elements into a kind of coherence. Their connection is made perspicuous. It was the lack of this coherence that made the initial explanation, although it was an explanation of sorts, unsatisfactory. By the same token, it was the fact of coherence, which emerged with the 'filling', in that made the subsequent explanation satisfactory.[6]

Here, though, we might consider briefly the other edge of the knife example. Most people, I suspect, would want to have the beliefs themselves of the 'savage' or the 'peasant' – the beliefs described in the little story – explained. Some of these beliefs are strange to us; and we might want to go, in the manner of an anthropologist, to the 'cosmology' and so on of the people who hold the beliefs in order to explain them. Doing this, however, is a different matter from explaining the original action; it takes us into another dimension altogether. For we are here explaining, rather than using, *beliefs*; and explaining these has only a tangential bearing on, and perhaps even a different logic from, the explanation of the original action.

This distinction between 'filling in' on the agent's beliefs and the *explanation* of these beliefs, after they are 'filled in', is an important one. The first of these two things is *part* of the explanation of the agent's action, while the second is the explanation of a particular *belief*, usually by reference to other beliefs. This second order of explanation could, actually, take any of several form: we could trace the agent's belief to some social convention, or even attribute it to a mistake in inference; but most often, and most interestingly, we would locate its genesis in some very general system of belief, in the *Weltanschauung* of a particular culture.

There is, moreover, a curious regress in the explanation of a *belief*: We refer the belief to be explained to a second and that in turn to another, and so on. This feature of regress is lacking when we *use* a belief in the explanation of an *action*. What 'fixes' the belief here, and prevents the regress, is the place that belief has in the explanatory story it is part of. In short, the explanation of an action may require the insertion of certain beliefs – and here, I think, is the sound point in von Wright's thesis – but we don't need also to be able to explain that belief in order to *use* it in the explanation of the action. (But see *EU*, 98–100.)

Of course, a historian might want to explain, e.g., Brutus's 'relevant' purpose; he might want to explain why Brutus was attempting to preserve the republican constitution of Rome. Or he might want to explain why Brutus *believed* that he needed to join the conspiracy of Cassius. And here certain further features of Brutus's thought, or features of Roman thought, might be called in as the explanatory beliefs behind Brutus's original intention or his means/end belief. But this is a different matter from explaining Brutus's *action* of joining the conspiracy of Cassius, which is what the historian in von Wright's example wanted to explain. Indeed, the subsequent explanation of Brutus's intention or his means/end belief here actually presupposes the truth of the explanation of his *action* that the historian gave.

By the same token the anthropologist might want to explain the magical notions or the primitive ideas about causation that functioned in the knife/wound story. But his subsequent explanation of these things actually *presupposes* the truth of his original explanation of the 'peasant's' action. This is, I think, an important feature in the logic of the relationship that holds between *using* a belief in the explanation of an action and the explanation of that same belief.

The main point I want to stress here, though, about the knife example is that the explanation when offered on the von Wright principle, where merely citing the agent's *belief* as to a means/end relationship is thought to suffice, is not a satisfactory explanation. If it had been, then there would have been no need, or no need felt, to supply the additional information which did fill in the picture satisfactorily. Even if the difference here is only that between understanding and understanding *better*, the fact remains that what we had on the von Wright principle was not up to a certain standard. The proof is in the pudding: Can we say upon 'filling in' that we understand the agent's *action* better than we did at the point where we knew only that he believed a means/end relationship to hold between what he did and what he intended? If so, I think we have in this example a sort of counter-instance to the thesis of von Wright.

The general moral we can extract from this is perhaps less clear. We need to determine here the margin of difference that exists between the understanding we have upon 'filling in' and the understanding we have on the von Wright principle alone. The proposition that seems to catch this difference is this: If the putting together of an agent's intention and his deed, considered here as a means to that intention or as part of accomplishing it, is not itself perspicuous, then the mere citing of the fact that the agent *believes* a means/end relationship to hold between deed and intention is not sufficient to make it perspicuous. And if the putting together of an agent's deed with his intention is not intelligible, then using these facts in an explanatory way lacks force. In fine, the mere insertion of an agent's sincere belief to the effect that *A* is a means to *E* is not sufficient in such a case to make the explanation of *A* go through.[7]

What does make an explanation go through, then, is that the putting together of the agent's intention with his deed *is* perspicuous. Or, better, we should say that this *allows* the explanation to go through. For it is not from the facts alone, even when their connection is perspicuous, that an explanation derives its force but, rather, from these facts *when placed within their proper logical framework or context*. The proper framework for these facts is provided by von Wright's schema for action-explanations.

In short, the idea or form of a teleological explanation is given in the schema itself; it is here, at this level, that we perceive the basic outline of the explanation. For it is in terms of the schema that we exhibit our basic conception that action is being considered as a means to an intended end, or as part of accomplishing it. But for facts to count as satisfying that schema in a given explanation we must not only see them as instances of the elements in the schema but also see them as exhibiting, materially, an intelligible connection.

My claim, then, is that the warrant for using certain particular facts in an explanatory way is simply that their connection, where they are taken as standing in the deed-intention relationship, is a perspicuous or intelligible one. Basically, this means that two facts, one a deed and the other an intention, are, under given descriptions, coherent one with the other; it means that their conjunction, under those descriptions, is not in itself opaque or problematic. We *have* this warrant, then, insofar as the explicit material connection provided – as when we bring together Brutus's joining the conspiracy with his alleged intention to save the Roman constitution – is a sound or plausible one. But if the investigator could not validate his connective reasoning at this level – at the level of a particular material connection – then his bringing these details under the schema of practical inference would be an empty gesture. For, if the connection between a deed and its alleged intention is an opaque one, as it was in the knife-wound example, then we cannot bring that deed and that intention, as described, under the formula: we cannot really use *these* facts as substitution instances of the 'variables' of the formula.

Or, to be more precise, if we use these facts in an explanation we are using facts we don't really understand. And here I think a *distinction* between explanation and understanding begins to emerge. The explanation so achieved is a minimal one; it is an explanation of sorts but it fails to satisfy the standard of *understanding*. This standard does not rule out the explanation as an explanation; it does, however, indicate a deficiency in an important respect. For an action-explanation should tell us something more than that the facts cited satisfy the basic schema. An explanation should yield understanding: it should provide a factual narrative that we can follow, an account of action that we can 're-enact' (the term is Collingwood's; see *Idea of History*, 215).

There is even a sense in which von Wright's point about means/end *beliefs* concedes the logic of my argument. For, after all, if there were in every case a perspicuous connection between the two basic facts – the fact of the agent's intention and of his action (considered here as a means to the end intended) – then there would be no need to insert yet a third fact – this one the fact of the agent's *belief* as to means/end. Hence, the very fact that we sometimes do need to call in the agent's *belief* (and, presumably, it will be a somewhat odd or alien one) indicates that initially the desired connection does not hold between the agent's deed and his intention; and the very point of inserting what the agent believes in the matter is to effect this connection. By the same token, where the insertion of that fact does not render the initial connection perspicuous, we do not consider the agent's action satisfactorily explained. The point of crucial difference between the inadequate attempt at an explanation and the adequate one, then, is not marked by the insertion of the agent's means/end *belief* but, rather, by the establishment of an intelligible connection between the elements of the agent's action (his deed, his intention, and his means/end *belief*).

Now that I have made this point, it might prove valuable to turn to the only extended historical example that von Wright provides in his book. I think this example sheds an interesting, and unexpected, light on the argument I have been conducting. Von Wright's example concerns the causal role played by the assassination of the Austrian archduke at Sarajevo (July 1914) on the outbreak of the First World War. (See *EU*, 139–43.)

Von Wright's point is that the Sarajevo incident was *causal* not with respect to some putative general law but, rather, in the sense that it supplied a *motivation* for further action. Indeed his main point is that the causation (the *explanans*) of the outbreak of the war was a complex series of actions, each one of which provided a motivation for the next action and ultimately for the last in the series, the outbreak of the war itself. As he put it (*EU*, 142–43):

We have a sequence of independent events: the assassination, the [Austrian] ultimatum, -----, the outbreak of the war. The events are linked, we said, through practical syllogisms. But *how*?.... The following schematic picture could be used to illustrate this – a dotted arrow meaning that a fact affects the premises of a practical inference and an unbroken arrow that a new fact emerges as a conclusion [i.e., as an action] from the premises [and here von Wright provides us with a diagram]:

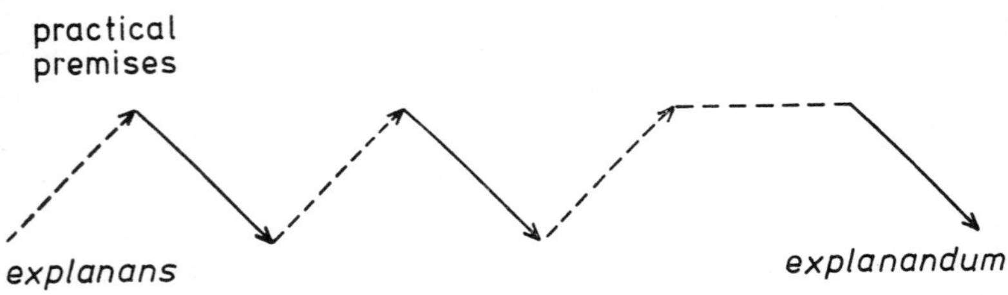

Two points of considerable theoretic interest emerge from von Wright's example. First, the explanation offered (of the outbreak of World War I) does *not* conform to the teleological model of explanation which he had offered us, at the beginning of his book, as *the* model for explanation in history and the social sciences (see *EU*, 27). The teleological or practical inference schema moved from premises to a conclusion (i.e., that an action is done); whereas the model we are considering here moves from situational facts or 'motivation background', in von Wright's phrase (*EU*, 141), to the practical premises, where the agent's operative intention is affected by his perception of these facts, and thence to the action done. Indeed, von Wright is aware that the Sarajevo example does not fit the teleological model and says, "To call the explanation 'teleological' would certainly be a misnomer, although teleology essentially enters into the practical inferences which link the *explanans* to the *explanandum*" (*EU*, 142). He proposes, instead, that it be called a "quasi-causal historical explanation" (*EU*, 142–43; see also 85–86).[8]

Nonetheless, this first point is not too telling against von Wright since, as he points out, teleological explanations are still required to move us from a given 'motivation background' (e.g., the assassination at Sarajevo) to a particular action, which then constitutes a new situation (or 'motivation background') for yet another practical inference culminating in a further action, and so on. However, it is worth noting, and this is my second point, that von Wright's idea of a situational fact or 'motivation background' is nowhere mentioned *explicitly* in his schema for action-explanations or practical inference. Now this is an important point; so, perhaps, we would do well to introduce that notion explicitly into our discussion of his schema.[9]

Let us return at this point to the argument I was advancing earlier, to determine the bearing of that argument on this proposal for a second base line of explanatory connection, this one to be drawn between the agent's situational motivation and his formed intention (see the dotted line in the schematic picture above and also *EU*, 140–41, in particular). It could be alleged in this case, analogously to the claim that what counts is simply the agent's means/end belief, that we need only take account of his *belief* that a certain end-in-view or purpose, if achieved, would satisfactorily take care of whatever it was (i.e., his situational motivation) that moved him to act in the first place. This belief, in fine, is *all* we need to possess in order to complete the required connection between the agent's situational motivation and his operative purpose. And, again, it could be replied that merely being in possession of this belief is not enough. For we have then failed to rule out those opaque, problematic connections which could exist in this relationship, just as they did in the intention/deed one.

The point here is the same one we met with earlier. It is the point that unless we can understand how someone having a particular situational motivation could believe that achieving his intention would resolve the situation he thought himself to be in, then we lack the essential standpoint from which we could *understand* his action and, hence, offer a full-fledged explanation of it.

To see that this is so, suppose we had evidence to show that the agent had a certain purpose and that he was moved to action by a certain situation and that he believed, or apparently believed, that this purpose 'fit' his situation. The question is whether the mere command of these facts (all of them supposed to be true) is sufficient to yield an adequate explanation. Again, an example will help us decide.

A reporter who was covering President Nixon's historic trip to Peking describes his own visit with a factory manager in China:

> On the walls are numerous injunctions to study the thoughts of Mao, and Mr. Ching [the factory manager] indicates that, thanks to such study, production has recently been doubled, from five thousand jeeps per year to ten thousand. I try a couple of times to discover precisely how Mao's thoughts have helped to raise output. On one occasion when I pose the question, I am shown how front-end grilles are picked up by an overhead belt and dipped into a bath of paint. Formerly, it is explained, they were sprayed one by one. When I pose the question again, I am shown a West German machine for making wire. Giving up on that line of research, I ask Mr. Ching what impact the Cultural Revolution had on the plant (Joseph Kraft, 'China Diary', 105).

Now it should be apparent that the manager's resolve 'to study the thoughts of Mao', which was being urged on the workers as well, was not thought to be his only relevant intention or even his ultimate one. But it was, clearly, *one* of his resolves respecting the situation that he envisioned (as involving, presumably, a need to 'raise' production). And it was precisely because the manager believed this resolve to 'speak to' his situation that the reporter asked the questions that he did. For the point is that the reporter could *not* 'see' the connection of intention with situation conception on the manager's part despite the manager's belief that there was a connection. The connection, then, was the very point he was questioning.

Should the reporter, then, accept the claim that the factory manager's conception of this situation did connect with his resolve to study the thoughts of Mao *simply* on the basis of the manager's express belief that there was a definite connection there? Would the reporter 'see' the connection or allege its intelligibility simply on this basis? Could he justifiably claim to understand the action(s) he wanted explained? Evidently not.

This shows, I think, that the agent's sincere belief *alone*, as to there being a connection between his situation as perceived and his formed resolves, is not enough. If it were, then we would have to accept 'crazy' beliefs, ones that are askew or even mad, as affording the desired connections. But this is exactly what we won't do, and can't do, in an explanation. For we need to know not only that a belief is supposed to connect the agent's purpose with his situational motivation but also that it does in fact fit in, and that using it does provide or help provide an intelligible connection.

I am not, obviously, suggesting that the investigator should discount the beliefs of agents and look only for 'intelligible connections'. Rather my claim is that if the putting together of an agent's situational motivation and his purpose, considered here as something which, if accomplished, would satisfy his impulsion to act, is not itself perspicuous, then the mere citing of the fact that the agent sincerely *believes* his purpose to be 'responsive' to his perceived situation is not sufficient to make it perspicuous.

What is necessary, then, is that the putting together of the agent's situational motivation with certain of his beliefs and with his purpose be perspicuous. We require that the explicit material connection provided, as when we bring together Brutus's perception of the threat posed by Caesar (*i.e.*, Brutus's 'motivation background') with his alleged intention to save the republican constitution of Rome, be an intelligible one. If it is not, then we 'fill in' our picture of the deed so as to exhibit these elements, the situational motivation and the agent's purpose as originally described, more fully. We do this by providing additional information, which in effect redescribes these elements, so as to bring them into a kind of coherence.

In the case of the factory manager, for example, we could get some purchase on the role played by his resolve 'to study the thoughts of Mao', if we were to 'fill in' a bit on the picture the reporter has provided. The manager perceives a deficiency in the level and quality of production, and he feels moved to raise production; this is his situational motivation. The action taken consists in the various measures he had instituted on the assembly line. The question, though, is one of intention or purpose. It is likely that the end or purpose behind the factory manager's actions is in fact a comprehensive national program for production, and one that is socialist in nature. Here, then, we begin to see how the thoughts of Mao could fit into the reporter's picture and how, by the same token, the manager's resolve to study these thoughts would be ingredient in his overall intention – how, in short, the manager's resolve *might* be responsive to his perception of the situation.

In sum, the investigator in an explanation is working with several rather precise relationships, the ones we are familiar with from our discussion of the expanded version of von Wright's schema of practical inference; and in giving an explanation, the historian is in effect asserting that the facts cited *can* stand in the required relationships to one another. This means that for each of the designated relationships the facts must, in order to fulfill the relationship, exhibit an intelligible connection. Such intelligible connection is required both for the facts that are taken as satisfying the terms of the situational motivation/purpose relationship and for the facts that are said to fulfill the relationship of deed to operative intention. There must be a 'coherence' between each pair of facts in order for the facts to serve as end points on a line of the teleological angle and, hence, when all the points are covered, to provide the explanations we are seeking:

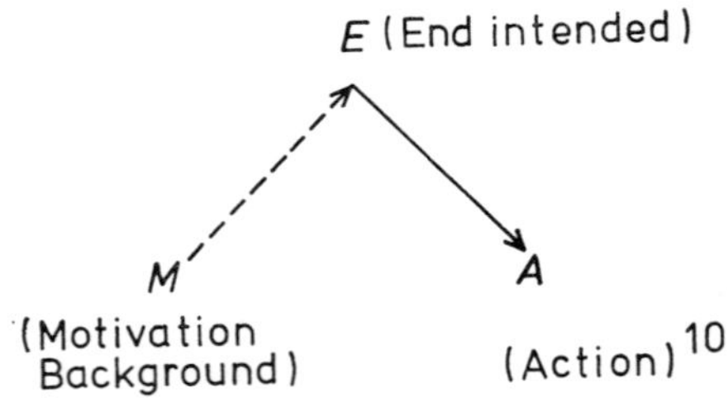

Now it might be well to add at this juncture what no one would deny, that the facts in question are 'established' by going to the evidence. For, after all, it is by reference to our body of evidence that we would decide that *E*-1 (Brutus's attempting to save the constitution) *is* a fact (a true statement of the agent's intention) and that *E*-2 (his getting rid of an envied political rival) is not. Even so, we should beware of drawing a faulty conclusion here. For a problem can arise with the implicit metaphor of 'plugging in' facts (or with the somewhat analogous one of 'valuing' the 'variables').

If the metaphor is taken too literally, then one could imagine that *all* you need do is show, by going to the evidence, that *E*-1 is a fact. The point is that there are *two* distinct relations involved and not just one: the relation of facts to evidence as true/false and the relation of facts to one another as perspicuous/opaque. Hence, it is one thing to say that the relationship of facts in a given historical account is a plausible one and another to say that the facts cited are evidentially well supported and, hence, in that sense, true.

Clearly, we want in an explanation to satisfy the relationship of factual statements to evidence, so that the statements we use are evidentially well grounded or supported (true). But that does not negate the need for 'intelligible connections' between facts, such that facts cited *can* fulfil the terms, say, of the intention/deed axis.

If my argument up to now has been sound, then this business of simply 'going to the evidence' to 'value the variables' of the schema won't do, contrary to what von Wright suggests (see esp. *EU*, 97). The angle metaphor at least has the virtue of indicating that the facts are always linked and can't be 'plugged in' simply as discrete individual units. For we can plug facts in only after we have established that a 'fittingness' holds between, say, a given action (like joining the conspiracy) and a particular end-in-view intended (like wanting to save the constitution).

Moreover, if my argument has been a sound one, then it also follows that we must add a criterion to those von Wright has been prepared to acknowledge. For the argument I've conducted has shown that it is not enough to require simply that we consistently substitute the same facts throughout, in the places where they belong under the schema, in order to have an explanation. It is not enough even when the facts themselves are evidentially well grounded. The argument has this thrust precisely because it requires that there be a perspicuous or intelligible material connection between a set of facts in order for the facts to serve as end points when superimposed on a line of the metaphorical angle of teleological explanation and, hence, for them to count as fulfilling of conditions in von Wright's schema.

What I am claiming here is that there are *several* criteria for using the schema in any given case of explanatory reasoning. It is not just that the facts be evidentially well supported, or that the substitution of these facts be a consistent one throughout, but also that the facts cited do, as we follow the lines of the angle, exhibit coherence. These three criteria – evidential support, univocal substitution, intelligible or understandable connection – are distinct and they are equally necessary, for each one tells us something essential about the correct application of the schema to a piece of behavior in a given case. We might, then, describe those criteria as 'application principles' (the term is Braithwaite's) for von Wright's schema. When these criteria are satisfied, and the conditions of the schema are in effect fulfilled, then the facts cited have explanatory force. We just *see* these facts, the ones that state the agent's action and the factors of his thought, as an instantiation of the basic schema for teleological explanations.

These criteria are not, it should be evident, directed at the schema itself; their job is not to specify additional conditions in the formula. Rather, they are directed at the facts. They indicate when, in a given case, certain individual facts can justifiably be brought under the schema for teleological explanations. Their function is to validate the use of those particular facts in an explanatory way. The logical role of understanding, then, is simply that it is a criterion for *applying* the schema: an understandable, or intelligible, material connection of this fact to that one, where they are taken as satisfying the terms of one of the lines of the teleological angle, is required in order for these facts to be treated as fulfilling of conditions in the schema for action-explanations. This is *one* criterion that must be met by the facts.

Perhaps my point here can be made clearer if it is put in rather more familiar terms, in terms that von Wright himself has employed. Let us, in our general account of teleological explanations, distinguish between 'formal' and 'material' validity (see *EU*, 120–21). A given explanation is *formally* valid if it provides us with facts that satisfy, at least in effect, the conditions or 'premises', as he calls them, of von Wright's schema. Now, of course, if any one of the facts provided is incorrect, that is, *not so*, then the explanation is said to be *materially* invalid. As von Wright comments of one of his own examples.

> This explanation can be 'materially invalid' (false, incorrect) in the sense that the reason why *A* pressed the button was in fact different. But it is 'formally valid' (correct) as an *ex post actu* construction of premises to match a given conclusion (*EU*, 120).

What I am contending is simply an extension of this distinction. I am proposing that an explanation in which the relationship of facts in either of two dimensions, motivation/intention or intention/deed, is not intelligible be regarded as *materially* invalid – as the explanations were, for instance, in the Mao (motivation/intention) and in the knife/wound (intention/deed) examples. In other words, when we cannot understand *facts* said to be standing in a certain relation to one another, even though we can comprehend the relationship itself, then we have a case of *material* invalidity; and this is so whether the facts are judged on other grounds to be true or false. Hence, beside invalidity due to falseness I would set invalidity due to implausibility. What we want in an explanation, otherwise formally valid, is both plausibility *and* truth. Materially valid explanations are both, and only those should we let count as full-fledged or wholly adequate teleological explanations of action.

Indeed, now that I have made this second kind of material invalidity explicit – the kind we call *implausibility* – it is possible to find mention of it, here and there, in von Wright's own text. For instance, he tells us that "we cannot understand or teleologically explain behavior which is completely alien to us" (*EU*, 114–15) and the context indicates that the failure here is at the *material* level, in our sheer inability to penetrate an alien form of life and thereby to understand its characteristic modes of behavior. (See also *EU*, 29, on this point.) And, again, in terms that have a strong resonance with the argument I have been developing, von Wright asserts, "In order to qualify as *fact*, one could say, the material at hand must already have passed a test of explicability" (*EU*, 166). Now, admittedly, von Wright's test here can be read narrowly as requiring merely that we must *tell what some group or person is doing* (and in that sense 'interpret' their behavior, putting it under some specific action-description); but his remark is susceptible of a wider reading, as it has in his own gloss on the passage cited, where he says, "I.e., it [the behavioral datum] must have been shown to be intelligible as action" (*EU*, 206 n. 27).

So, my criticism of von Wright should not be taken as suggesting that he has totally ignored the *material* invalidity induced into 'formally valid' explanations by unintelligible or implausible (material) relationships between the facts cited. Rather, I am principally concerned to show that he did not accord this factor *explicit* status and thereby failed to incorporate it into his account of teleological explanation. I am concerned, as I have already said, to show the inadequacy of von Wright's contention that *all* we need in order to 'mediate' an explanatory connection between the agent's end-in-view and the action he performs is his *belief* that this action is a means to that end. (Recall: "What the agent *thinks* is the only relevant question here" [*EU*, 97].) This may be all that is needed under normal conditions in order for the action to be *done*, but it is not all that is needed in order for the action to be *understood.*

Von Wright has provided no real role for *understanding* in his account of the logic of action-explanations. Although he has found a place for evidence and for *true* statements within the conceptual space occupied by his schema for teleological explanation, he has found no corresponding place for understanding and for *plausible* statements.

Lest my point be misunderstood, let me emphasize that von Wright does have a positive doctrine of understanding in his book. He says, "Before explanation can begin, its object – the *explanandum* – must be described.... If we call every act of grasping what a certain thing is 'understanding', then understanding is a prerequisite of every explanation, whether causal or teleological" (*EU*, 135). But his emphasis here is unmistakable: "[I]t seems to me", he says, "clearer to separate this step from explanation proper, and thus to distinguish between the *understanding of behavior* (as action) and the *teleological explanation of action* (i.e., of intentionalistically understood behavior)" (*EU*, 124). Clearly, then, von Wright not only distinguishes understanding from explanation but also *separates* them into two distinct operations. (See also *EU*, 26, 123, 132, 134, and 152.) They relate to one another only *externally*; there is no penetration, no internal role for understanding *within* the teleological explanation of actions.

Von Wright has failed to *integrate* understanding into his account of the teleological explanations of actions. And this, I think, is the foremost shortcoming in his book *Explanation and Understanding*.[11]

Now that this very basic point has been made, let us return to the theme with which this essay began, to von Wright's contention that the schema of practical inference provides the model for explanation "in history and the social sciences" (*EU*, 27). I think the concept of *understanding* that I have been discussing in this essay fits the case of history particularly well. For I have in effect contended that understanding consists in the ability, given a particular set of facts, to construct an unforced narrative. Or to put the matter differently, the essence of understanding does not reside in some datum of *experience*, say, an 'aha!'- or 'hat doffing' experience (the latter phrase is F. Waismann's) or a 'seeing' of the light or a 'perception' of any sort; rather it resides simply in the telling and following of a plausible story, the factual details of which can be displayed as instantiations of the elements of von Wright's schema for practical inference.

I think this may well represent what von Wright had in mind when he explicitly contrasted a 'psychological' construal of understanding (as *Einfühlung*) to what he called the 'intentionalistic' construal or "semantic dimension of understanding" (see *EU*, 6) and contended that understanding "is a semantic rather than a psychological category" (*EU*, 30). And this idea of understanding, as the capacity to tell or follow a particular story, certainly fits in nicely with his *aperçu* that "behavior's intentionality is its *place* in a story about the agent" (*EU*, 115).

If my reading of von Wright here is sound, or even suggestive of a tendency in his thought, then he is much closer to the *verstehen* tradition, or at least to Collingwood, than he thought. (See *EU*, 33.) For Collingwood put great emphasis on the category of action with its attendant features of motivation background and purpose (cf. von Wright's schema of practical inference) and his idea of re-enactment can be taken simply as a peculiar way of saying that we understand a deed when we can put it in place in a story about the agent's motivation background and his purpose.

Be that as it may, I am not so much concerned with von Wright on understanding or Collingwood on understanding as with the claim that understanding is the 'narrative dimension' in a teleological explanation (*i.e.*, an explanation that conforms to von Wright's schema). And I do want to contend that an integrated model of explanation, where von Wright's schema is applied in accordance with the criterion of narrative understanding, is peculiarly well adapted to serve as the model for explanations *in history*.[12]

So, one half of von Wright's basic contention is solidly made; indeed, it is enhanced when his idea of the schema of practical inference is supplemented by the notion of narrative understanding – a supplementation which is required in any case, as I have tried to make clear. But how do things stand with the other half of von Wright's basic contention, with his claim that the schema also provides the model for explanation *in the social sciences*? Here the matter is not so clear.

I do think the logical interrelationship of the practical inference schema and the notion of narrative understanding has interesting implications for von Wright's theory of the social sciences.

(a) The range for using the schema marks off the human-from the natural-events category. We say a process or event is natural when the basic schema is *in principle* not applicable. It is a category mistake for the terms of the formula, such as intention or situation-motivation, to be applied to *non-rational* causes or events. This is a way of stating von Wright's distinction between nature and agency. (See *EU*, 160–61.)

(b) Within agency, moreover, the possibility of 're-enacting' the elements that make up a particular instantiation of the schema marks off the specifically rational category from the *irrational*. We say an activity or an action is irrational when the schema itself, with its peculiar terms, such as intention, is applicable but where the instantiating elements are *not* re-enactable.

We are now in a position to provide a thumbnail account of psychology on the basis of these distinctions. Actually, there ought to be *two* types of psychology, and not just one, in this account. First, there should be a naturalistic or physiological psychology. It is a natural science as per (a) above, in that it studies the *non-rational*. It would be a causal science. The scientific study of human perception might be an example of such a science. Second, there ought to be a psychology that studies purposive but *irrational* behavior. This science is *not* a natural science, in the sense of (a) above, in that the basic terms of the schema for teleological explanation are broadly applicable; but neither is it a *verstehende* science, in the sense of (b) above, precisely in that it studies things not re-enactable. Or, to be exact, the connections of instantiating facts are *in principle* not perspicuous: they are rationally inappropriate or unintelligible. The study called psychoanalysis, especially in its Freudian form, might be an example of such a science. In this latter case, where the distinction rational/irrational is drawn, we have the conceptual foundation for drawing a sharp distinction between history (which has the dimension of narrative understanding) and psychoanalytic psychology (which does not) while at the same time retaining the schema itself as the model for explanation in both of them.[13]

There are, of course, alternative ways to formulate the notion of psychoanalysis as a science – and here I am considering just the descriptive and theoretical element in psychoanalysis – and it is quite possible that von Wright would prefer one of these to the one I have just elaborated.[14] But the point is that the alternative formulation, if it is to conform to von Wright's basic contention, must preserve the schema in its role as providing the model for explanations in that science. This aside, it is still open to the critic to respond that history and psychoanalysis and perhaps even psychology are not *social* sciences and that for von Wright's contention to make good it must hold up in the case of recognizable *social* sciences, *i.e.*, the sciences of economics, sociology, and cultural anthropology. And these are hard cases for his basic thesis to cover.

It is difficult, indeed, to see how von Wright *could* make his point here. He does not appear to regard an economic system or a society or a culture as a supra-individual agent, with a life and destiny of its own, of whom it would be appropriate to posit intentions, a motivation background, and so forth. In fact, he seems to regard this view, when it is expressed straightforwardly, as it was by Hegel, as a metaphysical excess. (See *EU*, 154, 156 and *PI*, 40.) Moreover, he appears to regard much talk of, say, a mode of behavior as a *function* of some social end or a social practice as *purposeful* as being quasi-intentional in character and as coming ultimately, not under the teleological model, but under the *causal* model of explanation. (See *EU*, 59–60, 80–85, 153, and esp. 156 and 160.)

On the other hand, von Wright's Sarajevo example provides an interesting possibility for extending his thesis to the hard-core social sciences. Von Wright's use of this example is quite instructive. Most philosophers of history, certainly those that are defenders of the teleological or 'intentionalistic' model of explanation, tend to emphasize as their favored examples cases of explaining the actions of *individuals* (*e.g.*, the Brutus example). But von Wright is almost unique in having fixed upon an example of a complex or macro-event (the outbreak of a war) which was the doing of many individual agents and which was, conceivably, not intended by any one of them. We might describe the outbreak of the war here as itself a complex and unintended action. I am not, of course, suggesting that historians don't normally write about such things; it is rather that philosophers of history don't normally cite them as examples or take them particularly seriously.[15] And the remarkable thing is that von Wright has managed to bring this kind of example – despite the misleading description of it as 'quasi-causal' – into the framework of teleological explanation. (See also *EU*, 157–59.)

If this is so, perhaps the economist's explanation of a particular depression or the sociologist's explanation of the impact of a given technological change on the consciousness and way of life of a particular class can be brought under von Wright's schema for teleological explanation. Indeed, we could even imagine the development of social scientific *theories*, say, of economic depressions in general, in which the "statistically correlated generic features" (*EU*, 164) of these macro-events can be expressed as the 'quasi-causal' outcome of the actions of a multitude of *anonymous* individuals acting and reacting under *typical* motivations. (See *EU*, 163ff.) Here social science theories (*i.e.*, explanatory models or 'laws') would themselves be conceived as conforming, ultimately, not only to von Wright's teleological schema but even, in a special and peculiar way, to the criterion of narrative understanding or intelligibility.[16]

On the other hand, it may well be that all or some of the *theories* of social science are causal in nature – are general laws or, at least, general laws in principle – and that the proper explanation of an individual macro-event, e.g., a particular economic depression, is by reference to these laws and, hence, causal and non-teleological. If that proved to be so, then von Wright's schema would have a highly restricted application in the explanation of macro-events and probably only a marginal role at best in the explanations offered by social scientists. But even there, if we were to decide that the characteristic patterns of explanation in social science do *not* come under the teleological schema, it would still be necessary that we have a clear conception of what such a schema is and how it could be deployed in order to decide that. So, von Wright's analysis, even though it does not really address the issue of explanation in the social sciences, does point in a theoretically interesting and fruitful direction.

At the same time, as we move in this direction, the question will inevitably arise as to the role of narrative understanding *in particular* in the social sciences. As I have already indicated, such understanding appears to have a role in *historical* explanations but, possibly, not to have a role in *psychoanalytic* ones – even though both kinds can be regarded as deployments of the teleological schema. If we are to determine the degree to which explanations in the *social* sciences also fall under this schema, we will need to consider its deployment in both the integrated form, where narrative understanding is a condition of its application (as it is in history) and the unintegrated form, where narrative understanding is *not* a condition (as it is not *ex hypothesi* in psychoanalysis).

This leaves two large pieces for von Wright to fill in in his basic thesis. He will have to consider my argument that narrative understanding is a criterion of applying the teleological schema in characteristically *historical* explanations. And then he will have to address, directly and with extensive argument, the question whether the schema is, especially when enlarged to include this narrative dimension, the principal model for explanations in the *social* sciences. Only then can von Wright claim to have established the thesis that the 'practical syllogism' provides the 'explanation model' for both history *and* the social sciences.

University of Kansas

NOTES

* This paper was written during the spring and summer of 1973 while I was a Fulbright Research Scholar in the Institute of Philosophy of the University of Helsinki. An earlier

version was read to the Research Seminar there; the present version owes much to philosophical conversations I had with Raimo Tuomela and, especially, with G. H. von Wright. I am also indebted to my colleague Donald Marquis for his helpful criticisms and to my wife's brother David Mel Paul for translations from the Swedish.

[1] See also *EU*, 160. Cf. von Wright's claim in a subsequent essay that "as a schema of explanation, the practical inference pattern holds a position in the human and social sciences similar to that of the deductive-nomological [or *D-N*] inference pattern ('the covering law model') in the natural sciences" (PI, 39: author's summary by von Wright)

In this paper I will cite and refer to von Wright's writings, in my text and in the notes, according to the following convention: initials followed by page number(s), e.g. *EU*, 160, where that means p. 160 of his book *Explanation and Understanding*. At the end of this paper, in a bibliographical note, I have provided a key to these abbreviations together with full bibliographical information on these and the other sources used in my essay.

[2] My own statement of the schema is substantially like von Wright's. For von Wright's "final formulation" see *EU*, 107 and also *PI*, 47. Since he discussed his schema at considerable length in *EU*, 96–107 and 125, in *PI*, 45–49 and 52, and in *DS*, one could incorporate more of this qualifying discussion into the composite statement of the schema than I have done here, especially at the point of so-called normal conditions. These conditions are the ones that obtain "in normal cases [where] we start from the fact that an action has been accomplished and can thus take it for granted that the agent also 'set himself' to do it" (*EU*, 119; see also *PI*, 50–51).

[3] Von Wright is here quoting Kant's *Groundwork of the Metaphysic of Morals* (tr. H. J. Paton; London, 1948), 84–85. He quoted the same passage in *VG*, 170 n (there the passage is quoted in German; see the Prussian Academy edition, IV, 417). And Donagan uses the same passage from Kant, to make the same point. He says ('Alternative Historical Explanations and their Verification', 76):

> I take it ... that if a man unconditionally intends to bring about a certain end *E*, and judges that his doing *A* is required for bringing about *E*, he must intend to do *A*. As Kant remarked, "who wills the end, wills (so far as reason has decisive influence in his actions) also the means which are indispensably necessary and in his power. So far as willing is concerned, this proposition is analytic."

It is, I think, worth remarking that von Wright, like Kant and Donagan, regards the "transmission of intention from ends to means" as exhibiting an analytic, or as he calls it, a "logical" connection (see *PI*, 45).

[4] In his book von Wright did not use the 'transmission of intention' argument (or the Kant quotation). Rather, he relied on the much weaker claim that "the intention or the will to do a certain thing cannot be *defined* without making reference to its object, *i.e.*, its intended or willed result ...," He adds, however, that "the logical dependence of the specific *character* of the will on the nature of its object is fully compatible with the logical independence of the *occurrence* of an act of will of this character from the realization of the object" (*EU*, 94; see also *EU*, 67–68). We might call this second, or weaker, argument von Wright's 'logical connection' argument, for he expressly acknowledges his debt for the two passages cited to Stoutland's 'Logical Connection Argument', 125.

[5] Many of von Wright's examples in *EU* are wholly unproblematic, commonplace examples – like pushing buttons to ring bells or turning handles to open doors/windows or crying for help when drowning – which are germane to the points he is making but

are of no real interest to working historians. The Brutus example, which is also one of von Wright's (see *FH*, 8ff; *SA*, 191ff; *HS*, 320ff), is, of course, of interest to historians, although it is, like those others, of unproblematic character. Interestingly, the Brutus example is one of Donagan's stock examples (see his 'Popper-Hempel Reconsidered', 150–54).

[6] Cf. Davidson: "[W]hen we explain an action, by giving the reason, we do redescribe the action; redescribing the action gives the action a place in a pattern, and in this way the action is explained" ('Action, Reasons, Causes', 692).

[7] The term 'intelligible' is drawn from Gellner; in his case it refers to "a *conceivable* reaction of human beings to circumstances" ('Holism versus Individualism', 492 [italics added]; see also p. 494). The term is also used by Donagan ('Popper-Hempel Reconsidered', 155), but without explication:

> Considering what human history has been, an historian would be in a pretty pass if he were obliged to assume that the only actions he may succeed in understanding were rational. They must, indeed, be intelligible; but that is another thing.

Along these same lines, the translators of Wittgenstein's 'Remarks on Frazer's 'Golden Bough' use the term 'perspicuous'; Wittgenstein's own word is *übersichtlich* (See p. 35 and 35 n). Wittgenstein's term has the sense of lucid or distinct, i.e., in that things are clearly arranged; his word suggests a synoptic or connective vision: hence, literally, a seeing of connections. In a similar vein Louch uses the term "transparent" (*Explanation and Human Action*, 163; and on pp. 120–21 he uses "intelligible", with the same meaning).

In each case the point is the same: we are concerned with the intelligible (perspicuous, coherent, transparent) connection of elements in a designated relationship. Here it is useful to note Scriven's notion of "parts whose relation we understand" (review, 502–03); this is the crux of what I mean by 'intelligible connection'.

[8] Von Wright provides one other important historical example in his book, that of the economic recovery of Poland in the 14th century under Casimir the Great (see *EU*, 153–55). But since this example has the same logical features as the Sarajevo one, I have not elected to give it any independent discussion.

[9] In *DS* von Wright explicitly introduces situational change into what he calls the logic of events. And, rather than the static graph of zig-zag lines, we might introduce the metaphor of a wheel that *moves*, presumably through the locus of points on those zig-zag lines. The metaphor would be an apt one. But the important point is that we can take the explicit introduction of situational change, in the context of his argument in *DS*, as representing von Wright's introduction of situational motivation, or 'motivation background', as he calls it, into his schema of practical inference. It would, presumably, be an additional premise, or condition, of his schema; and it would rank with the other thought-factor premises. Hence, we would have three such premises: (1) an intention premise, (2) a cognitive or 'epistemic attitude' premise (the agent's means/end belief), and (3) a motivational one; these three premises, together with (4) normal conditions, would entail the *doing* of an action. (The term 'epistemic attitude' used in [2] above is found in *PI*, 50.)

[10] We could, of course, direct the dotted line from the agent's motivation background to his means/end belief (as von Wright suggests on *EU*, 143). Our schematic representation would then look like this:

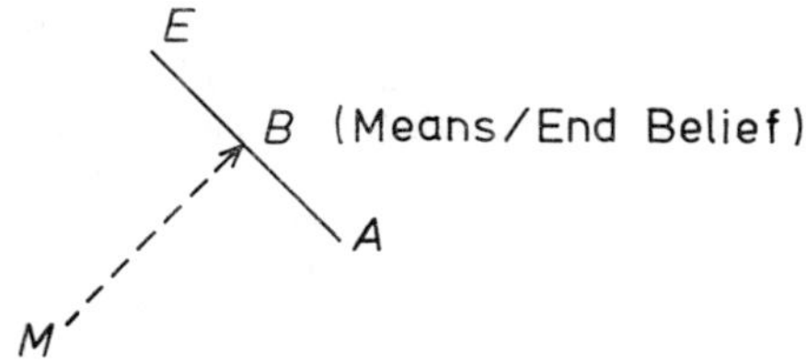

Or we could keep the representation as it is in my text and rename one of the points as *P* (standing for Premises – *i.e.*, the intention premise and the 'epistemic attitude' premise – of Practical Inference):

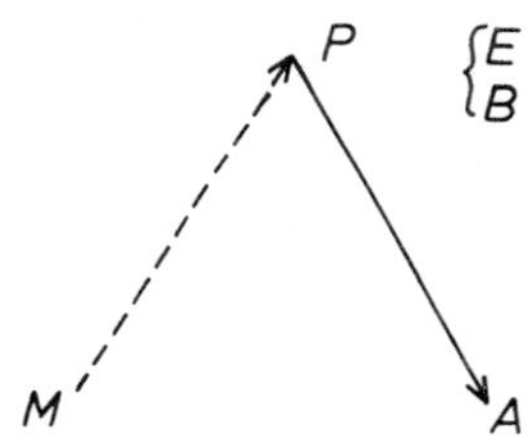

But these different schematic renderings do not affect von Wright's basic contention that the motivational background operates by affecting and thereby altering either or both of the original premises of the inference. Nor do these different renderings affect the logic of my point. For the point still stands that if we require a connection between the agent's motivation background and his means/end belief, then that connection is mediated by the agent's perception of the "requirements of the situation" (the phrase is von Wright's *EU*, 141) on the action he must take to achieve his intended end. And, again, we would require that this perception of the "requirements of the situation" provide an intelligible material connection of the agent's motivation background with his means/end belief. If we do not have this then we have a failure to understand at this point.

[11] In the 'Postscript' to his subsequent essay *PI*, von Wright disputes the idea that we normally 'interpret' behavior as action (see *PI*, 51–53), and he thereby, implicitly, disputes his own account of understanding in *EU*. Nonetheless, the point I am making still stands, for even in *PI* von Wright never really *integrates* the themes of explanation and understanding. But to make this point stand I'd need to discuss the positive doctrine of *PI* at some length, and this I can't do in the present essay.

[12] I am using 'narrative' in the sense specified by Danto (see his *Analytical Philosophy of History*, esp. 123–29). My underlying claim is that the model for the notion of understanding is that provided by story-telling and, more precisely, by *discourse* itself. I think it of some interest to note that the Greek word *historia* originally meant 'research' or 'inquiry'; thus Herodotus' book of that name was his book of 'investigations'. But it is, perhaps, not accidental that by the time of Aristotle the term *historikos* (*i.e.*, historian) had taken on the primary sense of 'story teller'. This shift reflects my point that the nature of understanding in history is exemplified in narrative, in story telling.

Of course a full account of the nature of understanding would require extended treatment. I have advanced elsewhere the claim that an explanation on the order of

"Caesar invaded Britain because he intended to conquer the tribes there" is supported, as a special case, by the judgment that 'intending *E*' (wanting to conquer certain tribes is an intelligible reason for 'doing *A*' (invading their lands). In short, there is an implicit susceptibility to generalization in all such explanations, that is, if they are truly explanatory. For if Caesar's reason is a plausible or understandable reason for what Caesar did, then it must be a plausible reason for *any* man to do it. Moreover, it is, I have argued, a necessary condition for such a generalization to be used, or asserted, in an explanation that there be *some* degree of regularity – even if it amounts only to a very loose correlation – between intending *E* (e.g., wanting to conquer) and doing *A* (e.g., invading).

These elements – generalizability and grounding in regularity – are *logical* features of the analysis of understanding. There are also, it seems to me, important semantic or pragmatic features. Among them I would include (a) that the community of inquiry is addressed by and must support these generalized judgments of plausibility and (b) that such judgments have cross-cultural application. One might want to argue, even, that they be in principle *acceptable* to persons of diverse cultures and historical periods. There are, no doubt, other features; but these suffice to illustrate my point.

13 One question that arises naturally from my account is why one should treat appropriateness or intelligibility as an application criterion rather than as a further condition of the schema. One reason is that this allows for teleological *explanations* that don't exhibit narrative understanding; if intelligibility were a condition in the schema, there could be no such explanations. Moreover, there does seem a virtue to retaining the formal/material distinction. This allows us to count a given explanation as false and another as implausible (though each is still an explanation of sorts, i.e., in 'form'). And since I have already shown that von Wright's schema allows for just this sort of thing, the way I have drawn the formal/material distinction here would certainly hold for *his* version of the schema. In any event, it seems obvious that no matter how much we elaborated or added to the basic conditions of the schema, there would always have to be some rules (semantic rules) for *applying* the schema to data. The so-called semantic dimension is indispensable. The only question is whether a certain criterion belongs there or, alternatively, in the schema, as one of its conditions. And, unless we wholly *identify* understanding with explanation, there will always be some elements of understanding that belong in the semantic dimension. My own position is that they ought to be distinguished – and here I agree with von Wright – but that they should be integrated, not separated, in an account of the logic of explanations in history and social science – and here we disagree.

14 Some theorists might be inclined to disagree with my account of psychoanalysis (see, for example, Apel's account in 'Communication and the Foundations of the Humanities,' esp. 25–26). But Apel's concern is principally with *therapy*; so it is not clear that there would be any significant disagreement between us on the way I have characterized the descriptive and theoretical side of psychoanalysis. On the other hand, I suspect a theorist like R. D. Laing would disagree with my characterization.

15 Nilson also has provided an interesting discussion of the relationship of the Sarajevo assassination to the outbreak of World War I, a discussion that takes as its framework, not the schema of practical inference, but the logic of conditions. (See his 'On the Logic of Historical Explanation', esp. 76–79, and for the logical point, 70–74.)

16 Allardt has pointed out, convincingly, the pervasive degree to which standard sociological theories of the structural sort presuppose and incorporate the idea of purpose and of purposive behavior (see his article 'Structural, Institutional, and Cultural Expla-

nations'). This has the effect of providing impressive support for von Wright's basic thesis. The notions of the *anonymous* agent and of *typical* human dispositions and patterns of motivation have been taken from Watkins (see his 'Historical Explanation in the Social Sciences', 513–14).

At this point, we have entered into what is called in the literature the dispute about 'methodological individualism'. This is a deep and murky subject, and one that von Wright has hitherto been cautious about entering. Certainly it would not be of value to enter into it in this essay. I do feel, though, that much light could be shed on the subject by shifting it from its current focus on the individual/collective distinction and on the possibility of reducing terms at the one level to those of the other. For I think that one undisclosed and principal locus of the problem of 'methodological individualism' is simply this issue of the nature and role of *narrative understanding* in explanations of the sort historians and social scientists give.

BIBLIOGRAPHY OF WORKS CITED

Works by G. H. von Wright (with Citation Code)

EU *Explanation and Understanding*, Cornell University Press, Ithaca, New York, 1971.

DS 'Determinism and the Study of Man'. (Unpublished manuscript of paper delivered, 3 July 1973, at the Jyväskylä Arts festival in conjunction with the meeting of the International Conference on the History and Philosophy of Science in Jyväskylä).

FH 'Om förklaringar i historie-vetenskapen' ['On Explanations in Historical Science']. (Article in *Historiallinen Arkisto*, **63** [1968], 1–14).

HS 'Historiallisista selityksistä' ['Historical Explanations']. (Article in *Historiallinen Aikakauskirja*, No. 4 [1967], 311–28. There is a brief summary in English, by von Wright, at the end of this essay, pp. 327–28. *HS* is a Finnish version of the Swedish article *FH*.)

PI 'On So-called Practical Inference'. (Article in *Acta Sociologica*, **15**, No. 1 [1972], 39–53.)

SA 'Historiallisista selityksistä', in *Suomen akatemia puhuu* [*Finnish Academy Addresses*]. Helsinki: WSOY, 1968, 179–200. (This article is a reprint of *HS*.)

VG *The Varieties of Goodness*, Routledge and Kegan Paul, London, 1963. In 1959 and 1960 von Wright gave the Gifford Lectures in the University of St. Andrews. This book is substantially the same in content as the second series of lectures.)

Other Books and Reviews of Books

Danto, Arthur C.: 1965, *Analytical Philosophy of History*, Cambridge University Press, Cambridge.

Louch, A. R.: 1966, *Explanation and Human Action*, Basil Blackwell, Oxford.

Scriven, Michael: 1966, Review of A. C. Danto's *Analytical Philosophy of History*, in *Journal of Philosophy*, **63**, 500–504.

Other Articles in Books and Periodicals

Allardt, E.: 1972, 'Structural, Institutional, and Cultural Explanations', *Acta Sociologica* **15**, 54–68.

Apel, Karl-Otto: 1972, 'Communication and the Foundations of the Humanities', *Acta Sociologica* **15**, 7–26.

Davidson, D.: 1963, 'Actions, Reasons, and Causes', *The Journal of Philosophy* **60**, 685–700.

Donagan, A. H.: 1969, 'Alternative Historical Explanations and Their Verification', *The Monist* **53**, 58–89.

Donagan, A. H.: 1966, 'The Popper-Hempel Theory Reconsidered', in *Philosophical Analysis and History* (ed. by W. H. Dray), Harper and Row, New York, 127–59. (Originally published as 'Historical Explanation: The Popper-Hempel Theory Reconsidered', *History and Theory* **4**, [1964], 3–26.)

Gellner, E.: 1959, 'Holism versus Individualism in History and Sociology', in *Theories of History* (ed. by P. L. Gardiner), Free Press, Glencoe, Illinois, 489–503. (Originally published as 'Explanations in History', in *Proceedings of the Aristotelian Society* **56** [1956].)

Kraft, Joseph: 1972, 'China Diary', *The New Yorker*, March 11, 100–113.

Kroeber, A. L.: 1917, 'The Superorganic', *American Anthropologist* **19**, 163–213.

Nilson, S. S.: 1970, 'On the Logic of Historical Explanation', *Theoria* **36**, Part 2, 65–81.

Ryle, Gilbert: 1950, 'If', 'So', and 'Because', in *Philosophical Analysis: A Collection of Essays* (ed. by Max Black), Cornell University Press, Ithaca, New York, 323–40.

Stoutland, F.: 1970, 'The Logical Connection Argument', in *Studies in the Theory of Knowledge* (ed. by N. Rescher), Monograph No. 4 in the *American Philosophical Quarterly Monograph Series*, Basil Blackwell in cooperation with the University of Pittsburgh, Oxford, 117–29.

Watkins, J. W. N.: 1959, 'Historical Explanation in the Social Sciences',in *Theories of History* (ed. by P. L. Gardiner), Free Press, Glencoe, Illinois, 503–14. (Originally published: *British Journal for the Philosophy of Science* **8** [1957], 104–17.)

Wittgenstein, L.: 1971, 'Remarks on Frazer's 'Golden Bough', tr. by A. C. Miles and R. Rhees and introd. (pp. 18–28) by R. Rhees, *The Human World* **1**, 18–41. (Originally published in German: *Synthese* **17** [1967], 233–53.)

ILKKA NIINILUOTO

INDUCTIVE EXPLANATION, PROPENSITY, AND ACTION

> I am, then, to define the meanings of the statement that the *probability*, that if a die be thrown from a dice box it will turn up a number divisible by three, is one-third. The statement means that the die has a certain 'would-be': and to say that a die has a 'would-be' is to say that it has a property, quite analogously to any *habit* that a man might have. Only the 'would-be' of the die is presumably as much simpler and more definite than the man's habit as the die's homogeneous composition and cubical shape is simpler than the nature of the man's nervous system and soul. (Charles S. Peirce, 1910.)

1. INTRODUCTION

The inductive aspect of scientific explanation is often ignored in philosophical and methodological studies in sociological, psychological and historical explanation. In particular, many critics of the deductive covering law model of explanation seem to implicitly assume that their arguments *mutatis mutandis* apply to inductive explanations as well. A more sophisticated position is held by G. H. von Wright, who does not discuss inductive explanations in his work *Explanation and Understanding* – except for brief and interesting comments in the introductory chapter. There von Wright explicitly states his reasons for the intentional omission of these kinds of explanations from the rest of his book: he thinks that inductive-probabilistic explanations in Hempel's well-known model are not genuine explanations at all, but only instances of reason-giving argumentation.[1]

The main purpose of this paper is to argue for the relevance, or even indispensability, of the theory of inductive explanation to the problems of explanation and understanding in the social sciences and the humanities. The typology introduced in Section 2 illustrates that there are many

Manninen and Tuomela (eds.), Essays on Explanation and Understanding, 335–368.

different kinds of inductive explanations which have not yet been studied as much they would deserve. In Section 3, certain arguments against the covering law model of explanation are discussed, and they are shown to be based on a misinterpretation of the questions asked in an explanatory situation. The nature of inductive-probabilistic explanation in general is discussed in Section 4, with the aim of showing that there are 'genuine' – neither epistemically relative nor reason-giving only – explanatory patterns of inductive kind which may be indispensable for scientific explanation. A propensity theory of explanation, or a theory of dispostitional inductive explanation, is outlined in Section 5, and its applicability to explanation in the social science and to the explanation and the understanding of human action is illustrated in Sections 5 and 6.

My account of the nature and the applicability of inductive-probabilistic explanation in this paper leaves many problems unanswered and many details to be filled in later. Still, I hope that it serves the purpose of putting the covering law model of explanation, and its inductive variant, into a perspective which helps to clarify its advantages and its difficulties.

2. Types of Inductive Explanation

A number of preliminary definitions and distinctions are introduced in this section. The resulting typology enables us to give a compact summary of the present state of the theory of inductive explanation. The contextual or pragmatic features of explanation are abstracted here for a moment (cf. Section 3 below), so that explanation is viewed as a two-place *inductive explanatory relation* E holding between an *explanans* x and an *explanandum* y.

Explanandum y may be either a factual statement or a set of factual statements. In the former case an explanation having y as its explanandum is *local* with respect to the explanandum; in the latter case it is *global* with respect to the explanandum. The statement, or the statements, forming the explanandum y may be either singular or general. In the former case we speak of explanation of *particular facts* or *events*; in the latter, we speak of explanation of *laws* or of *generalizations*. To the latter case we include not only deterministic laws but also probabilistic laws and statistical generalizations. A *statistical generalization* is a statement which specifies the proportion (relative frequency) of instances of kind G in a reference class (population) F, i.e., a statement of the form $rf(G, F)=r$. A probabilistic statement of the form $P(G/F)=r$ is a *probabilistic law* only if 'P' designates a physical probability and the class determined by 'F' is (potentially) infinite (cf. Section 5 below).

Explanans x is a set of statements which may contain both singular and general statements (but not necessarily both). If x essentially contains probabilistic statements, such as probabilistic laws, we speak of *probabilistic* explanation, and otherwise of *nonprobabilistic* or *deterministic* explanation. At least some statements in explanans x typically contain concepts that do not occur in the corresponding explanandum. If these new concepts are theoretical, we speak of *theoretical* explanation, and otherwise of *nontheoretical* explanation.[2]

Explanatory relation E is at least an inducibility relation, i.e., xEy holds only if y is inducible from x in either of the following senses: x gives *inductive support* to (confirms) y, or y is rationally *inductively acceptable* on the basis of x.[3] Relation E is supposed to be weaker than the deducibility (entailment) relation in that necessary preservation of truth is not required. Relation E may be characterized either qualitatively or by regarding explanation as a matter of degree, that is, by giving a definition of the *explanatory power* of x with respect to y.[4] A measure of explanatory power, and a corresponding explanation, is said to be *global* with respect to the explanans if it emphasizes the explanatory strength of the explanans with respect to a class of potential explananda. It is said to be *local* with respect to explanans, if the explanans is intended to account for some particular explanandum y (local or global) as well as possible. This distinction corresponds to what Hintikka (1968) has called global and local theorizing: while in global explanation we seek an explanans having a maximal explanatory appeal to a variety of phenomena which include the explanandum y as a special case only, in local explanation we seek an explanans that gives the maximal amount of information about the particular explanandum y.

The theory of inductive explanation has for long been concentrated on such explanations of particular facts that are non-theoretical, probabilistic, and local with respect to the explanandum. Hempel's theory of inductive-probabilistic explanation and Salmon's theory of statistical explanation are both of this type.[5] The crucial difference between these two accounts can be expressed in terms of the inducibility relation in-them: while Hempel requires the inductive probability of the explanandum relative to the explanans to be high (close to one), Salmon emphasizes the idea of statistical relevance of the explanans to the explanandum. It seems that this difference reflects a deeper point of disagreement between Hempel and Salmon: while in the former's account inductive explanations are local with respect to the explanans, in the latter's account they are rather global with respect to explanans.

Inductive-probabilistic explanations which are global with respect to the explanandum have been studied by Greeno (1970, 1971). In their accounts of inductive systematization, Hempel (1958) and Scheffler (1963) have studied examples which can be regarded as theoretical inductive-deterministic explanations of particular facts. Similarly, Lehrer (1969) has provided an example of theoretical inductive-probabilistic explanation of particular facts.[6]

The emphasis on the explanation of particular facts is misplaced, if one's aim is to develop a systematic theory of scientific explanation. The explanandum of explanations in natural science as well as in the social and behavioural sciences is often not a statement describing some particular fact or event, but rather a universal or statistical generalization. This is true even for many explanations encountered in our every-day life, as I argue in Section 3 below. Still, there are so far very few attempts towards systematic accounts of the inductive explanation of laws and generalizations – for example, of the logical form and the methodological role of inductive-probabilistic explanations of statistical generalizations (see Section 5 below). Theoretical inductive-deterministic explanations of universal generalizations have been studied by Niiniluoto and Tuomela (1973) – the methodological relevance of their approach is illustrated by explanations of disposition-laws, How Possible? – explanations in archaelogy, teleological or purposive explanations in biology, and partial explanations in psychology.[7]

3. Explanation and Why-Questions

Explanations can be regarded as (complete or partial) *answers to why-questions*. This immediately suggests that explanation can be viewed as a four-place relation 'A explains y to B by x'; here explanans x is an answer to a why-question, 'Why is it the case that y?', that the explainer B has put to the explainee A. This *pragmatic* notion of explanation is employed in this section to argue that certain objections against the simple covering law model of explanation of particular facts are misplaced and, when properly interpreted, show only the need to extend this simple model so as to apply to the explanation of laws and generalizations as well. The argument given here concerns both deductive and inductive explanations, and its relevance to the explanation of action is also illustrated.

Consider the following two explanatory arguments:

(1) Question: Why is this bird black?
Answer: It is a raven, and all ravens are black.

(2) Question: Why is this man trembling?
Answer: He has caught malaria fever, and all who have caught malaria fever do tremble.

While argument (2) seems intuitively satisfactory, the adequacy of argument (1) as an explanation has been denied. Thus, von Wright argues that "a law stating the universal concomitance of the two characteristics ravenhood and blackness" is not strong enough to have explanatory power (see von Wright, 1971, p. 19). By this von Wright wants to defend the reasonable view that generalizations used in explanation should be *lawlike*, and that mere universality (i.e., constant conjunction) is not sufficient for lawlikeness. This remark applies to argument (2) as well. But von Wright also suggests that (1) does not 'really' explain why the bird is black, since

> We should like to know why ravens are black, what it is about them that 'is responsible for' the color which, so we are told, is characteristic of them all.

An obvious difference between explanations (1) and (2) is the fact that while the event of cathing malaria fever can be regarded as the cause of the event of trembling, being a raven is not a cause of being black. But as many scientific explanations are not causal explanations (cf. Hempel, 1965, p. 352), it is not correct to assume that an answer to a why-question of the form (1) should always refer to the cause of the state of affairs described in the explanandum. In particular, argument (1) represents a legitimate explanatory pattern in which the *specific property* of blackness is explained by appeal to a *natural kind* term 'raven' (for this terminology, see Achinstein, 1973, pp. 538–539). In brief, I think it is wrong to assume that all explanation-seeking why-questions are *cause-seeking* why-questions. But it is equally wrong, I will argue, to claim that the answer (1) is unsatisfactory for the reason that it does not explain *why ravens are black*.

Recent studies in the so-called 'logic of questions' have made it clear that what counts as a potential answer to a question is dependent upon the knowledge (or beliefs) of the questioner.[8] Thus, we may expect that a questioner who knows neither what ravens are nor that they are black would not be satisfied with the answer given in (1). On the other hand, a questioner who knows that ravens are black, but does not know that the bird in question is a raven, might very well find the answer given in (1) perfectly satisfactory – similarly as the questioner in (2) who knows enough about malaria and its symptoms, but does not know that the poor man is a victim of this disease. But a questioner who already knows that the bird referred to in question (1) is a raven might respond – with von Wright – to the given answer by saying: "That I know already. What I want to know is why ravens are black". Thus, it seems that if there is a difference in the adequacy of arguments (1) and (2), it can be accounted for, in pragmatic terms, by assuming that in (1), but not in (2), the questioner knows already before asking the singular premise of the answer (i.e., the initial condition of the explanans). However, this does not yet carry the analysis of the situation far enough.

In general terms, we may say that the problem of giving conditions for an explanation to be *appropriate* or *satisfactory* to the explainee is analogous to the problem of 'answerhood' studied in the logic of questions. However, what counts as an answer to a given question depends, not only upon the knowledge of the questioner, but also upon *what is asked* in the first place. As different questions usually call for different answers, there seems to be more 'logic' in explanation than the above analysis of (1) and (2) would suggest. The pragmatic presuppositions concerning the initial condition of the explanans are as it were built in the corresponding why-question – so that the questioner who is dissatisfied with the answer given in (1) has simply asked a wrong question.

To see this, suppose you are watching or studying a black bird the species of which is unknown to you. Then an appropriate question to ask in this situation is the following:

(3) Why is this *bird* black?

Suppose then that you are watching a black bird of which you know that it is a raven. Now an appropriate question to ask is not (3) but rather

(4) Why is this *raven* black?

If 'B' designates the property of being a black bird and 'R' designates the property of being a raven, then the explanandum of an explanation answering to the why-question (3) is of the form

(5) Ba.

The explanandum corresponding to the why-question (4) is not

(6) Ba & Ra

(note that (6) would correspond to the question 'Why is this bird a black raven?'), but rather

(7) Ba, if Ra.

In many cases, the questioner asking (4) regards 'this raven' as a representative of all ravens – i.e., 'a' in (7) plays the role of a free individual variable rather than that of an individual constant. In these cases, (7) has the same logical strength as the generalization

(8) $(x)(Rx \supset Bx)$.

In other cases, explanandum (7) takes the form

(9) $Ra \supset Ba$

instead of (8). (For an illustative example, see scheme (19) in Section 5 below.) This is the case especially if the corresponding why-question has the form

(10) Why is *this* swan black?

where the emphasis on 'this' has a particularizing effect.

This analysis may be summarized by saing that argument (1) provides an adequate (potential) answer to question (3), but not to question (4).[9] Therefore, there is nothing wrong in the explanatory pattern exemplified in (1): even if (1) is trivial as an explanatiton why 'this' *raven* is black (i.e., why ravens are black), it can be used to explain why 'this' *bird* is black.[10]

The difference between questions (3) and (4) is similar to the difference between questions

(11) Why is this man trembling?

and

(12) Why is this man who suffers from malaria fever trembling?

While argument (2) gives an adequate answer to question (11), a questioner asking (12) would normally like to know, in all physiological detail, why trembling is a symptom of malaria fever. It is easy to find similar examples of inductive-probabilistic explanatory arguments.

We have seen that many, though not all, explanation-seeking why-questions are implicit requests for explanations of laws or of generalizations. These explanations in science are typically theoretical and often inductive only. For example, the explanation of the generalization 'all ravens are black' refers to the 'hidden' causal factors responsible for the manifest property, that is, to the physico-chemical or physiological constitution of the members of the species in question. The fact that this generalization has a theoretical explanation is one indicator of its *law-likeness*. At the same time, this generalization can be used as a covering law in explanations of type (1). Naturally, argument (1) does not explain why ravens are black. No explanatory argument explains its own *premises* – this is the task of another 'higher-order' explanation which has a different why-question corresponding to it.

In a similar way one may argue that certain kinds of questions which may seem to require an explanation of a particular *act* are, in fact, requests for an explanation of the *premises* of an argument explaining this act. If you ask why a savage who has accidentally inflicted a knife-wound on his own leg is *cleaning his knife*, a satisfactory (potential) answer to this question can be given by telling that the man wants his wound to heal and believes that cleaning his knife is the means to this end. But if you ask why this savage is *trying to cure the knife-wound by cleaning the knife*, you already implicitly suppose that he believes cleaning the knife to be a means to curing the wound. What you are then really asking for is an explanation of the curious *belief* of the poor savage. The scheme of practical syllogism for the explanation of action is applicable to questions of the former kind, but not to those of the latter kind. It seems to me that this gives at least a partial answer to Rex Martin's criticism of practical syllogism (see this volume, pp. 305–334).

4. Explanation, Begründung, and Epistemic Relativity

Many, perhaps most, inductive explanations are *incomplete* in the sense that they are only *partial answers* to the corresponding why-question. Such explanations are inductive because of some imperfection in our knowledge: if we only could supply the missing true, but yet unknown, premises to them, they could be turned into deductive explanations. If this were true of all inductive explanations, then inductive explanations would be at best *methodologically indispensable* in science. It is argued in this section that there are deeper, *ontological* grounds for the indispensability of inductive explanations. There may exist inductive-probabilistic explanations which are 'genuine' in the sense that they are as complete as any true answer to the given why-questions my be; a variant of such explanations is provided by maximal *causal* inductive explanations. These genuine explanations are not epistemically relative in Hempel's sense, nor are they only instances of *Begründung* or reason-giving argumentation.

The simplest form of the deductive-deterministic explanation of particular facts is the following:

$$(13)\qquad \frac{\begin{array}{l}(x)(Fx \supset Gx)\\ Fa\end{array}}{Ga}$$

(It is assumed, even if our extensional way of writing (13) does not explicitly show it, that the generalization in (13) is lawlike.)

By weaking the universal premise in (13) to a probabilistic law, and the deductive relation in (13) to an inductive one, a scheme that Hempel calls the basic form of inductive-probabilistic explanation is obtained:

$$(14)\qquad \begin{array}{l}P(Gx/Fx) = r\\ Fa\\ \hline\hline Ga\end{array}\quad [r]$$

Here r in square brackets is the inductive probability of the explanandum 'Ga' relative to the explanans $\{P(Gx/Fx)=r, Fa\}$.

A potential explanation of the form (13) is called *true*, if the explanans is true. ((13) is *well confirmed*, if the explanans is well-confirmed relative to the total evidence available. For the non-epistemic notion of true explanation, which interests us here, it is of course irrelevant whether we are justified in believing in the truth of the explanans.) A true explanation of this form may be said to explain why 'Ga' is true by showing that it must be true in view of its deducibility from the true explanans. According to Hempel (1965), the nature of inductive explanations of the form (14) is drastically different, however. Hempel noticed that, unlike in the deductive case, there may be true inductive explanations of the form (14) with logically incompatible explananda. The notion 'true inductive-probabilistic explanation' thus suffers from *ambiguity* which, according to Hempel's view, can be resolved only by restricting the class of predicates 'F' appropriate to the probabilistic law '$P(Gx/Fx)=r$' in the explanans. This restriction, i.e., Hempel's *Requirement of Maximal Specificity*, makes a reference to a knowledge situation K as represented by a class of accepted statements at a certain time (cf. Hempel, 1965 and 1968). Hempel thinks that the reference to K is 'unavoidable', and calls this phenomenon the *epistemic relativity* of inductive-probabilistic explanation. In this view, the notion of true inductive explanation is meaningless, since an argument of the form (14) may qualify as an explanation relative to a knowledge situation K, but may fail to qualify as such relative to another, enlarged knowledge situation K. In short, unlike the logical notion of deductive explanation, the notion of inductive explanation is *epistemic*.

In spite of the alleged epistemic relativity, Hempel accepts (14) as a model of explanation. von Wright (1971), pp. 13–14, argues that arguments of the form (14) cannot be used to *explain* what happens, but only to *justify* certain *expectations* and *predictions*. He says that while deductive explanations do genuinely explain what happens, i.e., why the explanandum event occurred, and therefore secondarily show why these things were to be expected, inductive-probabilistic explanations tell us only why things which happened were to be expected. Similarly, Stegmüller (1973) claims that arguments of the form (14) do not answer to what Hempel calls explanation-seeking why-questions, but rather to reason-seeking or epistemic why-questions ('What reasons are there for believing that y?'). Accordingly, Stegmüller reinterpretes and reformulates Hempel's model (14) as a model of *Begrüdnung*. These arguments of von Wright's and Stegmüller's can, indeed, be supported with reference to Hempel's rationale for his own model. Hempel assumes that an inductive explanation has to show that the state of affairs described by the explanandum *was to be expected*: he interpretes the inductive probability r associated with (14) as the degree of *nomic expectability* of the explanandum, and requires that r should be high (i.e., close to 1). Hempel (1968), p. 121, distinguishes explanation from prediction by saying that

the point of an explanation is not to provide evidence for the occurrence of the explanandum phenomenon, but to exhibit it as nomically expectable;

therefore the Requirement of Maximal Specificity is not a substitute for the Requirement of Total Evidence which applies in the case of prediction. Still, exhibiting the explanandum event as nomically expectable seems to amount in Hempel's view to showing that we could have justifiably *predicted* its occurrence. Thus, Hempel's model seems to be applicable primarily to the justification of expectations in a given knowledge situation, i.e., to *Begründung* rather than to explanation. This impression is reinforced by the fact that in explanatory situations Hempel's problem of ambiguity creates no real problem at all: it has a trivial solution in that we do not have any problem in choosing the appropriate one of two rivalling explanations with incompatible conclusions, since it is always assumed that the explanandum of an explanatory argument is true (cf. Coffa, 1974). It is also interesting to note that von Wright, in one of his early studies in probability, has formulated a requirement according to which one should choose as the basis of *prediction* a class which is a minimal maximally specific reference class in the given knowledge situation (see von Wright, 1945).

As the ambiguity of inductive arguments is a genuine problem only in a context of prediction, the Requirement of Maximal Specificity is not an indispensable part of the notion of inductive explanation. This requirement is needed only to characterize the class of 'genuine' or *complete* inductive-probabilistic explanations. Moreover, it is argued below that this requirement can be formulated in nonepistemic terms, so that Hempel's thesis of the epistemic relativity fails for complete inductive-probabilistic explanations.

In Hempel's view, the epistemic relativity of inductive-probabilistic explanations is a consequence of the fact that an explanation of the form (14) has to satisfy – in order not to suffer from ambiguity – the following Requirement of Maximal Specificity:

(RMS) If $\{F_1 a, (x)(F_1 x \supset Fx), P(Gx/F_1 x) = r_1\} \in K$, then $r_1 = r$ unless '$P(Gx/F_1 x) = r_1$' is a theorem of probability theory.

In Salmon's (1970) terminology, RMS requires that the reference class F (i.e., the class that is the extension of the predicate 'F') should be *epistemically homogeneous*, that is, we should not know of any property F_1, other than F & G and F & $-G$, which determines an infinite subclass F_1 of F such that the probability of G in F_1 differs from r (the probability of G in F).

But cannot RMS be formulated without mentioning the knowledge situation K at all, simply by speaking of *true* statements instead of *known* or *accepted* ones? The claim that reference to K in RMS is unavoidable – i.e., the thesis of epistemic relativity – means that a subclass F_1 of F effecting a statistically relevant partition of F can fail to exist only with respect to our limited knowledge or our ignorance. In reality, so to speak, such a subclass F_1 will exist for any G and F; otherwise the maximal specificity of F could be characterized without reference to a knowledge situation. This assumption thus implies the following *Density Principle* for reference classes:

(DS) For any probabilistic law $P(Gx/Fx) = r$, where $0 < r < 1$ and 'Fa' is true, there exists a property F_1, other than F & G and F & $-G$, such that '$(x)(F_1 x \supset Fx)$', '$F_1 a$' and '$P(Gx/F_1 x) = r_1$' are true and $r_1 \neq r$.

As Coffa (1974) has plausibly argued, it is essentially this principle which has led Hempel to his thesis of the epistemic relativity of inductive-probabilistic explanations – a thesis which is virtually equivalent to DS.

Principle DS is obviously satisfied, if the extensions of our predicates are allowed to be chosen among *all* subclasses of the universe of discourse. In technical terms, this would be tantamount to adopting a *standard interpretation for second-order logic.*[11] The thesis of epistemic relativity is thus a simple consequence of the presupposition that the reference classes of probabilistic laws 'live' in a standard model of second-order logic. But in a context where laws of nature are at issue, there does not seem to be any justification for operating with standard models; instead, a suitable non-standard model should be chosen. Parallelling Church's mathematical definition of randomness, this non-standard model could be specified by including into it only infinite recursively enumerable subsets of the class F.[12] However, in this context even this choice would be *ad hoc*: the *physical properties* of the objects in consideration should be let to restrict the choice of classes to be included in the model. (This sound idea underlies Reichenbach's theory of probability and Salmon's account of statistical laws and statistical explanation.)

In a non-standard model, where only a subclass of the class of all subsets of the universe of discourse is included, principle DS need not hold, so that RMS can be formulated in non-epistemic terms: the model itself has the same restrictive function (in excluding the existence of F_1) as the knowledge situation K in Hempel's formulation of RMS. Relative to such a model, the notion of *physical randomness* (of G in F) can be defined. It is to be remembered that such a model represents *reality conceptualized* in a certain way. In view of this, it may be said that even if the notion of (complete) inductive-probabilistic explanation is not epistemically relative (i.e., relative to a knowledge situation), it still is *conceptually relative* (i.e., relative to a conceptual system). In this respect, the difference between inductive-probabilistic and deductive explanations is not a qualitative one, but only a matter of degree.

It was seen above that the principle DS is valid with respect to some interpretations of second-order logic, but nonvalid with respect to some other interpretations. Is DS valid with respect to that non-standard interpretation (conceptual system) which is the 'correct' or 'best' representation of reality? This question has an intimate connection with the philosophical doctrines of determinism and indeterminism. A determinist will answer affirmatively to this question: principle DS, and hence the thesis of the epistemic relativity of inductive-probabilistic explanations, is a consequence of the determinist view that all probabilistic laws of nature are ultimately reducible to deterministic ones. (Cf. the so-called hidden variable theories in quantum mechanics.) Conversely, if DS is not valid, then there are irreducible probabilistic laws of nature, just as an indeterminist is willing to contend.

Alberto Coffa (1974) has recently argued that "to accept Hempel's thesis of epistemic relativity amounts to accepting the claim that there are no inductive explanations". If principle DS is valid, Coffa argues, then "in the limit, God would find no inductive explanations relative to his knowledge situation: He knows too much". Salmon (1974) accepts Coffa's argument, with the conclusion that Hempel's "claim that inductive-statistical explanation *must be* relativized to the knowledge situation thus seems to yield a commitment to determinism". This Coffa-Salmon argument thus amounts to the thesis that principle DS implies determinism. It seems to me that this thesis as such is not valid, but it holds if one additional – and problematic – assumption is made.[13]

Let $P(Gx/Fx)=r$ be a probabilistic law with $0<r<1$. It follows from principle DS that there exists an infinite subclass F_1' of F, determined by a property other than F & G and F & $-G$, such that $P(Gx/F_1'x)=r_1'$ and $r_1'\neq r$. Choose $F_1=F_1'$, if $r_1'>r$, and $F_1=F-F_1'$ otherwise. Then F_1 is an infinite subclass of F such that $P(Gx/F_1x)=r_1>r$. If $r_1<1$, then the same argument can be repeated. Either $r_i=1$ for some finite i or we obtain by repeated applications of DS a sequence $\langle F_i \mid i<\omega\rangle$ of infinite subsets of F and a sequence $\langle r_i \mid i<\omega\rangle$ of real numbers such that $F_{i+1}\subseteq F_i$, $r_i<r_{i+1}$, and $P(Gx/F_ix)=r_i$ for all $i<\omega_i$ Let

$$F_\omega = \bigcap_{i<\omega} F_i$$

and

$$r_\omega = \lim_{i\to\infty} r_i$$

be the limits of these sequences, respectively.

There is no guarantee that the limit r_ω has the value one. What is more, even if $r_\omega=1$, the limit set F_ω may fail belong to the non-standard model M we are operating with. In such a case, it does not follow from DS that the law $P(Gx/Fx)=r$ can be reduced to a deterministic law by selecting a suitable narrower reference class.

But suppose that the non-standard model M satisfies the following principle requiring that M is *closed under limits*:

(CL) If $F_1\supseteq F_2\supseteq\ldots$ is decreasing sequence of sets belonging to M, then the limit F_ω of this sequence also belongs to M.

If now $r_\omega<1$, then CL justifies the application of DS to F_ω, to obtain an infinite subset of F_ω with a corresponding probability (of G) higher than r_ω. If M satisfies a suitable closure condition like CL, then you cannot stop repeating the application of DS until you have shown the existence of a property H, other than F & G, which determines an infinite subclass H of F such that $P(Gx/Hx)=1$.

The above arguments show that DS *alone is consistent with indeterminism*, so that the Coffa-Salmon thesis is not correct. On the other hand, DS *together with* CL implies that particular form of determinism which claims that all probabilistic laws of nature are ultimately reducible to laws of the form

$$(15) \qquad P(Gx/Hx) = 1\,.$$

A stronger form of determinism follows from DS and CL only for those interpretations of probability which make (15) equivalent to the deterministic law

$$(16) \qquad (x)\,(Hx \supset Gx)\,.$$[14]

In other words, Hempel's thesis of the epistemic relativity of inductive-probabilistic explanations yields a commitment to a weak form of determinism *if* the model M which 'correctly' represents reality satisfies the principle CL. Whether this additional requirement is valid or not is another problem to which I have no answer.

It is interesting to note that even if DS and CL together implied a strong form of determinism, claiming that all probabilistic laws are reducible to deterministic ones, this would not yet imply that all inductive-probabilistic explanations can ultimately be replaced with deductive-deterministic explanations. The reason is that the above argument does not guarantee that the individual a mentioned in the explanation (14) has the property (i.e., belongs to the appropriate subclass) which effects the reduction of the probabilistic law in (14) to a deterministic one. This observation provides another complication to the arguments of Coffa and Salmon.

The above remarks notwithstanding, I agree with the core of Salmon's and Coffa's criticism of the Hempelian view of inductive-probabilistic explanation – i.e., with their view that there are 'objectively' random processes in nature and, correspondingly, genuine or complete inductive-probabilistic explanations which are not epistemically relative. In an indeterminist's view, randomness is not an epistemic notion: the indeterminacy of certain phenomena resides in nature itself, not in our mind or in our ignorance. This is the reason why explanation by means of probabilistic laws constitutes a mode of explanation *sui generis*.[15] If F is a reference class which satisfies some non-epistemic version of RMS with respect to G, then no increase in our knowledge can lead us to a better explanation of 'Ga' than (14): there are no new and true premises to be added into the explanans which could increase the probability r. This explanation is then, though inductive, as *complete* as any true explanation of 'Ga' may be. The point of such an explanation is not to exhibit the explanans as nomically expectable, or to justify any expectations, but rather to identify those features of reality that are nomically, but yet non-deterministically, relevant to the explanandum. As the case may be, the probability r associated with such an explanation may be low, so that this explanation had not been applicable for reasonable prediction, or *Begründung* of any kind. As Jeffrey (1969) and Salmon (1971) have contended, inductive-probabilistic explanations need not be *arguments* showing that the explanandum can be inferred or 'detached' from the given premises.

Stegmüller (1973) has argued that this idea does not give rise to a new model of statistical explanation, but only to a mode of *statistical depth analysis* (statistische Tiefenanalyse). He is not willing to say that such a depth analysis 'explains' anything, even if the requirements of *positive relevance* (i.e., the explanans increases the probability of the explanandum) and *coherence* (i.e., the explanandum is more probable than its negation relative to the explanans) were satisfied. On the other hand, Salmon (1970) assumes that a complete depth analysis always gives a genuine explanation, independently of such additional conditions as positive relevance and coherence. My view lies between these two extremes: I should say that a complete depth analysis may, or may not, amount to an explanation, depending upon whether the explanans gives information about the explanadum. In other words, a complete depth analysis satisfying at least the condition of positive relevance is what I take to be a genuine inductive-probabilistic explanation.[16] An important special case of such explanations is provided by maximal causal inductive-probabilistic explanations.

Let F and G be two generic events (i.e., event-types), and let f_t and $g_{t'}$ be two particular events of type F and G, respectively, such that f_t occurs at time t and $g_{t'}$ occurs at time t'. A probabilistic law of the form '$P(G/F)$

$=r$' expresses a non-deterministic connection between the generic events F and G, so that particular events f_t and $g_{t'}$ instantiating this law bear to each other a nomic, but non-deterministic relation. Following Suppes (1970), event f_t is said to be a (potential) *prima facie cause* of event of $g_{t'}$ just in case

(i) $t < t'$
(ii) $P(G/F) > P(G)$.

The cause f_t does not here *nomically bring about* the effect $g_{t'}$; f_t is rather *nomically relevant* to $g_{t'}$ (cf. Coffa, 1974). If the law '$P(G/F)=r$' is irreducible in the above sense (i.e., F satisfies RMS with respect to G), f_t is said to be a (potential) *maximal cause* of $g_{t'}$. In actionist terms, it may be said that by doing f_t we can produce changes in the possibility that $g_{t'}$ will show up. However, nomic relevance is basically an objective notion, a weaker counterpart of an objectivist notion of a 'deterministic' cause (cf. Tuomela, this volume). Statistical (positive) relevance is an indicator of such nomic relevance.

Explanations of particular events which refer to probabilistic laws and to earlier events that are *prima facie* causes of the explanandum event are *causal inductive-probabilistic explanations.* Such explanations are *maximal* if they refer to maximal causes. These explanations are genuine in the above sense.[17]

5. Propensity Theory of Explanation

According to the *frequency interpretation* of probability, a probabilistic law of the form

(17) $P(G/F) = r \quad (0 \leqslant r \leqslant 1)$

states that the relative frequency of instances of property G in the infinite reference class F is equal to r. In another version of this interpretation, law (17) states that the relative frequency of occurrences of event of kind G in a sequence of occurrences of event F would converge to value r if this sequence would be extended to an infinite random sequence of occurrences of F. In this view, probabilistic laws are simply lawlike statistical generalizations or 'statistical laws'.

In the frequency interpretation, probability is primarily regarded as an attribute of a generic event relative to its occurrence in a random sequence. In the *propensity interpretation*, outlined already by Charles Peirce[18], probability is viewed as a dispositional property of an 'experimental arrangement' or 'chance set-up'. Hacking (1965) characterizes a *chance set-up* as "a device or part of the world" on which one or more *trials* may be conducted. For example, tossing a coin and observing the result is a trial on a chance set-up constituted by a coin and a tossing device. Through a series of trials, a chance set-up may generate a sequence of events. Particular probabilities are properties of chance set-ups; they manifest themselves in the sequences of events that chance set-ups may give rise to. In this sense, probability is a quantitative 'physical' concept, like mass and length; moreover, it is a theoretical concept which is not explicitly definable in terms of observational concepts, such as observable relative frequencies.

According to the *long-run propensity interpretation*, a statistical law of the form (17) states that an experimental arrangement or chance set-up of kind F possesses a dispositional tendency to generate an outcome of kind G with the relative frequency r in the long run. This view was adopted by Hempel as the interpretation of statistical laws used in inductive-probabilistic explanations.[19] From this point of view, Hempel's model of inductive-probabilistic explanations (see (14)) amounts to a rule for applying knowledge of long-run frequencies to a single case, so that his RMS has essentially the same function as the rules for choosing the appropriate reference class for probability statements in the frequency sense. It also follows that inductive-probabilistic explanations should be regarded as a special sort of *dispositional explanations*.

A dispositional explanation of the breaking of a glass can be given by stating that it was hit by a stone, that it is fragile, and that all fragile objects break when hit by a stone. The logical structure of this explanation is the following:

$$(18)\qquad \frac{\begin{array}{l}(x)\,(Tx \supset (Fx \supset Gx))\\ Fa \,\&\, Ta\end{array}}{Ga}$$

where the universal premise is lawlike. (Cf. Hempel, 1965, pp. 457–462.) The covering law of this explanation is a *universal dispositional law*; it states that if x is subjected to a test of kind T and if x has the property F (fragility), then x manifests behaviour of kind G. The premise 'Fa' ('a is fragile') is a *singular dispositional statement* which attributes a dispositional property to a single individual. The premise 'Ta' ('a was hit by a stone') describes the event that is the *cause* of the explanandum event.

There is another type of dispositional explanation which is not a kind of causal explanation, as (18) is. Explanations of this type are answers to questions of the form 'Why did a manifest behaviour of kind G under the test conditions of kind T?'[20] The structure of such an explanation is the following:

$$(19)\qquad \frac{\begin{array}{l}(x)\,(Fx \supset (Tx \supset Gx))\\ Fa\end{array}}{Ta \supset Ga}$$

where the universal premise is lawlike. (Cf. Section 3.) Explanation (19) employs the same universal law as (18), but its initial condition does not refer to any event. (Having a disposition does not belong to the category of events.)

Isaac Levi (1969) has argued that inductive-probabilistic explanations viewed as dispositional cannot be covering law explanations, since dispositional statements play the role of initial conditions of dispostitional explanations. However, probabilistic laws should not be compared to singular dispositional statements, but to universal dispositional laws (cf. Fetzer, 1974). The dispositional law in (18) and (19) can be expressed by saying that all objects of kind F possess a dispositional tendency of *universal strength* to produce, under test conditions of kind T, an outcome of kind G. Fetzer (1974) suggests that probabilistic laws should be interpreted as weaker counterparts to dispositional laws of this kind.[21] For this purpose, he adopts the *single case propensity interpretation* of probability (cf. also Mellor, 1971; Fetzer, 1971; and Giere, 1973) according to which a statement of the form (17) expresses that a chance set-up of kind F has a dispositional tendency of strength r to produce an outcome of kind G on a *single trial* of that set-up. While in the long-run propensity interpretation probability is essentially regarded as a universal disposition to produce infinite sequences of a certain kind, in the single case propensity interpretation probability is a metrical disposition, or a disposition to a degree, to produce outcomes of a certain kind. A probabilistic law corresponding to the law in (18) and (19) will then state that if x is a chance set-up of kind F, then x has the dispositional tendency of strength r to produce, if subjected to a test of kind T, an outcome of kind G, i.e.,

$$(20) \qquad (x)\,[P(Gx/Fx \,\&\, Tx) = r].$$

In other words, the propensity probability of an outcome of kind G is r for any chance set-up of kind F which is subjected to any test of kind T.

It follows immediately from the law (20) that there cannot be any subclasses F_1 and T_1 of F and T, respectively, such that $P(Gx/F_1x \,\&\, T_1x) = r_1 \neq r$. Therefore, a probabilistic law of the form (20) must be irreducible in the sense of Section 4. In the single case propensity interpretation of probability, *genuine* inductive-probabilistic explanations of particular facts have thus the following form:

$$(21) \qquad \begin{array}{l} (x)\,[P(Gx/Fx \,\&\, Tx) = r] \\ Fa \,\&\, Ta \\ \hline\hline Ga \end{array} \quad [r]$$

where it is required that F & T is positively relevant to G. If T is positively relevant to G relative to F[22], then the event that a is subjected to a trial of kind T is a maximal probabilistic cause of the event that a produces an outcome of kind G. Inductive-probabilistic explanations in this sense have a *universal non-deterministic covering law* (20) which is applied to express the strength of the dispositional tendency of the chance set-up described in the initial condition to produce the outcome described in the explanandum. A crucial difference to Hempel's model is the fact that, in (21), it is not knowledge of long-run frequences, but rather theoretical knowledge of the single cases themselves, that is applied to a single case. Here we seem to be justified, for a good reason, to assume that the inductive probability r associated with the explanation is exactly the same as the propensity probability given in the explanans.

It is important to notice that the law (20) used in (21) is not a *statistical* law: it is not a statement about relative frequencies but about the strength of dispositions. Of course, single-case propensities and relative frequencies have a probabilistic connection through Bernoulli's and Borel's theorems (which are special cases of the weak and strong laws of large numbers, respectively). Relative frequencies are thus relevant for *testing* the law (20), but they do not constitute a part of the *meaning* of (20). For this reason, it is preferable to call (21) a pattern of *probabilistic* explanation rather than statistical explanation. – Incidentally, I think that one may speak of 'statistical explanation' in a derivative sense only: statistical generalizations are lawlike, and may be used in explanation, only if they are 'backed up' by propensities.

In those cases where a probabilistic law of the strong form (20) cannot be found, it may be possible to replace it with a weaker law, and use that law in explanatory arguments closely similar to (21). For example, there may be a *lower bound* r_1 for the propensity probabilities $P(Gx/Fx \,\&\, Tx)$; then (20) can be replaced with

(22) $(x)\,[P\,(Gx/Fx \,\&\, Tx) > r_1)$.

Or there may be a lawlike 'statistical' statement to the effect that

(23) The average of probabilities $P(Gx/Fx \,\&\, Tx)$ is r_2.[23]

As far as the values r_1 and r_2 are greater than the probabilities $P(Gx)$ or $P(Gx/Fx)$, (22) and (23) can be used as premises of inductive-probabilistic explanations.

The propensity theory of probability is still beset with many philosophical problems that require clarification. But for a scientific realist and for an indeterminist propensities are not suspectible or occult entities as such – even if a strict empirist would challenge their epistemological status, and a determinist would deny their ontological status. In any case, we have seen that the propensity theory of probability gives rise to an interesting *propensity theory of inductive-probabilistic explanation* in which essentially theoretical probabilistic laws are assumed to have real or genuine explanatory power.

As an example of the applicability of the propensity theory of inductive-probabilistic explanation let us consider Dray's famous example of the unpopularity of Louis XIV – which von Wright (1971), p. 25, takes to lead to "a complete rejection of the covering law model". Suppose we state that Louis XIV died unpopular, because he pursued politics detrimental to the national interests of France. It is obvious that this argument cannot be represented as a *true* deductive-deterministic explanation, since it is simply false that all rulers who pursue such politics become unpopular. Dray argues that a true covering law needed for this argument would require so many specifications that it would be limited to one instance only, viz. Louis XIV. Such a statement would not be a law any more, and since it would be logically limited to one instance only, it should be rejected on the basis of the sound methodological principle, introduced already by William Whewell in the 1840's, that it should in principle be possible to support an explanans independently of the explanandum in question.

One could suggest that the required covering law should be expressed as a statistical generalization stating that *most* rulers who pursue politics detrimental to the national interests of their country become unpopular. This statement would surely have other instances than Louis XIV. But it may be doubted whether such a frequency statement would be lawful enough to serve its purpose. It seems more promising to seek a suitable dispositional probabilistic law. If we define

Fx = 'x is a ruler pursuing politics'
Tx = 'x pursues politics detrimental to the national interests of his country'
Gx = 'x becomes unpopular',

then (22) and (23) become probabilistic laws which are quantitative counterparts to the qualitative dispositional statement that a ruler pursuing politics detrimental to the national interests of his country *risks* unpopularity, or becomes *liable* for unpopularity. The propensity probabilities r_1 and r_2 in these laws are thus measures of the liability of the ruler with property T for unpopularity. If r_1 or r_2 here has a value greater than the prior propensity probability of unpopularity, $P(Gx/Fx)$, then an inductive-probabilistic explanation for the unpopularity of Louis XIV is obtained. Moreover, it may be possible to improve on this explanation by sharpening the condition T, but there seems to be no reason to suppose that the covering probabilistic law of the corresponding complete explanation (21) would be trivial.

As another example of this kind, the causal role of the assasination of the Austrian archduke at Sarajevo in the outbreak of the First World War may be mentioned (cf. von Wright, 1971, pp. 139–143). In modern peace research, the national states with conflicting interests are often regarded as constituting a system which has, relative to certain kinds of conditions, a dispositional tendency to 'explode', or to change from the state of peace to that of war. Relative to the historical situation, the shots at Sarajevo increased the strength of this dispositional tendency. This event can thus serve as an initial condition of a causal (nonmaximal) inductive-probabilistic explanation of the outbreak of the war.

These examples can easily be multiplied. Especially in the social and behavioural sciences it is common to formulate laws in terms of tendencies and dispositions. As examples the following laws, mentioned in Nowak (1970), may be given: 'The lower middle class on taking the power in society reveals a tendency to disintrested punishment', 'People with very high incomes tend to be conservatively-minded', 'Economic deprivation breeds the tendency to change the political system', 'Thoughts which breed anxiety tend to vanish', 'People tend to accept opinions which justify the legitimacy of the privileges they have'. The propensity theory gives a plausible account of the structure of explanations using laws of this kind.

The propensity theory of explanation outlined above can easily be extended, in a straightforward manner, to *theoretical inductive-probabilistic explanation of statistical generalizations.* The explanandum is here of the following form:

(24) $rf(Gx/Fx \,\&\, Hx) = p$,

i.e., the relative frequency of instances of G among instances of F and H is p. Explanandum (24) is neither a probabilistic statement nor a statistical law; it expresses a statistical 'fact' of the constitution of a *finite* reference class.

Examples of such explanations are often met in biology and social science. Suppose, for instance, that you want to know why the proportion of boys among the children born during the years 1947–1967 in Finland was 0.51. It is well-known that a biologist can give an *ex post facto* deductive explanation of the sex of a newly born baby. A finite set of such explanations, explaining deductively the sex of each member of the reference class $F \,\&\, H$, would not be appropriate here, since it would not explain why the *proportion* of instances of G in this class was 0.51. To explain this, one may assume that all babies in H were conceived under conditions of kind T, and that there is a probabilistic law of the form '$P(Gx/Fx \,\&\, Tx)=p$'. (Examples with laws of the forms (22) and (23) can be treated similarly.) This amounts to treating the parents of a baby as an experimental set-up for producing male or female offsprings, so that the strength of the dispositional tendency of producing a male offspring on a single trial in this set-up depends upon the relevant physiological test conditions. The structure of this explanation can then be expressed as follows:

$$\text{(25)} \quad \frac{\begin{array}{l}(x)\,(Fx \,\&\, Hx \supset Tx) \\ (x)\,[P(Gx/Fx \,\&\, Tx) = p]\end{array}}{rf(Gx/Fx \,\&\, Hx) = p}$$

The statistical fact described in the explanandum is here subsumed under a probabilistic law concerning a potentially infinite 'reference class'. The explanandum is not deducible, but at best inducible, from the explanans. If it is further assumed that the class F & H is a random sample from the population F & T, then p, given in the explanandum, is the most probable value of the relative frequency of G in F & H given the explanans. Class T in an explanation of the form (25) is typically determined by a theoretical concept.

This example is perhaps over-simplified, but is serves to remind us that probabilistic explanations of statistical generalizations concerning particular finite populations are *inductive* in their character. Ernst Nagel (1961), pp. 509–520, has argued that the formal structure of explanations of statistical generalizations in the social sciences is always deductive. This is true in the sense that explanations of *probabilistic laws* usually follow the deductive-probabilistic pattern, and it is the explanation of laws that Nagel has in mind. But such explanations concern the question why the *premises* of explanations of the type (25) are true.

Explanations in the propensity model are subsumptive explanations of a certain kind, so that the applicability of this model to exhibit the structure of sociological and historical explanations may be said to reinforce the claims of the covering law theorists of explanation. The relevance of this model of the explanation of action is discussed in the next section.

6. Inductive Explanation of Action

When Peirce introduced the idea of propensity probabilities, he compared them to habits that men may have – in his view, both probabilities and habits have a dispositional character. This analogy suggests that the explanation of human behaviour could be regarded as a special kind of dispositional explanation. A detailed defence of this view has been given by Ryle (1949), who carefully analyses the dispositional nature of human reflexes, instincts, habits, hobbies, mannerisms, skills, qualities of character, beliefs, and motives.[24] It is, indeed, very easy and quite natural to apply the schemes (18) and (19) of deductive dispositional explanation to exhibit the structure of explanation of reflex behaviour, instinctive behavior or behaviour performed 'automatically' from the force of habit. But all habits do not entail uniform dispositions to behave in the same way in similar circumstances – I do not smoke after every meal in spite of my deeply rooted habit of doing so. This is even more true of intentional behaviour or of action done from motives. It is argued in this section that the thesis of the dispositional character of explanation of action can be defended if and only if *inductive* dispositional explanations are taken into consideration.

By action we mean behaviour understood in intentionalistic terms; explanation of action is therefore assumed to refer at least to the beliefs (convictions, knowledge, cognitive attitudes) and to the intentions (aims, wishes, wants, preferences, volitional attitudes) of an agent. The agent's beliefs and intentions are assumed to entail dispositions, inclinations, tendencies or propensities to act in a certain way in certain kinds of situations. (Of course, this may be only a part of a proper analysis of beliefs and intentions.) In other words, an agent with certain kinds of beliefs and intentions is supposed to have a dispositional tendency of some degree to act in a certain way in certain kinds of circumstances. Thus, an agent may be regarded as a chance set-up for producing different acts relative to his beliefs and intentions and to the external circumstances. If the connection between the agent's situation, his system of beliefs and intentions, and his behaviour can be expressed as a dispositional probabilistic law (of the form (20), (22), or (23)), then his particular acts can be explained in the propensity model of inductive-probabilistic explanation. To work out this suggestion would require a discussion of the ontological aspects of acts and a justification for the assumption that propensities to act are probabilities.[25] These considerations are beyond the scope of this paper; I only try to illustrate by simple examples the general viewpoint outlined above, and to point out some of its philosophical implications.

Suppose that an agent a intends to bring about a certain state of affairs S. Then a has a disposition to act in a way which would favour, according to his beliefs, the attainment of the end S. It is a plausible assumption that the strength of a's dispositional tendency to perform an act A depends upon his views of the relevance of A as a means to the end S. A case in which the strength of this dispositional tendency might be taken to be the strongest possible has been described by von Wright in his 'final formulation' of the scheme of *practical syllogism* (see von Wright, 1971, p. 107). The following statements characterize this situation:

F_1a = 'a is an agent who from now on intends to bring about S at time t'

F_2a = 'a considers (believes) from now on that he cannot bring about S at time t unless he does A no later than time t', and that he from now on is able to do A'

Ta = 'a is not prevented, from now on, from doing A'

In the scheme of practical syllogism, it is assumed that the conclusion

a sets himself to do A no later than
when he thinks time t' has arrived

follows with 'practical necessity' from the premises F_1a (a's intentions), F_2a (a's beliefs), and Ta (a's being under 'normal conditions'). We may now suggest that any agent finding himself in a situation characterized by F_1, F_2, and T has a very strong dispositional tendency to set himself to do A. (Note that this is not a statistical statement of the agent's mode of behaviour, but a theoretical attribution of a set of single-case propensities to him.) If 'Ga' is a statement which describes the conclusion above, then the scheme of practical syllogism can be converted into an argument of the form

$$(26) \quad \begin{array}{l} (x)\,[P(Gx/F_1x \,\&\, F_2x \,\&\, Tx) = r] \\ F_1a \,\&\, F_2a \,\&\, Ta \\ \hline\hline Ga \end{array} \quad [r]$$

where r is close to one.[26] This argument conforms to the scheme (21) of inductive-probabilistic explanation with a probabilistic covering law.

The question whether practical syllogism is logically conclusive reduces now to the question whether r takes the value one in (26). (Cf. note 14.) Von Wright's arguments suggest that r should be less than one (cf. von Wright, 1971, pp. 116–117). If $r=1$, however, then the (extensional part of the) covering law of (26) reduces to

(27) $(x)\,[Tx \supset ((F_1x \,\&\, F_2x) \supset Gx)]$

or perhaps even

(28) $(x)\,[Tx \supset ((F_1x \,\&\, F_2x) \equiv Gx)]$.

This corresponds closely to Tuomela's reformulation of the scheme of practical syllogism (see this volume, pp. 183–205).

The explanation (26) is *self-evidencing* in the sense that a's doing A in the situation T may constitute an indispensable part of the available evidence for attributing to him the intention F_1 and the belief F_2.[27] There is noting peculiar in this, however, since this is a feature shared by all dispositional explanations. Still, it is not unplausible to regard the covering law of (26), with $r=1$, as a conceptual or *analytic* truth rather than as a factual statement (cf. Tuomela, this volume, p. 183–205) – this is an analogue of Aristotle's view that in the search of premises for explanatory syllogisms we finally end up with 'laws' which have the nature of definitions. The intention F_1 and the belief F_2 are characterized here in so specific terms that they do not have what Hempel calls a *broadly* dispositional character (cf. Hempel, 1965, p. 460). While laws concerning broadly dispositional properties do have factual content, laws concerning specific dispositions which can manifest themselves in one way only serve to define these dispositions.

The situations to which practical syllogism is applicable can be viewed as special, or limiting, cases of the range of a general explanatory scheme (26) with a covering law of the form: 'Any agent of kind F_1 & F_2 has a dispositional tendency of strength r to perform an act of kind G under conditions of kind T', where F_1 and F_2 characterize the system of intentions and the beliefs of the agent. Such laws may be broadly dispositional, and consequently they may have factual content. The initial condition of such an explanation gives an *overall* characterization of the agents intentions and beliefs. It may therefore be possible to apply the same explanans to explain inductively several acts of the agent, so that the explanans may be testable independently of the particular explanandum. In other words, while practical syllogism is local – and extremely so – with respect to the explanans, the general inductive scheme has a more *global* character. Arguments instantiating this general scheme help us to *understand behaviour* as action and serve to *explain* action 'teleologically'.

The scheme of practical syllogism is limited to situations where the agent considers an act *a* to be *necessary* for an end *S*. The inductive scheme (26) is free from this restrictive assumption. The agent's preferences may be multi-dimensional, and they may fail to be connected – in such cases there need not be any particular *S* which he prefers to all alternative ends. Even when such an end *S* exists, the agent will normally act in accordance with the *strategy* that he considers *best* for attaining his end. There is no guarantee that such a strategy can always be reduced to a number of steps involving actions necessary to the intermediate ends. For example, the agent may face a choice between several acts which are *sufficient*, but not necessary, for bringing about *S*, or between acts which may bring about the desired consequence *S* only with different *probabilities* (cf. von Wright, 1971, pp. 101–102). The behaviour of an agent in such situations – which have been extensively discussed in the quantitative theory of rational deliberation, viz. decision theory – cannot always be *understood* by means of practical syllogism. Still, it seems possible to give an account of such behaviour in the inductive scheme (26).[28]

As other examples of the applicability of the propensity theory of explanation we may mention the explanation of action by broadly dispositional motives or characters of personality – such as jealousy, politeness, and conformity to laws of society or to social norms. The covering law of such explanations is of the form (20), (22), or (23); the initial condition contain a singular dispositional statement and a description of the situation which occasioned the explanandum action.

As we have seen above, dispositional explanations may be causal explanations: if you strike by hammer a fragile object, then the event that the object is struck by hammer is the cause of its breaking in parts (cf. scheme (18)). It was argued in Section 4 that inductive-probabilistic explanations may be (non-deterministic) causal explanations. If particular acts can be explained in the dispositional inductive-probabilistic model, as we have argued, it means also that causal explanation of action is possible, and that acts may have causes. To be sure, some propensity explanations of action are noncausal in that they do not refer to any event that could serve as a cause of the explanandum. Just as dispositional properties like 'fragility', propensities to act – entailed by beliefs and intentions, motives, characters of personality – are not causes. The causes of action are typically changes in the agent's situation: sighting red light is the cause of stopping at a traffic light, the neighbour's request is the cause of passing milk to him at table. But relative to otherwise stable situation, a change in the agent's system of beliefs, intentions, and motives may play a causal role in his acting; for example, the event that the agent comes to think that he is able to do an act, or the event that he is persuaded to choose bringing about S as his end. Even if a disposition is not a cause of anything, the *onslaught* of such short-termed dispositions as beliefs and intentions may be one (cf. Davidson, 1963).

There is nothing in the above account of action that would suggest that particular acts may have *humean* (deterministic) causes. The causes of action that are discussed above are probabilistic or non-deterministic, possibly maximal causes: they are not nomic sufficient conditions of the effect, they do not nomically bring about the action. Such a weak notion of causality does not presuppose the manipulability of the causal relation for 'technological' purposes, control, and prediction. This gives a refutation to the view of some representatives of the 'hermeneutic-dialectical' school who claim that a 'technological interest' is always involved in the covering law theory of explanation (cf., for example, Apel, 1972). Moreover, it is interesting to note that the inductive propensity account of action leaves, in a sense, more room for the *freedom* of an agent than the theory of practical syllogism. Referring to this scheme, von Wright (1971), p. 165, says

> If an action can be explained teleologically, it is in a sense determined, viz. determined by certain intentions and cognitive attitudes of men. If every action had a teleological explanation, a *kind* of universal determinism would reign in history and the life of societies.

The propensity model, even if it were applicable to the teleological explanation of all actions, does not have any kind of determinism as a corollary; on the contrary, if some actions can be genuinely explained in this model, a kind of indeterminism reigns in the realm of human behaviour.

University of Helsinki

NOTES

1 See von Wright (1971), pp. 13–15, and Section 4 below.

2 For theoretical concepts, see Tuomela (1973).

3 For a brief discussion of inducibility and its variants, see Niiniluoto (1972).

4 Several measures of explanatory power are discussed in Niiniluoto and Tuomela (1973).

5 See Hempel (1962, 1965, 1968) and Salmon (1970, 1971). Stegmüller (1973) has argued that Hempel's and Salmon's theories have different explicanda, neither of which should be called 'explanation'. This problem is discussed in Section 4 below.

6 The notion of inductive systematization and the examples of Hempel, Scheffler, and Lehrer are discussed in Niiniluoto (1972). See also Niiniluoto and Tuomela (1973).

7 The following framework is used in Niiniluoto and Tuomela (1973) for the study of theoretical inductive-deterministic explanations of universal generalizations. Let L_o be an observational language and let L be an extension of L_o obtained by adding a number of new, theoretical predicates to L_o. Let e be a singular evidence statement in L_o, g a universal generalization in L_o, $\mathbf{D}$ the set of constituents of L_o, and T a theory in L employing theoretical predicates. Applying the theory of semantic information, different measures of the explanatory power of theory T and evidence e with respect to generalization g, or set $\mathbf{D}$, are defined. With the help of Hintikka's system of inductive logic, these measures are studied in the special case where L_o and L are monadic first-order languages.

8 A question of the form 'Why is it the case that y?' may be interpreted as a request 'Bring it about that I know why it is the case that y'. Hintikka (1974) has shown quite convincingly that 'a' is a satisfactory (potential) answer to a WH-question 'Who (what, etc.) is such that ——?' only if the questioner knows who (what, etc.) a is. Cf. also Hempel (1965), pp. 426–427.

9 If I know that all ravens are black and that a is a black raven, then I may be said to know already why a is black: 'Ba' is a trivial logical consequence of my knowledge which includes '$(x)\ (Rx \supset Bx)$' and 'Ra'. But if I don't know that a is a raven, then the inclusion of 'Ra' into my corpus of knowledge *brings it about* that I know why a is black.

10 Radnitzky misses this point when he argues that covering law explanations look "like a logistic travesty of idle-running small talks: A says 'Look, this piece of ice floats on water' and B rejoins with a resignated shrug on the shoulders: 'Sure, they *all* do'" (Radnitzky, 1970, p. 169). Radnitzky mistakenly supposes that, in the covering law model, the fact *that this particular piece of ice floats on water* is explained by the generalization that all pieces of ice float on water. An argument of the form

$$\frac{(x)\ (Fx \supset Gx)}{Fa \supset Ga}$$

does not qualify as an explanation in the covering law model (cf. the non-comparability condition in Tuomela's (1973) DE-model of explanation). In Radnitzky's example, the questioner *A* would normally request for information explaning why ice floats on water, i.e., why all pieces of ice float on water. This generalization, which itself may be explained by other laws in the covering law model, may function as a part of the explans in the following argument: *A* says 'Look, this object is floating on water', and *B* rejoins with 'Sure, don't you see that it is a piece of ice'. Cf. also Apel (1973), p. 24, where the same mistake is repeated.

[11] For the difference between standard and non-standard interpretations of second-order logic, see for example Robbin (1969).

[12] These are the classes that can be selected from *F* by means of a recursive function.

[13] When I wrote the first draft of this paper, I thought that the Coffa-Salmon thesis is valid. Professor von Wright pointed out to me a serious flaw in my attempted reconstruction of this argument, and helped me to realize that something like the limit condition CL, to be formulated below, would be needed to support this thesis.

[14] In the frequency interpretation of probability, (15) does not imply (16), but in the single case propensity interpretation this implication holds true (cf. Section 5 below).

[15] This is the view of Hempel, too (see Hempel, 1965, p. 418). We have seen that the relation of this view to Hempel's account of inductive-probabilistic explanations as epistemically relative is a very complex one – even if it is not an inconsistency as Coffa and Salmon argue.

[16] Salmon's model admits of explanations in which the explaning property is probabilistically irrelevant to the explanandum. The view that an explanation, as an information-providing argument, has to satisfy the requirement of positive relevance has been defended in Niiniluoto and Tuomela (1973).

[17] It should be noted that, even in the case of deterministic causes, there may exist two different *true* causal explanations of the same event, only one of which is the *correct* explanation. For example, suppose that both '$(x)\,(Fx \supset Gx)$' and '$(x)\,(Hx \supset Gx)$' are true causal laws, and that $t'' < t' < t$. Then both the occurrence of $f_{t''}$ at t'' and the occurrence of $h_{t'}$ at t' may serve as initial conditions of *true* causal explanations of the occurrence of g_t at t. Which one of these explanations is the *correct* one depends upon which one of the events $f_{t''}$ and $h_{t'}$ was really productive in the case concerned. For example, it may happen that event $f_{t''}$ does not have enough time to realize its effect g before cause $h_{t'}$ brings about g at t. Similar remarks apply to causal inductive-probabilistic explanations. It follows that a cause which is *spurious* in Suppes's sense (cf. Suppes, 1970, p. 23) may be the correct cause of the given event.

[18] Cf. the quotation from Peirce that opens this paper (see Peirce, 1932, §664).

[19] See Hempel (1965), p. 378. Later he has returned to the customary frequency interpretation of probability (cf. Hempel, 1968).

[20] Ryle (1949), pp. 86–87, has observed that to explain why a glass broke is different from explaining why a glass broke when it was hit by a stone. Cf. Section 3.

[21] Rozeboom (1973) argues that dispositions should be analyzed in terms of a 'base' relation *B* over attribute triples. Using this relation, the universal dispositional law to the effect that all objects of kind *F* possess a dispositional tendency of universal strength to produce, under test conditions of kind *T*, an outcome of kind *G* is expressed simply by

$$B(F, T, G).$$

Probabilistic laws can be treated similarly in terms of 'base' relations weaker than *B* above.

[22] Note that T is positively relevant to G relative to F just in case

$$P(Gx/Fx \,\&\, Tx) \;>\; P(Gx/Fx),$$

that is, the dispositional tendency of a chance set-up of kind F to produce outcomes of G is greater relative to trials of kind T than relative to trials of kind $-T$. Recall that the above inequality is equivalent to

$$P(Gx/Fx \,\&\, Tx) > P(Gx/Fx \,\&\, -Tx).$$

[23] Cf. Giere's example in which "each person is regarded as a chance set-up which at each moment of time has a definite propensity to contract cancer under certain condition", and "on the average, a man's propensity to contract cancer is greater if he smokes than if he does not" (Giere, 1973, p. 480).

[24] Ryle regards explanations by motives as analogous to a dispositional explanation which explains why the glass shivered when struck by stone (see Ryle, 1949, pp. 86–88). It is argued above that the logical structure of such explanation – in the case of universal dispositions – is given by scheme (19). In Ryle's view, the explanans of this explanation states a "general hypothetical proposition about the glass" i.e., for any t, if the glass a is struck by stone at time t, then it shivers at t. The structure of this explanation is then something like the following:

$$\frac{(t)\,(Tat \supset Gat)}{Tat_1 \supset Gat_1}$$

This is not a genuine explanation, however. (Cf. Section 3 and note 10.) The pattern

$$\frac{\begin{array}{l} Fa \supset (t)\,(Tat \supset Gat) \\ Fa \end{array}}{Tat_1 \supset Gat_1}$$

where 'Fa' states that a has the disposition F may qualify as an explanation. But this pattern (just as (19)) presupposes that dispositional terms are referring terms, which is incompatible with Ryle's instrumentalism.

[25] Cf. Suppes's (1973) representation theorem showing that the propensities involved in radioactive decay are probabilities.

[26] The covering law in (26) could, of course, be replaced with a law of the form $(x)\,[P(Gx/F_1x \,\&\, F_2x \,\&\, Tx) \geqslant r]$.

[27] The notion of self-evidencing explanation is discussed in Hempel (1965) pp. 371–373.

[28] An agent facing a choice between alternative acts with uncertain consequences is often unable to order the acts, even if he could order the consequences according to their preferability, without implicitly assuming such a coherence in his system of beliefs and preferences that his beliefs and preferences are representable by means of a subjective probability measure and a utility function, respectively. In such a case one might suggest that the agent's propensity of doing a is proportional to the *expected utility* of a. This would give us, for example, a method of understanding and explaining *randomized acts*. The imaginary agents of decision theory and game theory are often supposed to toss a coin between two alternative acts. The real-life agents may do a similar thing, i.e., a randomized act, by deliberately doing an act the consequence of which depends upon chance or upon acts of other agents. (An example is provided by one of my colleagues who occasionally, but intentionally, orders in the restaurant the same drink as the person sitting on his left-hand side when the waiter is approaching him from the right.)

BIBLIOGRAPHY

Apel, K. O.: 1972, 'Communication and the Foundations of the Humanities', *Acta Sosiologica* **15**, 7–26.

Apel, K. O.: 1973, 'Programmatische Bemerkungen zur Idee einer "Transzendentalen Sprach-Pragmatik"', in *Studia Philosophica in Honorem Sven Krohn, Annales Universitatis Turkuensis* **B 126**, Turku, 1973, pp. 11–35.

Coffa, J. A.: 1974, 'Hempel's Ambiguity', *Synthese* **28**, 141–163.

Davidson, D.: 1963, 'Actions, Reasons, and Causes', *The Journal of Philosophy* **60**, 685–700.

Fetzer, J. H.: 1971, 'Dispositional Probabilities', in R. C. Buck and R. S. Cohen (eds.), *Boston Studies in the Philosophy of Science*, Vol. 8, D. Reidel Publishing Company, Dordrecht, 1971, pp. 473–482.

Fetzer, J. H.: 1974, 'A Single-Case Propensity Theory of Explanation', *Synthese* **28**, 171–198.

Giere, R. N.: 1973, 'Objective Single-Case Probabilities and the Foundations of Statistics', in P. Suppes *et. al.* (eds.), *Logic, Methodology and Philosophy of Science IV, Proceedings of the Fourth International Congress for Logic, Methodology and Philosophy of Science, Bucharest, 1971*, North-Holland Publ. Co., Amsterdam, 1973, pp. 467–483.

Greeno, J. G.: 1970. 'Evaluations of Statistical Hypothesis Using Information Transmitted', *Philosophy of Science* **37**, 279–294.

Greeno, J. G.: 1971, 'Theoretical Entities in Statistical Explanation', in R. C. Buck and R. S. Cohen (eds.), *Boston Studies in the Philosophy of Science*. Vol. 8, D. Reidel Publishing Company, Dordrecht, 1971, pp. 3–26.

Hacking, I.: 1965, *Logic of Statistical Inference*, Cambridge University Press, Cambridge.

Hempel, C. G.: 1958, 'The Theoretician's Dilemma', in H. Feigl, M. Scriven, and G. Maxwell (eds.), *Minnesota Studies in the Philosophy of Science*, Vol. 2, University of Minneapolis Press, Minneapolis, 1958, pp. 37–98. (Reprinted in C. G. Hempel, *Aspects of Scientific Explanation and Other Essays in the Philosophy of Science*, The Free Press, New York, 1965, pp. 3–46.)

Hempel, C. G.: 1962, 'Deductive-Nomological vs. Statistical Explanation', in H. Feigl and G. Maxwell (eds.), *Minnesota Studies in the Philosophy of Science*, Vol. 3, University of Minneapolis Press, Minneapolis, 1962. pp. 98–169.

Hempel, C. G.: 1965, 'Aspects of Scientific Explanation' in C. G. Hempel, *Aspects of Scientific Explanation and Other Essays in the Philosophy of Science*. The Free Press, New York, 1965, pp. 331–496.

Hempel, C. G.: 1968, 'Maximal Specificity and Lawlikeness in Probabilistic Explanation', *Philosophy of Science* **35**, 116–133.

Hintikka, K. J.: 1968, 'The Varieties of Information and Scientific Explanation', in B. van Rootselaar and J. F. Staal (eds.), *Logic, Methodology and Philosophy of Science III. Proceedings of the 1967 International Congress*, North-Holland, Publ. Co., Amsterdam, 1968, pp. 151–171.

Hintikka, K. J.: 1972, 'On the Ingredients of an Aristotelian Science', *Nous* **6**, 55–69.

Hintikka, K. J.: 1974, 'Questions about Questions', in M. K. Munitz and P. Ungar (eds.), *Semantics and Philosophy*, New York University Press, New York, 1974.

Jeffrey, R. C.: 1969, 'Statistical Explanation vs. Statistical Inference', in N. Rescher *et al.* (eds.), *Essays in Honor of Carl G. Hempel*, D. Reidel Publishing Company, Dordrecht, 1969, pp. 104–113.

Lehrer, K.: 1969, 'Theoretical Terms and Inductive Inference', *American Philosophical Quarterly*, Monograph series, Studies in the Philosophy of Science, Monograph No. 3, pp. 30–41.

Levi, I.: 1969, 'Are Statistical Hypotheses Covering Laws?' *Synthese* **20**, 297–307.

Martin, R.: 1975, 'Explanation and Understanding in History', this volume, pp. 305–334.

Mellor, D. H.: 1971, *The Matter of Chance*, Cambridge University Press, Cambridge.

Nagel, E.: 1961, *The Structure of Science*, Routledge & Kegan Paul, London.

Niiniluoto, I.: 1972, 'Inductive Systematization: Definition and a Critical Survey', *Synthese* **25**, 25–81.

Niiniluoto, I. and R. Tuomela: 1973, *Theoretical Concepts and Hypothetico-Inductive Inference*, D. Reidel Publishing Company, Dordrecht and Boston.

Nowak, S.: 1970, 'The Logic of Reductive Systematizations of Social and Behavioural Theories', *The Polish Sociological Bulletin* **22**, 79–102.

Peirce, C. S.: 1932, *Collected Papers*, Vol. 2. (ed. by C. Hartshorne and P. Weiss), Harvard University Press, Cambridge, Mass.

Radnitzky, G.: 1970, *Contemporary Schools of Metascience*, Akademiförlaget, Gothenburg.

Robbin, J. W.: 1969, *Mathematical Logic. A First Course*, W. A. Benjamin, New York.

Rozeboom, W. W.: 1973, 'Dispositions Revisited', *Philosophy of Science* **40**, 59–74.

Ryle, G.: 1949, *The Concept of Mind*, Hutchinson, London.

Salmon, W. C.: 1970, 'Statistical Explanation', in R. G. Colodny (ed.), *Nature and Function of Scientific Theories*, University of Pittsburgh Press, Pittsburgh, 1970, pp. 173–231. Reprinted in W. C. Salmon, *et. al.* (eds.), *Statistical Explanation and Statistical Relevance*, University of Pittsburgh Press, Pittsburgh, 1971, pp. 29–87.

Salmon, W. C.: 1971, 'Explanation and Relevance: Comments on James G. Greeno's "Theoretical Entities in Statistical Explanation"', in R. C. Buck and R. S. Cohen (eds.), *Boston Studies in the Philosophy of Science*, Vol. 8, D. Reidel Publishing Company, Dordrecht, 1971, pp. 27–39.

Salmon, W. C.: 1974, 'Comments on "Hempel's Ambiguity" by J. Alberto Coffa', *Synthese* **28**, 165–169.

Scheffler, I.: 1963, *The Anatomy of Inquiry: Philosophical Studies in the Theory of Science*, Alfred A. Knopf, New York.

Stegmüller, W.: 1973, *Personelle und Statistische Wahrscheinlichkeit, Probleme und Resultate der Wissenschaftstheorie und Analytischen Philosophie*, Springer-Verlag, Berlin, Heidelberg & New York.

Suppes, P.: 1970, *A Probabilistic Theory of Causality* (*Acta Philosophica Fennica* **24**), North-Holland Publ. Co., Amsterdam.

Suppes, P.: 1973, 'New Foundations of Objective Probability: Axioms for Propensities', in P. Suppes *et. al.* (eds.), *Logic, Methodology and Philosophy of Science IV, Proceedings of the Fourth International Congress for Logic, Methodology and Philosophy of Science, Bucharest, 1971*, North-Holland, Publ. Co., Amsterdam, 1713, pp. 515–529.

Tuomela, R.: 1973, *Theoretical Concepts*, Springer-Verlag, Wien.

Tuomela, R.: 1975, 'Explanation and Understanding of Human Behavior', this volume, pp. 183–205.

von Wright, G. H.: 1945, *Über Wahrscheinlichkeit. Eine Logische und Philosophische Untersuchung*, Acta Societas Scientiarum Fennicae, Nova Series A, Tom. III, No. 11, Helsinki, 1945.

PART IV

REPLIES TO COMMENTATORS. SECOND THOUGHTS ON EXPLANATION AND UNDERSTANDING

GEORG HENRIK VON WRIGHT

REPLIES

The papers in the present volume can be divided into two groups, *viz.* those which deal directly with some aspects of my thoughts and those which do not. The line of division between the two groups is sharp. In my comments I shall confine myself exclusively to those essays which relate to my own work.

This decision means that here I shall not comment on the essay by Bubner on problems of 'hermeneutics' nor on Hintikka's suggestive paper on intentionality. Aleksandar Kron's article on the formal aspects of the theory of causality I shall also leave aside. Its topic is one to the clarification of which I myself have tried to contribute in various writings. But Kron's method of treatment and the techniques he employs are different and do not call for comments from me.

I shall use this opportunity for replying to the papers somewhat egotistically, partly to make further efforts to clarify my thoughts and partly to remove some misunderstandings which have sometimes arisen, I am afraid, due to my own shortcomings. The length or shortness of my comments on a particular essay should not be taken as an indication of the importance or value which I attribute to that essay.

It goes without saying that I am deeply grateful to the organizers of the colloquium in Helsinki in January, 1974, and to the contributors to the present volume for this opportunity of having my views debated and of contributing myself to their further progress.

From the point of view of subject-matter the essays on which I shall be commenting can be divided into five groups:

(i) Teleology. This is the topic of the paper by Makai and of Riedel's essay on causal and historical explanation.

(ii) Causality. To this group belong the papers by MacIntyre and Winch and the first half of the paper by Tuomela. (The paper by Kron I shall not comment on. See above.)

(iii) Formal theory of action. The paper by Kenny.

Manninen and Tuomela (eds.), Essays on Explanation and Understanding, 371–413.

(iv) Problems connected with intentionality, action, and the so-called practical syllogism. This group includes the papers by Hertzberg, Kim, Stoutland, and the second half of the paper by Tuomela.

(v) Related to the essays in the fourth group, but of a somewhat broader scope, are the papers by Martin on explanation and understanding in history and by Niiniluoto on inductive explanation.

I propose to deal with the papers in the order indicated by the above division into groups.

I. TELEOLOGY

Makai and Riedel comment critically on my treatment of teleology in *Explanation and Understanding* (henceforth abbreviated as *E & U*). Particularly the first author finds a great many faults in it. In fact, she thinks that "the specific teleological aspect" (p. 27) is missing from my account of action – and attributes this defect to my "purist" interpretation of finality which ignores the "ontic (causal) components" involved in action (p. 31).

I have little to say in defense against this criticism. The treatment of teleology in *E & U* was never intended to be complete. As a matter of fact it is even more defective than I realized at the time of writing the book. It is also very unsystematic. The scattered nature of my remarks may have made it difficult for some readers to see that in the background of the book there are several *different* aspects of teleology awaiting further exploration. Let me briefly recall three of them and the essentials of my position:

(i) The ideas, first put forward in the renowned paper (1943) by Rosenbluth, Wiener, and Bigelow on the use of the notion of negative feedback in accounting for purposive behaviour. Of these ideas I have wanted to say that they represent a 'causalization' or 'mechanization' of the notion of teleology. I think that they have important applications for the understanding of purposive mechanisms in biology. And I said I believed that, in biology, teleology is really 'quasi-teleology' and teleological explanations replaceable by (complex forms of) deductive-nomological, 'covering law' explanations. *Teleonomy* would perhaps be a suitable name for teleology thus subsumed under the "reign of natural law". (Cf. *E & U*, p. 86.)

(ii) Ideas concerning the key-role of the 'practical syllogism' as a pattern of explanation in the human sciences. Of this explanation model I maintain that it is not 'causalist' but genuinely 'teleological'.

(iii) Various ideas about explanations which I called 'quasi-causal' in history and social science. I tried to show that quasi-causal explanations in spite of a certain affinity with explanations of the 'covering law' pattern, contain genuinely teleological components. In my opinion, the applications of various cybernetic and system-theoretic ideas and formalisms to the explanation of social phenomena also fall under the heading of quasi-causality. Thus, if I am right, the applications of cybernetics and kindred theoretical tools in the social sciences are more unlike their applications in biology, not to speak of engineering, than a superficial resemblance on the level of formalism may suggest.

At the time of writing *E & U* I did not realize the enormous range of the topic 'teleology'. In particular, I did not see the complex nature of the ideas grouped under (iii) above. I thought that quasi-causality could be treated as a combination of Humean causation and of (latent and manifest) patterns of practical inference. I underrated, among other things, the role which norms and social institutions play as determinants of the actions both of groups and individuals.[1] I believe that, when due notice is taken of this, as well as of natural causation and the intentionality of agents, one can give the more complete account of "the specific teleological aspect" with its "ontic components" which Maria Makai now finds missing from the treatment in *E & U*.

The only aspect of teleology with which I have dealt at any length, so far, is 'practical inference'. I think that my treatment, even if not entirely satisfactory, is substantially on the right lines. But it is quite certain that, in *E & U* and in other of my earlier publications, I over-rated the importance to the human sciences of this particular explanation pattern. The exaggeration is perhaps understandable in the light of the fact that my approach to teleology, as Professor Riedel acutely observes (p. 3), was from the platform of a philosophy of *human action*. I also regret my use of terminology. 'Practical syllogism' is not a good name, – for one thing because of its misleading associations with the notions of deciding and planning. (See my reply to Hertzberg, below p. 394.) Nor do I any longer wish to use the name 'teleological explanation' for the explanation pattern in question – although I think it is teleological explanation *of a kind*. The best name for it seems to me to be *intentionalist explanation*.

I have read Professor Riedel's clarifying account of teleology in Hegel and Kant with great interest and profit. I had been aware of the fact that Hegel's treatment of teleology in the second part of *Wissenschaft der Logik* was concerned with the means-end relation and therefore with the 'technical' aspect of goal-directed action. In this it closely parallels explanation of action through 'practical syllogisms'. What I did not see, however, was that the broader aspects of teleology, discussed for example in Hegel's writings on the history of philosophy and philosophy of history, vastly transcend the rather narrow frame set for 'teleology' in *Logik*. Yet in my own remarks on quasi-causality there are some incidental references to Hegel which ought to have made me aware of this. That I did not see it was surely due to my own exaggerated fondness for the practical inference schema as the basic pattern of *all* explanation in the human sciences.

Riedel points to Kant (in *Kritik der Urteilskraft*) rather than to Hegel as the 'renovator' of an Aristotelian tradition in our thinking about teleology. I find this, as well as Riedel's criticism of some of the shortcomings of Hegelian teleology, extremely interesting. Does Kant then provide modern philosophy of science with a workable alternative to the subsumption-theoretic view of explanation for the realms of human action and goal-directed processes in history and social life? Following the hints given by Riedel in his paper, I hope one day to be able to answer this question – at least for myself.

It still seems to me, however, that there is a difference in principle between teleology in biology and teleology in the human sciences. I do not think one can draw a sharp border between the two types of science themselves, nor between the types of phenomena they study. Psychology, for example, stands with one foot in biology and with another in a different province. But I believe there are *sharp* conceptual differences between the patterns of explanation which, on the whole, are to be credited as 'scientific' in biology and those which, on the whole, are accepted in history and social study. The tendency to blur or neglect these differences is detrimental to our understanding of the ways in which man has endeavoured to grasp in rational terms the various aspects of reality.

Maria Makai's essay outlines a theory of teleology and finalistic structures which is meant to be an alternative both to positivist reductionism and to the "purist" man-nature dualism which she finds in my work. I am not sure whether I have been entirely successful in my effort to understand her, owing to differences between us in general approach and traditional background. Part of her criticism is, I think, sheer misunderstanding and part stems from a failure to recognize the *restricted* aims of the analysis undertaken in *E & U*. With much of what she says herself about finalistic structure and the depth-analysis of socio-economic forces I find myself in sympathy. I am not sure, moreover, that a treatment of these topics, in order to be successful, must employ conceptual techniques and adopt a general point of view which is *basically* different from my own. It seems that Makai has somewhat exaggerated the "purism" of my position. I can see why one should call my analysis of intentional explanation ('the practical syllogism') by that name. This is the aspect of teleology most strongly emphasized, and perhaps overemphasized, in my work hitherto. But there is also, even in *E & U* sketchy mention of the *interplay* between nomic connexions and intentional action which I called 'quasi-causality' and which underlies the dynamics of history and social change. (Cf. above.) It is questionable whether this should be called 'teleology' at all. At least it is not, in Makai's sense, "purist".

II. CAUSALITY

Professor Winch's critical discussion of my concept of cause and its relation to the concept of human action gives me a welcome opportunity for trying to clarify some controversial points connected with the position for which I have been arguing.

Let me first, in rough outline, restate part of my view. – To say that *c* is cause of *e*, in the sense which I call 'Humean', is to maintain, not only that whenever *c* appears *e* will appear, but also that on those occasions, when *c* in fact did not appear, *e* would have appeared had *c* appeared. In the notion of cause an element of 'counterfactual conditionality' is thus involved.

If an agent, on some occasion, does or produces *c*, then *c*, which was not there, came to be 'thanks to the interference of the agent'. To say this is to imply that, had the agent not acted, *c* would, on that very occasion, have remained absent. Thus in the notion of action, too, there is an element of 'counterfactual conditionality'.

The counterfactual conditional statements which causal relationships warrant I shall call *causal* counterfactuals. They are only one among many different kinds or types of counterfactual conditional statement.

Further, I have tried to argue that the notion of a ('Humean') cause is, in a characteristic sense, dependent upon or secondary to the concept of an action. I shall not repeat the arguments here.

(a) Some critics of my position think that my view is in some vicious way circular. Causal relations entail counterfactuals. Action statements too imply counterfactuals. *If* the counterfactuals involved in action were *causal* counterfactuals, *then* clearly it would be circular to try to base the concept of causality on that of action.

Winch fears that there is this circularity and that it vitiates my argument. I do not think he is right – but it is certain that in *E & U* little was said to forestall the criticism. I hope I have done better in my Woodbridge Lectures, *Causality and Determinism* (1974). But I shall not deny that the matter may still be open to debate.

The essentials of my 'defense' are as follows:

Causal laws are *primarily* laws connecting *changes*. The appearance or emergence of *c* in a situation in which initially *c* was not there, causes *e*, which was also absent, to appear too. This is the basic pattern. There are also causal laws about how things are kept unchanged thanks to *counteracting* or *preventive* causes. These laws, however, are *secondary* to laws of the first type, in the sense that the operation of a counteracting cause presupposes that a cause working for a change is already in operation.[2]

It follows from the above that the counterfactuals which causal laws warrant are primarily of the following form: had *c* appeared on an occasion when it was initially absent and did, in fact, not subsequently appear, then *e* which was and remained absent would have appeared too. Not all counterfactual conditionals which have this form are causal – but all causal counterfactual conditionals which can be extracted from primary causal laws are of this form.

Productive action is the producing of a change on an occasion when no change would otherwise have occurred. Preventive action is different. It is a counterpart on the action side to counteracting causes on the causation side. We shall here ignore preventive action. The counterfactuals presupposed in (productive) actions are accordingly of the following form:

had *c* not been produced on the occasion when initially it was absent, it would have remained absent. The conditional links one counterfactual non-change with another, subsequent, non-change.

When things do not change, but remain as they are, it is *normally* not because there is some cause which keeps them from changing. ('Normally' here means: except when some *counteracting* cause prevents an impending change – and such cases are in a minority.) Causes are typically factors which disturb or interfere with a natural equilibrium or state of rest. This, incidentally, is also reflected in the etymology of causal words. In several languages known to me the word for cause carries a connotation of guilt, responsibility for some bad thing. Health, being the normal state of a living body, has no 'causes', but illness usually has. The literal meaning of 'aetiology' is 'science of causes' – the received meaning is 'science of causes of diseases'.

I shall not maintain that what I have said here proves my claim that the counterfactual conditionals presupposed in action statements are not *causal* counterfactuals. But I hope to have shown that they are of a quite different form from counterfactual conditionals normally presupposed in causal statements and to have stated my reason for thinking that they are not themselves causal. If the charge of circularity persists, I think the *onus probandi* is with the critic. He has to show that not only changes but also non-changes are *normally* the effects of causation.

(b) Another question which I want to raise *à propos* Winch's paper is whether intentional acting is compatible with the agent's awareness of the simultaneous operation of a cause which effects the result of his action. I have wanted to say that action is incompatible with *awareness*, although not with the *existence* of causes of the results of our actions. Winch gives what I consider a very good and fair restatement of my view on pp. 130–2 of his paper. He then raises some doubts about its correctness for basic actions – admitting at the same time that he finds himself "unable to decide whether or not von Wright is right about this" (p. 132). The question is indeed difficult and I appreciate Winch's caution here.

Winch asks (p. 133): "Can I not suppose – and might I not have reason for supposing – that the operation of such Humean causes will only be observed on those occasions on which I am also acting intentionally?" It is not *quite* clear to me what Winch means by "such Humean causes", but, to judge from the context, he means a Humean cause of the *movement* which is the outer aspect of my action – for example my arm's rising when I raise my arm.

Let us suppose then that there is such a cause of the movement and that the cause, moreover, is both sufficient and necessary for the effect. It may, of course, happen that the operation of this cause is actually observed only on occasions when I am acting intentionally. But it is surely perfectly thinkable that this cause sometimes operates – say as a result of an experimenter's interference – and that my arm goes up although I do not intend this to happen. Likewise we can imagine that I sometimes intend to raise my arm but that the arm does not rise, *e.g.*, because I am lame or because an experimenter has cut the nerve-connections.

In view of these possibilities, I think the following should be said in reply to Professor Winch's question above: The 'Can I not suppose'-part of the question has an affirmative answer. We can make the supposition. But it is difficult to see that we could ever have a good *reason* for making it. To have a reason for making it would require, I think, that we suppose that the Humean cause in question is a cause, not only of the *movement* (of my arm, say) but also of my *intention* in performing this movement. I can easily conceive of the movement having a neural cause, but can the *intention* have one? So that, for example, every time my brain is being (experimentally) stimulated in a certain way, I have (get) an intention to perform a specific action. Winch does not say that he believes this is possible – and I doubt that he believes it. But there are other philosophers who claim to be able to believe this; for instance, those who support the thesis of 'central state materialism', – a view which is to me, as it is to Winch (pp. 133–4), unintelligible.

Winch, however, seems to think that there are alternative ways (to materialism) of arguing for a causal relation between brain state and intention. He says (p. 134): "It might be supposed, for example, that a given intentional action, involving given bodily movements, is always accompanied by given neural processes." I too find this completely conceivable. The truth of this supposition is a logical consequence of the supposition made earlier that there is a sufficient and necessary causal condition (among brain states and neural processes) for the movement which goes with the action. If this is true for example of my arm's rising, then it is also true that whenever I raise my arm intentionally a certain thing happens in my neural system. But this is not all that Winch needs in order to prove his point. What he needs is the fiction of a Humean cause for *just* the intention – irrespective of whether an accomplished action ensues or not. I doubt whether Winch would say that he can imagine this fiction to be true. To me it is equally unintelligible as the thesis of 'central state materialism'.

If there are sufficient and necessary conditions in the neural system for specific movements of our limbs, then it is *contingent* whether those neural conditions are satisfied when we intend to move our limbs. (Cf. *E & U*, p. 129.) Of those movements, however, which we can perform (which are under the control of our 'will'), it would then also hold true that *on the whole*, when we intend the movements, the conditions are satisfied, and *vice versa*. This would be a very remarkable fact of human natural history. We could call it the 'naturalistic' basis for the possibility of (intentional) action. Whether action has this basis or not depends upon whether the above hypothesis about the causal conditions of bodily movements is true or not. And it may well be that it is true.

So much for the general question of Humean causes of intentions. We now go back to the more specific question, raised above (p. 377), whether it is possible to *observe* a Humean cause of the result of one's intentional action on some individual occasion. The reason I think this is *not* possible has to do with our confidence that, had we not done what in fact we did, the results of our actions would not have materialized (on those occasions). If we say with emphasis "I did it", we are also prepared to say "had I not done this, it would not have happened just *then*". If now we observe something operating which we think is a cause of the result (of an action), then what we can do is to *let* this result come about or, maybe, try to *prevent* it from coming about. But we cannot ourselves do what is being done *for us*.

If the above is accepted, how then does it agree with my view that actions are compatible with the existence of Humean causes for their results, whenever the actions are done? One might think that my view of the 'counterfactual element involved in action' ought to commit me to indeterminism. I don't think it does. For, thinking that there is a cause, not however observed by me, for the result of my action is fully compatible with claiming that, had I not acted, the result would not have materialized. What I am then claiming is that, had I not acted, the cause would not have been operating either. This would be a statement that two facts contingently *coincide*, *viz.* my action and the operation of that cause. But it would not be a statement that they are, causally or otherwise, *connected.* The coincidence in question is nothing but the "remarkable fact of human natural history" alluded to above (p. 379) which, *if* the results of all actions have Humean causes, must be accepted as a precondition of the possibility of action.

As mentioned earlier (pp. 371–2), Professor Tuomela's paper divides into two sections, one for each of the two main topics in *E & U*, *viz.* the theory of causation and the theory of the 'practical syllogism'. Both sections contain interesting criticisms of my position and independent observations of value.

(a) Tuomela wishes to make a contrast between my 'interventionist' or 'experimentalist' view of causation and his own "objectivistic entailment account" (p. 193). I do not find his characterization of my "interventionist" position quite accurate. Tuomela's statement of his own position is too brief, moreover, to make it possible to judge how much, if at all, it really differs from mine.

First some words on Tuomela's characterization of my position. – Tuomela says (p. 184) that "according to this (*sc.* vW's) interventionist view, causation consists basically of this idea: by his intervention in a system an agent can bring about changes which would not otherwise have occurred". Tuomela also says (*ib.*) that if an agent can produce an instantiation of an event-type '*p*' and thereby bring about an instantiation of another event-type '*q*', then according to the interventionist account "the first (generic) event is a cause of the second". But this seems to me to be a very weak claim to which anybody could agree, whether 'interventionist' or not. For who would deny that, if by doing so that *p* we can bring it about that *q* then an event of the first type *causes* an event of the second type? Surely, the 'interventionist' view must make a stronger claim than this.

Before proceeding to clarify wherein my 'interventionism' consists, let us have a look at the alternative account proposed by Tuomela. On p. 192–3 of his essay, he says: "Let thus *p* and *q* be singular events. Then we say that the statement '*p* caused *q*' is true only if there are suitable statements *P* and *Q* describing *p* and *q*, respectively, and if there is a nomic backing theory *T* such that *T* jointly with *P* (plus perhaps some context description) deductively explains *Q*."

This as it stands does not contain anything with which I disagree. It differs from my formulation in *E & U* in that it conceives of the causal relation as subsisting, primarily, between individual events and not between event-types. This I would regard as an improvement. The subsistence of the causal relation, however, depends, according to Tuomela's view as well as mine, on the existence of a nomic or law-relation between event-types, *i.e.*, generic features of the individual events. When is a relation between event-types *nomic* – as distinct from accidentally universal? This question is, in my opinion, the crux of the whole matter. In my efforts to answer it I was led to the view that the concept of cause presupposes the concept of action. Tuomela still has to show that he can give an alternative account of nomicity (lawlikeness) which substantially differs from mine. He is excused for not giving it in his essay. (P. 193: "Space does not here permit a fuller discussion".) But it seems to me that he takes it all too easily for granted that "an objectivistic entailment account" *can be given* which does not make the notion of cause dependent upon actionist notions.

The terms 'objectivist' and 'subjectivist' are used in many senses and are therefore easily misunderstood, if not duly qualified. There is *one* sense in which the account I try to give of the causal relation is *objectivist*. In my view the causal relation (*i.e.*, *that* causal relation which is now being discussed and which is not the only one) must satisfy the following three requirements:

(i) The relation holds between mind-independent events (or states) in 'nature'.

(ii) It holds, on the individual occasion, independently of whether the occurrence of the cause is the result of action, or not.

(iii) It holds independently of whether the cause *can* be (re)produced by action.

On these three features of the causal relation I would insist so strongly that, if a critic convinced me that something else which I say about causation is inconsistent with them, I would give up that other thing rather than my insistence upon these features.

For reasons, connected with the above, I gave (*E & U*, p. 70) the following 'semantic characterization' (p. 184–5) of the causal relation: p is a cause relative to q, and q an effect relative to p, if and only if by doing p we could bring about q or by suppressing p we could remove q or prevent it from happening. Whether we *can* do or suppress *p*, is another matter. And so, of course, is also the question whether, when *p* occurs, it is there as a result of action, and whether, when *p* fails to occur, it is absent as a result of being suppressed.

Now then: wherein does the 'subjectivism' of my view of causation reside? The answer is, of course, that the subjective element resides in the link which I think there is between the concepts of cause and action. What makes me think that there is this connection? Briefly, the following observations:

One can distinguish between accidental and nomic (law-like) generalizations. (There are philosophers who wish to deny this, but Tuomela is not one of them.) The differentiating feature is that nomic generalizations support causal counterfactual conditionals, whereas accidental generalizations do not. In order to discriminate between the nomic and the accidental character of a generalization we must therefore be able to test whether or not the generalization supports counterfactuals. To this end experiments, *i.e.*, interferences with what is thought to be the action-independent course of natural events, are needed. Most people, Tuomela among them (p. 190), would agree that interference or manipulation is important as a methodological or practical device for testing nomicity. But many are reluctant to follow me when I maintain that experimentation (manipulation) is *essential*, a requirement inherent in the *concept* of a nomic tie. We are, of course, at liberty to assume *ex hypotheso* that a connection is nomic even when we *cannot* do anything to manipulate the assumed cause-factor involved in the connection. But if we never learn to manipulate it, we cannot test *this* hypothesis, *viz.* the hypothesis that the connection is nomic and not accidental.

(b) There is an interesting discussion in Tuomela's essay (pp. 186–9) of possible "backward" (or retroactive) causation. Tuomela finds the idea "unacceptable" (p. 187). He is inclined to reject, *a priori*, any account of causation which leads "to a 'proof' of retroactive causation" (*ib.*). And he apparently thinks (p. 186) that I have wanted to give such a proof. This, however, is a misunderstanding.

On the question whether there exists "backward causation" or not, I have not wanted to say anything at all. My problem was this:

The causal relation has a characteristic asymmetry or directedness. If, on some particular occasion, *p* causes *q*, then it cannot be the case that, on that same occasion, *q* causes *p*. How is this to be explained? For reasons which I shall not repeat here, I do not think that the asymmetry of the causal relation can be derived from the asymmetry of the temporal before-after relation. If this is accepted, the problem whether the direction of causation could be opposed to the direction of time, becomes intriguing. What I wanted to show in *E & U* was that retroactive causation is *thinkable* (logically possible) and, *not* that it *exists*. To this end I made a *Gedankenexperiment* involving a 'hypothesis' about the relation between brain events and muscular movements. This hypothesis may be completely erroneous. But that is irrelevant, unless the hypothesis also happens to be *logically* false.

I am still of the opinion that the asymmetry of the causal relation is not a simple matter of 'before' and 'after'. But I am more inclined now than I was then to look for conceptual reasons *against* (the possibility of) retroactive causation. These reasons would, I think, have to do with the conceptual nature of *time* rather than with the conceptual nature of causation.

If retroactive causation existed, past time would have 'gaps' in it, be 'open' in the sense in which we think that future time is 'open', *i.e.*, speaking in topological terms, a branching and not a linear ordering of events. Is *this* 'thinkable'? All I can say is that I find the idea hard to accept, but that I cannot think of arguments to refute it. And I do not wish to label it unacceptable "*a priori*".

If a later event could cause an earlier one to happen, it must be possible to *conceive* of an experiment to test this. The test must be performed in a situation when the *earlier* event has not already *taken place*. For, the purpose of the test is to see whether it will take place – albeit before a certain later event occurs which we are going to produce and which we think is its cause. For this reason retroactive causation, if possible at all, could be efficacious only within the 'span' of an experimental situation. This excludes such absurdities as those mentioned by Tuomela (p. 187) about effecting the cooling of the earth billions of years ago by raising one's arm now.

(c) Another special point about causation which Tuomela discusses (p. 190) concerns the limitations of manipulative causation (in natural science). Malaria may seem to be caused by a mosquito-bite, but the 'real' cause is a certain amoeba transmitted by the insect and a still 'more real' cause perhaps 'some event or process in the organism' which the amoeba activates. From the acceptance of such 'underlying' causes it follows, Tuomela says (p. 190), "that the idea of potentially manipulable causes is not very helpful even for epistemological and methodological purposes".

I find this obscure. If we want to show that it is the virus and not the bite which causes malaria, we must find means of injecting the virus without the aid of the mosquito. This is thinkable. And if we want to show that the virus does not cause the fever directly, but that the fever is caused by some internal process which the virus activates, then we should have to think of other ways of activating those processes. The 'theoretical' character of the "unobservable and nonmanipulable 'underlying' causal entities" (p. 190) does not make them immune to experimental test. If it did, medicine would not be an empirical science.

(d) Tuomela also charges my account of causation with circularity (pp. 191–2). The circularity is alleged to consist in my reliance, when trying to explain nomicity, "on the *causal* notion of bringing about, which is left without sufficient clarification". I find it hard to follow Tuomela's argument. But I seem to detect in it an idea to which I must take exception. It is the idea of a "real causal force gluing p and q together" (p. 191). I do not believe in the existence of a "causal glue" (*ib.*) *over and above* the fact that, if p causes q, then, if p had been when in fact it was not, q would have been too. The 'causal glue' is the *truth* of the counterfactual statement, one could perhaps say. And it is in the idea of 'counterfactual truth' that, in my view, the concepts of cause and action meet.

My attitude to Alasdair MacIntyre's paper is somewhat divided. The essay contains a greater number of 'substantive' remarks on historical explanation than is usual in the writings of philosophers about history. With those remarks I find myself on the whole in agreement. Similarly, I can agree with many of the *critical* comments which MacIntyre makes, particularly in the beginning of his paper, on the present state of the philosophy of social science. But with a good many of his more 'theoretical' remarks on causality in general and of his own alternative suggestions for a sound methodology of history and social science, I must disagree.

(a) MacIntyre is critical of those who regard with suspicion the applicability of causal notions to the study of history and society. There can be no doubt, according to him (p. 138), that the social sciences yield "genuine causal knowledge". It is also clear to him that it is quite implausible to give "two different accounts of causality, one for nature and one for society" (*ib.*). In MacIntyre's view, the "genuine causal knowledge" which the study of history and society yields is of the same kind as the knowledge we have of causes operating in nature.

This, as it stands, is in good accord with a current 'positivist' or 'monistic' attitude in the methodology of the human sciences. But in one respect MacIntyre is strongly critical of this attitude. Causal knowledge, in the tradition of Hume and Mill, is knowledge of generalizations or of law-like uniformities in the sequence of phenomena. MacIntyre is anxious to detach (p. 144) "our knowledge of particular causes from our knowledge of generalizations". Causal relations are between *particulars* and can, often at least, be identified as such "without invoking law-like generalizations" (p. 144).

It is not clear to me how MacIntyre's position should be understood. One can take the view that the causal relation is, primarily, a relation between particulars – and yet think that an explication of the statement that the relation holds must make reference to a nomic or lawlike connection "in the background". This would answer to the positions of Tuomela (above p. 380) and Davidson and also to the position taken by Morton White in his book *Foundations of Historical Knowledge*. But MacIntyre wants to reject this position. "We do not know a priori", he says (p. 144), "that for every particular causal connection which we identify there is some law-like generalization awaiting to be advanced". As far as I can see, he thinks that *sometimes* there is an 'associated' generalization, sometimes not. This I think is correct. But after this point of agreement, MacIntyre and I part ways. MacIntyre is not, it seems, willing to make the fact that some causal connections can be associated with a general law, but others not, a ground for a dualistic (or pluralistic) account of causation – as I would make it. Furthermore, he does not think that the existence or non-existence of a 'nomic foundation' for a causal relation has anything to do with the question whether the relation subsists between events in nature or in the human world (p. 144).

"From the fact that a particular cause produced a particular effect nothing whatsoever follows about how in general that effect can be produced" (p. 147). As it stands, this is an exaggeration. If in a well-defined and well-controlled experimental situation we think that it was the occurrence of *c* which made *e* happen, then we are also ready to generalize and say that *ceteris paribus c* will (always) cause *e*. Were we not willing to generalize, we should not think that the particular occurrence of *c* caused that particular occurrence of *e*. But in situations of a different type we may be quite right in thinking that this particular cause produced that particular effect without wishing to commit ourselves to any general statement at all about how that effect can be produced. Situations, in which we look for the causes of actions, are, I should say, normally of this second type.

MacIntyre also says (p. 144) that the "detachment of our knowledge of particular causes from our knowledge of generalizations is of crucial importance in those areas such as history and the social sciences where many particular causal connections are open to view, but few, if any, genuine law-like generalizations". With this I completely agree. I would perhaps go a step further than MacIntyre and say that causal connections between the results of individual and collective human actions are never nomic or 'Humean'. This is not to say that 'Humean' causation plays no role in human contexts. Actions have causal consequences, connected with the results of the actions by genuine, and usually easily recognizable, nomic ties. For example: when a bomb, thrown at his carriage, kills the emperor. And the effects of causes in nature, for example the destruction of a town by an earthquake, create new situations to which people react (*not*, however, in general like Pavlov-dogs) by doing things they would not otherwise have contemplated.

So, everything considered, MacIntyre has not convinced me that a dualist (or pluralist) account of causation is not needed in order to come to grips with this troublesome but important notion. He has said nothing which would make me think that the *distinction* between particular causal connections with and without a nomic backing is not of the greatest importance; and he seems to agree with me that non-nomic causation is particularly prominent in history and social science. – I am somewhat surprised at the vehemence with which MacIntyre conducts his polemics against what he thinks are views contrary to his own.

(b) Another criticism which MacIntyre levels against the Hume-Mill tradition in the theory of causation is that "causality on this view is taken to be essentially a dyadic relation" (p. 142). A causal explanation, he says

(p. 147–8), needs at least four related terms, *viz.* the intervening cause, the state with which the cause interferes, the effect of the intervention, and "the outcome that would have prevailed but for the intervention". These observations about the four factors are pertinent – but I do not think they constitute a very forceful criticism of traditional accounts of the causal relation.

It is not quite clear to me what MacIntyre means by the second factor, or the "state of affairs which is interfered with". A cause is normally a *change*, say from a state $\sim p$ to a state p. Thus the description of the change (cause) embraces both the intervention and the state interfered with. This is trivial. But MacIntyre might also have had something different, non-trivial, in mind: circumstances which are logically independent of the intervention but relevant to its causal efficacy. He might have wished to say that a statement to the effect that c causes e is *elliptic* and short for the statement that c causes e under such and such circumstances. The relativity of causal relations to a 'frame of circumstances' has not remained unnoticed. But it has seldom been duly emphasized or its character more closely scrutinized. If this relativity is *essential* to every causal relation, or to every causal relation associated with a generalization, then it would indeed be right to say that the causal relation is (at least) *triadic* and not dyadic.

The fourth factor mentioned by MacIntyre is not, in my opinion, to count as a fourth term of the causal relation, nor is it always involved in a causal explanation. An effect can be 'over-determined'. Then the outcome that would have prevailed but for the intervention is the same as the one which prevails after intervention has taken place. When the effect is *not* overdetermined, the outcome in the case of non-intervention would simply have been the contradictory of the effected outcome. MacIntyre's fourth factor is thus an 'indicator', whether the effect is, or is not, overdetermined. To pay attention to it may be important. But I do not think it ought to be regarded as a term in the causal relation. For the 'value' of this term does not affect the question whether the causal relation, in a particular case, subsists or not.[3]

"A cause is what makes a difference", MacIntyre writes (p. 149). Yes – *usually* a cause makes a difference. But, unless we are determinists, I think we must agree that, sometimes, differences occur (appear) without a cause. And, sometimes, the absence of a cause which was there would not have made any difference to what actually happened, because there was also another cause operating with the same effect.

(c) An important role in Professor MacIntyre's account of causation is played by the notion of a *causal order*. By a causal order he means (p. 150) "an interrelated set of items, such as a planetary system, or a crystalline structure, or some type of educational system, where the relations between the items can only be formulated in terms of some type of generalization". The generalizations specify "a bond between the items of a certain type" (*ib.*). MacIntyre adds: "it is important that bonds themselves function as causes" (*ib.*).

I find this rather obscure. It is not clear to me whether the order is a concrete state of affairs or a system of relations. The three examples mentioned in the quotation are, moreover, a rather mixed bunch. A planetary system is 'held together' or 'governed' by generalizations which have the character of natural laws ('nomic connections'). An educational system is, I should have thought, governed by norms or rules. To call the rules 'generalizations' seems to me very misleading. But if the rules (norms) are efficacious, it may also be possible to make some generalizations about the conduct of people 'governed' by them. (MacIntyre's example on p. 152 "Social class determines educational opportunity" is *perhaps* a generalization which is in this way 'backed' by a system of social rules.)

MacIntyre also speaks of the 'strength or fragility" (p. 152) of a causal bond. I find it difficult to reconcile this with the talk of the bond (itself) as a cause. Causes are not more or less strong. But the scope of circumstances under which a causal relation subsists can be more or less restricted. The more restricted it is, the greater is the number of circumstantial conditions which have to be satisfied if the relation is to hold – and the more 'vulnerable', on the whole, does the relation become to interfering and counteracting factors. I am not sure, however, that this observation will apply to all the examples of interference with a causal order which MacIntyre has in mind.

Considering what I should like to think of as a far-reaching agreement between us on matters of 'substance', I regret that I have to express so much disagreement with Professor MacIntyre on matters of conceptual analysis. My comments may seem pedantic. But I think stronger requirements of precision must be met, if MacIntyre is to be able to defend his position on the nature of causality against those who profess allegiance to the tradition which he attacks. With his attack, as such, I am to a great extent in sympathy.

III. FORMAL THEORY OF ACTION

Kenny's paper is a valuable contribution to the clearing of an obscure corner of modal logic, *viz.* the logic of what Peter Geach originally named *dynamic* modalities. This, one could also say, is the logic of the word 'can' when it refers to an *ability* of doing things ("can speak German", "can read", "can open this lock", *etc.*). I introduced the topic as a possible field of logical study in my *Essay in Modal Logic.* But neither I, nor anybody else, has since done much to develop it. Later on, beginning with *Norm and Action,* I embarked on the venture of developing a Logic of Action. It is natural to take it as a starting point for a logical study of abilities, too. This is, implicitly perhaps rather than explicitly, what Kenny is doing in his paper.

Kenny sets himself the task of investigating (p. 209), "whether the dynamic modality of ability is subject to the same formal rules as the alethic modalities" formalized in such systems as my System *M* or Lewis's *S*4 (my *M′*). He arrives at a negative answer. Indeed, he wants to argue that "dynamic modality is not a modality" (p. 209), *i.e.*, that the logic of the 'can' of ability is not a modal logic. It is too unlike any known system of modal logic, Kenny thinks, to be subsumed under that heading. The paper ends (p. 231f.) on a somewhat pessimistic note, saying that "we are still as far from a satisfactory formalisation of the 'can' of ability" as we were a quarter of a century ago.

I should like to contest Kenny's pessimism – or at least to mitigate it to some extent. But let me first note some points where I think Kenny is right. Obviously he is right in thinking that the logic of ability cannot be *S*4-like (pp. 212–6). A potential ('acquirable') ability is not yet (a possessed) ability. I also have sympathy with Kenny's position (pp. 213–4) on the more controversial question whether the *ab esse ad posse*-principle holds for the dynamic modalities. The fact that some pope on some occasion delivered a gracious speech in English, does not show that he knew English. Nor does occasional success with a complicated feat, say jumping a fence, show that one ('really') *can do* the thing. Yet on that occasion one *could* do it. It is perhaps useful to make a distinction here between two concepts 'can do' – the 'can do' of *success* and the 'can do' of *ability*.[4] The former obeys, but the second does not obey, the *ab esse ad posse*-principle. Kenny is concerned with the second 'can do'.

The chief reason why Kenny thinks that the logic of ability is not even a modal logic (in any standard sense of the term) has to do with the law of disjunctive distributivity of the possibility operator.

Let us consider Kenny's counterexample – the picking of a card from a pack. Any card I pick will be either a red or a black one. Thus there is a sense in which it is uncontroversially true that, if I can pick a card, then I can pick a card which is either red or black. But, as Kenny rightly observes (p. 215), it does not follow that either I can (bring it about that I am) pick(ing) a red or I can (bring it about that I am) pick(ing) a black card. Does this mean that the 'can' of ability is not disjunctively distributive?

Contrary to Kenny's opinion (p. 229), I think that the action logic presented in my *Rechtstheorie* (1973) paper will be of decisive help here. In order to see this, let us first discard the rather artificial 'bring it about that I do'-terminology used by Kenny and shift to the 'do'-terminology of my paper. To bring about a state of affairs, say that a door is open, is a relatively clear notion and useful in some logical contexts. To bring about an action (that one does something) is a much less clear idea and one which can hardly be used for the formalization of action logic. Let us then consider which the *actions* are in Kenny's example. There is first of all the action of picking a card from a pack. That this is an action is beyond doubt. I can describe how it is performed; I could teach somebody else to perform it; I, normally, know what I am going to do when I set myself to do it. Is there also an action of picking a red card? The answer is not obvious. But even if, as a matter of actual fact, there is no such action, we could, I think, imagine that one existed. We could, *i.e.*, imagine an agent endowed with a peculiar sensitivity in his fingertips who can discriminate between the action of just picking a card and picking a *red* card. And the same for the action of drawing a black card. So let us assume that these two (remarkable) feats were actions. Then there would also exist a 'disjunctive action' which consists in picking a red or picking a black card. When would an agent be performing *it*? I think the answer is: he performs the disjunctive action when he does not omit the omission of both disjuncts, *i.e.*, performs at least one of the two actions, both of which he *can* do.

The fact that any card is either red or black and that therefore, if a card has been picked, a red or black card has been picked, is totally irrelevant to the question, whether there even *is* a disjunctive performance "picking a red or picking a black card". When used to describe an action, the locution "he picked a card which was red or black" sounds slightly bewildering since, as a description of action, it means just the same as "he picked a card". So why add "which was red or black"? Perhaps because we wanted to instruct somebody who was ignorant about the colours of cards.

I think we must agree that the *action* of picking a card is distinct from the disjunctive mode of behaviour of picking a red card or picking a black card, although the *state of affairs* that a card has been picked is the same as the state of affairs that a red or black card has been picked. (Note that equivalence of descriptions of resulting states does not warrant intersubstitutability in the action logic described in the *Rechtstheorie* paper.) The question now is: Shall we say that if a person can perform a disjunctive action, he can also perform at least one of the action-disjuncts? I think that in any reasonable interpretation of what a disjunctive action is, we must say this – indeed we must say that he can perform *all* the action-disjuncts, since the very 'point' about the 'disjunctive action' is that it involves a choice between alternatives.

I shall not here propose a system of a Logic of Ability. But I am much more optimistic about the prospects of this logic than Kenny seems to be. I shall only advance the following observation which is also relevant to the discussion of the disjunctive distribution principle:

I do not think that the notion of 'can do' involves a superposition of operators, one for 'can' and another for 'do' – as suggested by Kenny (pp. 227–8). The 'can' operates directly on the verb. Let us introduce the symbol $C_x p$ to mean "x can p". 'p' stands, for example, for 'read'. The negation-sign in front of an action-verb means omission. So, if 'p' stands for 'read', $C_x{\sim}p$ means "x can omit reading".

A fundamental law of ability-logic seems to be this

$$C_x p \leftrightarrow C_x{\sim}p.$$

Ability to do and to omit are reciprocal. (Cf. Kenny, p. 228.) This, be it observed, is not in conflict with the possibility that, on an individual occasion, an agent cannot exercise an ability and therefore may be forced to omit an action – or forced to do it and thus 'unable' to omit it. Prevention and compulsion do not annihilate the ability, only the exercise of it on a particular occasion.

Another law of ability, I would suggest, is this: $C_x(p \,\&\, q) \rightarrow C_x p \,\&\, C_x q$. If one can do two things conjunctively (on the same occasion), one can do each one of them. But not necessarily the other way round.

To perform a disjunctive action is to omit the omission of all its parts (disjuncts). Hence $C_x(p \vee q)$ is the same as $C_x{\sim}({\sim}p \,\&\, {\sim}q)$. By the principle *eadem est potentia oppositorum* the last is equivalent with $C_x({\sim}p \,\&\, {\sim}q)$ which entails $C_x{\sim}p$ and $C_x{\sim}q$, which are equivalent with $C_x p$ and $C_x q$ respectively. Thus, from the ability to perform a disjunctive action follows the ability to perform its component parts. The converse cannot be proved on the basis of the above principles alone. (Perhaps the converse is not even true.)

The principles mentioned above certainly do not suffice as a basis for a Logic of Ability. I shall not stop to consider here what else may be required. But I hope to have shown that the task is not hopeless and I therefore conclude on a more optimistic note than Kenny about the possibility of creating a Logic of Ability. But he is perfectly right in thinking that this logic will *not* be a modal logic, if thereby one means that the 'can' of ability obeyed laws analogous to those of the 'may' of possibility. That this should be so is, I think, somewhat unexpected and therefore also interesting.

Another key-notion, beside ability, in the logical study of action-concepts, is the notion of an *opportunity* (for action). It too answers to a species of 'can'. The phrase "this can now be done" often means that there is now an opportunity for doing this thing. (But sometimes it means that the agents concerned have acquired a new ability).

On the notion of opportunity Kenny also has interesting things to say (pp. 216–23). He makes some critical remarks about my use, in *An Essay in Deontic Logic*, of the modal possibility-operator to cover both ability and opportunity. This criticism is, I think, well founded. To remedy the defects would require a rather different and more complex logical frame for a theory of action than the one employed in the *Essay*. I shall not discuss the problem here.

Kenny seems to take a more optimistic view of the possibilities of accommodating the 'can' of opportunity within the frame of a modal logic than of accommodating the 'can' of ability (pp. 221–3). He *may* be right, but I should prefer to suspend judgement on the issue. I shall conclude my comments on his essay with the following observation, which I think relevant to any attempt to construct an adequate logic for action-opportunities:

Opportunities for action are definitely of two main types. First there are what may be called *logical* action opportunities. An action which effects or prevents a change in the world can, for reasons of logic, be performed only provided that a certain other change, 'counterfactually' associated with the first, would 'otherwise', *i.e.*, if action did not take place, happen or, as the case may be, not happen. For example: I can open a door only provided it is closed and does not, on that occasion, open 'of itself', *i.e.*, independently of *my* action. Or: I can prevent a door from opening only provided it is closed and would, on that occasion, otherwise open. – A fully developed action-logic will have to notice this interrelation of changes and not-changes which constitute the logical opportunities for any action. How this is to be done is not a simple question to answer.

Sometimes the logical opportunity is there and the action is within my ability (it is the sort of thing I, in the generic sense, *can do*) but nevertheless I cannot do it, because something *prevents* me from 'seizing the opportunity'. I am, say, busy doing something else which keeps me engaged, or somebody interferes preventively with my action. The rich man and the poor may have the same logical opportunities of doing things. But the poor man has to toil for his and his family's maintenance and he can therefore avail himself only of a few of these opportunities. Not all the logical opportunities, therefore, are 'real' opportunities. This notion of 'real opportunity' is of great importance in social philosophy. It seems that Kenny chiefly had it in mind when discussing opportunities. "I have an opportunity to ø" he says (p. 220), "iff nothing prevents me from ø'ing, *i.e.*, if nothing compels me not to ø".

The logic of this second notion of an opportunity could also be termed a logic of preventing and compelling. It constitutes another sub-province of action-theory awaiting penetration by philosophical logicians.

IV. ACTION AND INTENTIONALITY. THE PRACTICAL SYLLOGISM

I am in sympathy with the general tenor of Hertzberg's paper and I can agree with a good many of the things which he says about deciding and deliberation and intentionality. He stages his own position against the background of what he calls (p. 233) "a long-standing tradition in philosophy" concerning the relation between decision and rational calculation. He classifies some of my views in *E & U* as belonging to, or continuing, this tradition.

I think the tradition in question is not a straw-construction – although I am not sure whether Hertzberg's identification of some of its leading representatives (p. 246) is beyond dispute. But I doubt whether I should be placed in it, – even "in an indirect way" (p. 233). My reason for doubting this is that I never put forward any theory of deciding at all. There are scattered observations on choice and decision in various writings of mine, but I have nowhere dealt with these concepts at length. (Perhaps I ought to have done so and perhaps in future I shall.) Nor can I see that anything I have said about some other topics, for example intention and explanation of action, would commit me to the view about deciding which Hertzberg criticizes – or to any other particular view. I therefore think it slightly misleading when Hertzberg assimilates my account to what some other philosophers have said about "what it is like to make a decision" (p. 238) or when he says that "On von Wright's view, reaching a decision must be similar to making a discovery: the correct answer is already determined before we find it" (*ib.*).

I shall not maintain that these misunderstandings have arisen without any fault of mine. One of their roots may be an overemphasis, of which I acknowledge myself guilty, of the importance of the practical inference schema as a model for the explanation of action. Far from all human actions flow from the peculiar combination of an intention and a cognitive attitude which the schema embodies; though it seems to me undeniable that a good many actions do this and also that the role of the schema in action explanation is, somehow, *pivotal.*

Another root of the misunderstanding may have been an undue emphasis on my part on *inference* rather than on *explanation.* The term 'inference' suggests that the agent whose action is being considered, goes through a process of calculation or of deliberation before he acts on his intention and understanding of 'the requirements of the situation'. Sometimes such 'inference' takes place, perhaps more often not. But when we – the agent himself or an outsider – look at the action in retrospect, as a *fait accompli*, we can offer an 'inference' in explanation or in justification of the deed. – I rather regret that I made 'practical inference' or 'practical syllogism' a key term in my writings on action. I also regret the term 'teleological explanation', as used by me. The best label for the thing under discussion seems to me now to be *intentionalist explanation*. (See above pp. 373–4)

Hertzberg's paper, being an essay on deciding, has thus only a loose connection with my theory of 'practical inference'. But the paper is of interest in itself, and I feel called upon to mention here some points on which I agree with Hertzberg and some where I think we disagree.

Notwithstanding its title ('On Deciding') the best things Hertzberg has to say in the paper concern the notion of *indecision*. I think I agree both with what he writes about the 'presuppositions' of indecision, *viz*. that we normally neither deliberate nor hesitate about our actions, and with what he says about the 'resolution' of cases of indecision, for example about the way a decision reached after indecision may reflect a man's character rather than his skill at calculating the optimal means to a given end.

I see, however, the relation between indecision and decision somewhat differently from Hertzberg. Decision, I should say, is an aspect of the process which might be called *forming an intention*. (A 'theory' of deciding is therefore a part of a 'general theory' of intention-formation.) If I have decided to do a certain thing, I intend to do it. But not necessarily the other way round: not everything I intend to do have I also decided to do. To give an account of decisions is therefore to give an account of *a way* in which intentions originate. Of this way *deliberation* is characteristic. Therefore I would not wish to question the "long-standing tradition" according to which "we arrive at decisions though a process of reasoning called deliberation" (p. 233). I think it would have been helpful if Hertzberg had made a distinction here between *deciding* and *choosing*. Choosing something does not logically presuppose that we have deliberated – although it is compatible with deliberation.

There is one more point on which I must take issue with Hertzberg. It concerns his comments on the notion of 'practical necessity', particularly in the 'caviar-example' which he discusses on pp. 240–1 of his essay. Here I have several counter-comments to make. First, that it is important to keep the distinction clear between *intending* and *wanting*. The fact that I want some caviar may lead me to consider possible ways of getting it.

These considerations may either terminate in an intention (decision) to get some caviar or they may terminate in no such intention, *e.g.*, for the reason that on calling the shop by telephone I hear that there is no caviar available or that the price is exorbitantly high. But if my longing for caviar leads to an intention to get some, then I am also thereby bound (committed) to do what the situation requires of me to try to get it. And what the situation requires depends partly upon the nature of the intention, partly upon the particularities of the situation, and partly upon what I *can do*. If I intend 'at all costs' to get some caviar – I am, say, crazy about the stuff – then I must also be prepared to steal it, if I cannot get it by any other means. If, again, I am sitting on the banks of the Volga and have access to a boat and fishing equipment, I may consider trying to catch a sturgeon. Should that be the *only* way of getting caviar in those circumstances, I *must* go fishing. If I am a millionaire, I can afford paying a price which most people simply cannot afford. What is practically necessary thus depends upon the concrete character of the individual case. And what is necessary is usually not that just *one* definite action be performed, but that one out of several is chosen. The practical necessity is then a *disjunctive* action – to employ the technical terminology which I have found useful in the 'Logic of Action'. (See my reply to Kenny above p. 390.) My intention and understanding of the situation commits me to the disjunctive action and explains my doing of it *completely*, and my choice of the particular disjunct only *incompletely*. That the explanation is incomplete means that, in principle, there is room for the residual question "Why did you do it, *e.g.*, get yourself some caviar, *in this very way*?" And this question may have an answer – for example, that this was the way I thought easiest, or the way people normally get themselves caviar. Or it may have no answer. Not all our actions have reasons and not all of them can be completely explained.

I now proceed to Professor Kim's essay. – What gives explanatory force to the practical inference schema (*PI*-schema)? In the view for which I have been arguing, the relation between the "volition-cognition complex", to use Kim's term (p. 250), and the resultant action is conceptual or even, I have sometimes thought, deductive. In the view to which I am opposed, the 'causalist' view, the relation is one of cause to effect. A causal relation may be held to be explanatory *per se*. If the relation between volition and cognition on the one hand and action on the other hand is not causal, how can it *then* be explanatory? Kim suggests (p. 250) that I might have thought that the deductive character of the relation made it explanatory. But this I did not do. As Kim quite rightly points out "the presence of a logical relation between two events is not sufficient in itself to yield an explanatory relation between them" (p. 253). I never doubted this. And I absolutely agree with Kim that the question whether the *PI*-schema is logically conclusive, or not, is "wholly beside the point as far as the explanatory efficacy of *PI* is concerned" (p. 253).

So, how is it possible for the *PI*-schema to have explanatory force? I don't think anybody would seriously deny that it *has* – although one may wonder with Rex Martin in his essay (see above pp. 311–3 and comments below pp. 411–2) whether some 'filling in' of the schema is perhaps needed to make it 'fully' explanatory. Kim himself hints at what I think is the right answer when he suggest, but rejects as unsatisfactory, that we are here facing "a primitive fundamental fact about the way our concepts of explanation, action, and the like work" (p. 251). If *A* *x*'ed, then the fact that he intended to *y* and thought *x*'ing necessary for this, *explains* why he *x*'ed. It explains this quite independently of the further question whether the relation of the action to the volition-cognition complex is logical or causal. I don't think one can tell *why* the *PI*-schema explains action, anymore than one can tell *why* giving the cause is to explain the effect.

Sturgeon's counterexample. At a meeting of the Cornell Philosophy Club where I was reading a paper on action explanation, Nick Sturgeon put forward what looks like a counter-example to the thesis that a combination of intentions and cognitions *ipso fact* explains action. Indeed, if Sturgeon's argument is valid, then what I have just said about the inherent explanatory force of the *PI*-schema can not be true. I think the following is a fair way of paraphrasing the alleged counterexample: I intend (have decided) to go to Cincinnati to read a paper. Let us call this my *primary* intention. I plan (intend), when in C., to look up my in-laws there. This we shall call my *secondary* intention. Unless I go to C., I cannot realize either intention. Yet the fact that I intend to look up my in-laws, in combination with the clear cognition that, unless I go to C., I cannot do what I intend, certainly does *not* explain my going (setting myself to go) to C.

Why does the volition-cognition complex not explain the action here? Why does reference to the complex not answer the question, *why* I go to Cincinnati? This must be so because of the 'secondary' character of the intention. The *reason why* I go to C., is *not* that I intend to see my in-laws. But how shall we make it clear that the secondary intention does not count as a premiss of a practical inference schema?

I used to think that it was relatively easy to dispose of Sturgeon's counterexample. But subsequent thinking and now Kim's interesting and thorough discussion of it has convinced me that it must be taken seriously. I don't think the example invalidates the idea that intentions *per se* stand in an explanatory relation to actions. But it offers a starting point for further developments in the theory of intentionality.

Kim's paper is a contribution to this further development. I shall not comment on it directly. Instead I propose to do the following: first I shall mention two points on which I disagree with Kim; then I shall make some independent comments on Sturgeon's example. They are not meant to be a rival alternative to Kim's theory of the "genealogy of intentions". They do not, as far as I can see, contradict anything he said – but they *may* offer a simpler way than his out of the difficulties.

(a) My first criticism concerns Kim's restatement of the counterexample in schema (2) on p. 252 of his paper. It goes as follows:

> "I intend to visit my in-laws in Cincinnati.
> I believe that I cannot visit my in-laws in Cincinnati unless I go to Cincinnati.
> Therefore, I set myself to go to Cincinnati".

"This is Sturgeon's counterexample", Kim says. Surely it is not. For, if my intention can be correctly described as an intention to visit my in-laws in C., then, if I understand that I cannot do this without going to C., I am bound to go to C., – and the premisses *explain post hoc*, why I went there. But this is not a correct description of my intention as it occurs in Sturgeon's example. My intention is not just to visit my in-laws in C. It is to visit them *when I am in C.*, (which I intend to be). If the first premiss in (2) is correctly formulated, the conclusion does not follow. But when it is formulated as by Kim in (2) *the conclusion follows*.

I think this criticism must be accepted. But I am not sure that Kim's less happy formulation of the example substantially affects his subsequent discussion of it and attempted solution of the difficulty.

(b) My second criticism may have more serious repercussions on Kim's theory. It concerns his schema, Pattern A (pp. 258–62) for the generation of intentions. I have myself entertained similar ideas when I spoke[5] of the *transmission of intention* from a 'primary' to a 'secondary' intention in the *PI*-schema (Phase 1 in Kim's modification of the schema). But I can no longer adhere to my previous views which were inspired by a famous passage to which Kim also makes reference (p. 268) in Kant's *Metaphysik der Sitten* about the 'transfer' of the will from ends to means.

It simply is not *true* that, if I intend to do a certain thing and realize that in order to do it I must (first) do another thing, then I intend to do that other thing too. *Sometimes* this may be so, *e.g.*, in a situation in which I deliberate about my actions and come to think that *this* is what I must do. But generally no such new intention is generated by the original one. What is true, however, is that if I do something which I think is a means to an intended end, then I do this other thing *intentionally*. My doing this other thing, so to say ,'gets' its intentionality from my intention to do the first thing.

I think these observations on 'transfer of intentionality' are important and I am sure that some modifications, perhaps even big ones, are needed to make Kim's theory of intention-generation acceptable.

It seems to me, however, that one can deal with Sturgeon's case without invoking a theory of intention-generation at all. This way of coping with the difficulty makes a more *direct* use of the notion of a *conditional intention* (p. 260) than is the case in Kim's treatment.

An intention could also be called a (self-)*commitment* to action. As long as I intend to *x*, then I am committed to doing whatever is required of me if I am to *x*. The variable here stands for a verb-phrase denoting an action – for example "visit my in-laws in C.".

If the action is deferred, then I am committed *not* to do anything intentionally in the meantime which I think will prevent me from putting the original intention into effect.

Now replace the variable in "I intend to *x*" by the phrase "visit my in-laws in C., if I go to C.". In the normal way of understanding the sentence, it is a declaration or intention. I have this intention *now*, but the carrying of it into effect is conditioned by or *contingent* upon the coming true of something not now (not yet) true. I shall call this a *conditional intention.*

A conditional intention must not be confused with an intention, the execution of which is deferred. If I say that I intend to go to my country-place over the next week-end, I am not committed to doing anything right now which would count as a means of reaching the destination. But eventually I must do this, *independently* of what happens – unless I change my mind (abandon the intention).

A conditional intention likewise does not commit me to anything right now but whether it will commit me to anything eventually *depends* upon what happens. In particular, it does not commit me to trying to make true the thing upon the coming true of which my intention is contingent. Intending to see my in-laws, should I go to C., in no way commits me to going to C..

But I may already be committed to this or for some other reason believe that I shall be in C.. Then I can say, replacing 'if' by 'when', that I intend to see my in-laws when I am in C.. I am sure there *will be* a commitment, although I am not yet committed. The commitment is still only conditional. In order to believe that I shall be in C., moreover, I need not be committed to going to C.. Perhaps it is not at all the case that I intend to go there, but that I shall be deported there much 'against my will'.

The 'when'-case, too, is different from the case of deferred action. If I intend to go to my country-place for the coming week-end, I am committed not to do anything which I think will prevent me from spending my week-end there – for example not to set out on a journey by train to Peking in the meantime. But my intention to see my in-laws, when I am in C., does *not* commit me not to do anything which will prevent me from going to C.. If my going to C. is something which will happen independently of what I myself want and wish, I may indeed try to do something which will prevent my going there. This will not affect my conditional intention to see my in-laws.

I think the distinction between an unconditional and a conditional intention can be made in a clear and unambiguous way. Then all that needs to be said to show that Sturgeon's example is no counterexample to the explanatory force of the *PI*-schema is to point out that the intention which belongs in the schema is unconditional whereas the secondary intention which figures in the example is a conditional one.

I should like to say here that I think very highly of Professor Stoutland's essay. First of all, its author summarizes the essentials of my position better than I could have done myself. Secondly, he restates the strongest arguments for the causal theory of action so as to do it – for all I can judge – a maximum of justice. Thirdly, he criticizes these arguments in a manner which seems to me fair and convincing. Finally, he modestly leaves the door open: the theory may still be true, although reasons for thinking it true seem to be lacking (p. 300). His concluding reflections are pertinent: the reason why the causal theory is attractive is that it explains the possibility of action. The theory is probably not correct. But to argue for an alternative is extremely difficult. Kant tried to do this with the whole force of his genius – but we must, I think, admit that he failed.

There is one point on which I should like to take issue with Stoutland. It is marginal to the substance of his paper, but concerns a question which is important in itself. The question is how to characterize non-intentional acting. I agree with Stoutland's view, if I understand it correctly, that non-intentional action is secondary to intentional action, the first concept 'parasitic' upon the second. When shall we say that action is non-intentional? According to Stoutland, an agent can be said to have acted non-intentionally on an occasion only if, on that same occasion, he *also* did something intentionally (p. 273 and p. 294). Stoutland's example is a man who inadvertently turns off a light when he intends, say, to ring a bell. Had he inadvertently turned off the light by some movement of his which did not aim at anything at all, he could not, in Stoutland's view, be said to have acted non-intentionally.

This may give us a correct account of what it is to do things *by mistake*. But doing things by mistake is not the only type of non-intentional action. There is also the related category of actions due to *negligence*. I wave my arms with *no* intention and hit a vase which breakes. This, usually, is classified as action and not as reflex. If it could be shown that the movements of my arms had a stimulus outside my control, *then* we should not say that I acted and perhaps talk of reflex movement. But if there is no such stimulus to be pinpointed, we think that the agent *could* have controlled his movements and, in view of the 'importance' of what he brought about (the damage to the vase), *should* have controlled them. And these two facts (could *plus* should) in combination seem to be the reason, why we attribute to the agent an *action*, albeit a non-intentional one. We could reproach him saying: "look what you have done!" And we could perhaps teach him to be cautious and to take care not to do the same on a future occasion. So it seems that one can *act* non-intentionally, even when one is not doing anything intentionally.

What has just been said shows that there are complications connected with the notion of non-intentional acting which Stoutland's account does not meet. But this observation does not make void Stoutland's point "that the concept of intentional action is more basic than the concept of action" (p. 273). The example which I mentioned can, on the contrary, be used for strengthening his point. For, the unfortunate effects of our negligent conduct are called (results of) *actions* precisely because we think they are things genuinely 'in our power' to do, and therefore also to omit doing, *intentionally*.

The second half of Tuomela's essay starts with an account of the practical inference schema and my view of its conceptual character. The account is entirely fair and Tuomela's comments are not at odds with my own opinions (pp. 194–6).

On p. 196 Tuomela says that "the theory of the practical syllogism is concerned with the 'qualitative' philosophical foundations of rational deliberation and decision making". When viewed from this angle, he says, the theory is "somewhat unsatisfactory". With this last I can agree. My theory of the practical syllogism is not a theory of deliberation and decision making at all. I am afraid that Tuomela has misunderstood my position in somewhat the same way as Hertzberg seems to have done. But since two competent people thus misunderstand me, it cannot be without some fault on my part. As stated in my reply to Hertzberg, I think the unfortunate term 'practical inference (syllogism)' has caused unnecessary confusion.

By introducing the notions of deliberation and decision, Tuomela gives to his subsequent discussion of the 'practical syllogism' a slant which leads in a somewhat different direction from my theory. I find what Tuomela has to say interesting and, on the whole, I think I agree with him. A few comments:

On pp. 197–8 Tuomela presents what he calls an "amended version" of my *PI*-schema. The main feature of the amendment is the introduction of a clause about the *rationality* of the deliberating and decision-making agent. I have some doubts about how the rationality clause 'matches' the rest of the schema. I should have understood things better if Tuomela had dropped from the schema the words "from now on" (which occur in my original version of it). If these words were dropped, one could allow for a time-interval between the intention and cognitive attitude on the one hand and the execution of action on the other hand. In this time-interval, factors of an 'irrational' character ("emotional and cognitive disturbances", p. 197) may intervene and disturb the connexion between the initial volitional-cognitive factor and the action. *Then* assumptions about rationality would be needed in order to safeguard the validity of the 'amended version' as a semantic principle of deliberated action.

On p. 198 Tuomela proceeds to consider what happens when (part of) the rationality requirement is dropped. He relates the 'amended version', with and without the rationality requirement, to the linguists' distinction between competence and performance respectively. But again I could here have followed him better if he had first omitted the words "from now on" from the amended version and considered the situation with an 'open' time-gap between intention and action. For *then*, clearly, the dropping of the rationality requirements will leave us, not with a semantic principle, but with some kind of *empirical*, inductive rule or generalization. Tuomela seems in the first place to be considering the case, when this rule has the character of a 'covering law', as in nomological explanations (pp. 200–1). An alternative would be to regard the relation between the volitional-cognitive complex and action as probabilistic. This would give us yet a further 'amended version' of the *PI*-schema. It is this version which Niiniluoto studies in his essay under the heading "inductive-probabilistic explanation of action". (See below pp. 408–11.) I assume that Tuomela would find Niiniluoto's variation of the theme an entirely acceptable supplementation to his own.

V. EXPLANATION

(a) There are several items in Niiniluoto's paper that I should like to discuss. The first I am going to take up may seem trifling, but is, I think, of interest to the very idea of 'explanation'.

Let us here consider only explanations which answer *why*-questions – and ignore those which are concerned with *how*- and *what*-questions.

Niiniluoto presents (p. 339) two 'explanatory arguments' of this kind for discussion. I shall start with his argument (2).

Obviously it is an appropriate answer to the question why a certain man trembles, to say that he has malaria. If someone does not understand the answer, *i.e.*, does not grasp its explanatory power, it would not be helpful to add "and all who have caught malaria fever do tremble". This addition is quite irrelevant to the answer. The answer is explanatory, because malaria is a *cause* of trembling – and since it is a cause, it follows that perhaps not all but a great, great many men who have caught malaria, have begun to tremble. This last fact is also *evidence* for the causal relation.

Now compare Niiniluoto's argument (2) with his argument (1), the raven case. Under which circumstances would the question "Why is this bird black?" be raised. It is hard to imagine. If someone asks why is this bird limping? or why can it not fly? it is easy to see the point of the question. It cannot fly because – (giving a cause of incapacitation). Or it cannot fly until it has – (stating conditions for the capacity). Suppose that I keep a farm on which I breed a white kind of bird. One day I see a black, but otherwise similar bird among them. I am astonished. I want an explanation. I ask, for example, "How did this bird come here?" But if I learn that it was bred in the regular way on the farm, I might be curious about its colour and ask: "Why then is it black?". I should like to know the *cause* of blackness in this particular bird. The answer could be that the bird is an A (say, one of a kind with a rare but possible combination of genes) and that Aness causes blackness. This being so, every A is, or is highly likely to be, B (black) – but this, by itself, carries no explanatory weight. The fact, however, that so many A's in the past were found to be B may be important *evidence* for the causal statement.

With the above I have not wished to say that answers to *why*-questions always point to a cause of something. This is not the case, and here I agree with Niiniluoto (pp. 339–40). What I am questioning is whether (mere) subsumption under a hypothetical general statement *ever* 'explains', whatever the context happens to be when we ask for an explanation and whatever the questioner's state of knowledge or ignorance is. This may seem a bold thing to question, considering how widely spread the contrary opinion appears to be (among philosophers).

There is an age-old idea that to explain something is to subsume this thing under a (general) *law*. *This* idea I am not disputing. On the contrary:

I think that subsumption under laws *has* explanatory power. This holds both for laws of nature and for laws of the state. (It is often an entirely satisfying answer to the question why a person does a certain thing, to answer that the law bids every citizen to do this).

Laws of nature are, or entail, general statements about the concomitance of events and states in nature. But not every true general statement of this sort is *nomic*, is a *law*. Is it, for example, a 'law of nature' that all ravens are black? I do not know what to say – except that it does not 'sound' like a law. But if someone were to explain to me why birds with such and such a genetic 'set up' are bound to be black and that ravens have this set up, then he would have referred to something which we should not hesitate to call a law of nature (genetics). This law is *not that* all ravens are black. It is a law explaining *why* ravens are black.

If these comments on the raven case are correct and if they can be generalized, I think this would be of some interest. It would show that certain ideas about generality and universal regularity do not hold the central place in the theory of scientific explanation which, implicitly at least, recent philosophy of science has accorded to them. It would also add urgency to the task of clarifying the notions of natural necessity (law) and of cause with the aid of conceptual tools which are substantially richer than those of the (lower and higher order) functional calculus of formal logic and of the theory of probability.

(b) From among the many interesting points which Niiniluoto makes about inductive explanation patterns I shall here single out only two for comment. The first concerns the question of the epistemic relativity of these explanations. The second concerns their relation to the *PI*-model for action explanation.

On Hempel's view, inductive-probabilistic explanations which are *complete*, *i.e.*, satisfy what Hempel called the Requirement of Maximal Specificity (RMS, *cf.* Niiniluoto, p. 346), are epistemically relative, *i.e.*, they make reference to a knowledge situation. They *have to be* this, the argument goes, because "a sub-class F_1 of F effecting a statistically relevant partition of F can fail to exist only with respect to our limited knowledge or our ignorance" (p. 346). This again is thought to be a consequence of what is called the Density Principle, which, Niiniluoto says (p. 346), "is obviously satisfied, if the extensions of our predicates are allowed to be chosen among *all* subclasses of the universe of discourse".

The question of epistemic relativity is further related to the question of *determinism*. It is thought that, if we adopt a standard interpretation for second order logic, then there will necessarily exist a probabilistically (statistically) relevant partition of the reference class such that the probability that x is G, given that x belongs to the one or to the other of the classes effected through the partition, is either 0 or 1. As Niiniluoto rightly observes (p. 349), this is in the first place a commitment to extreme probabilities and not *quite* the same as a commitment to determinism. But we may say: *almost* the same.

Well aware that it may sound unorthodox I should myself wish to question the validity of the Density Principle *even* under a standard interpretation of second order logic. For, it is then being presupposed that to every sub-class of the extension of a given property (F) there corresponds a property through which a partition of the class can be effected. This presupposition I find debatable – but I shall not here try to argue against it.

For, I think Niiniluoto is clearly right, when he says (pp. 346–7) that "in a context where laws of nature are at issue, there does not seem to be any justification for operating with standard models". We shall have to impose some restrictions on the class of properties which are used for partitionings of the reference class. Or, to speak with Niiniluoto (p. 347), the reality within which the probabilistic laws 'live' must be (somehow) conceptualized. As far as I can see, the partitionings which may be relevant to inductive-probabilistic explanations are in practice always made on the basis of some empirically identifiable properties of the individuals of the reference class. By this I mean properties for which it can be decided whether an individual has them or not, when this individual, so to speak, 'turns up for examination'. I have in previous writings called such properties *physical properties* of the individuals.[6] The same term is also used by Niiniluoto and by others. It is not very precise and stands in need of further specification. But it may, I think, be safely assumed that the various authors including myself who have employed the term have used it with *roughly* the same intended meaning.

Let us assume then that the inductive-probabilistic explanations are 'conceptually relativized' (p. 347) to some such 'concept-space' as, for example, that of the physical properties. Then the question will arise, whether the Density Principle is valid for this concept-space. In particular, one can ask with Niiniluoto (p. 347), whether the principle is "valid with respect to that non-standard interpretation (conceptual system) which is the 'correct' or 'best' representation of reality"? If the answer is No, then insistence upon maximal specificity can be sustained without any commitment to determinism or to extreme probabilities. For, the notion of maximum specificity itself can then be defined in terms which do not make reference to a knowledge situation.

The answer to the above questions concerning the Density Principle depends, of course, upon further details of the definition of the concept-space in which inductive-probabilistic explanations are assumed to 'live'. How is the class of 'physical properties' delineated? What is the 'best' representation of reality? Given that these questions have a satisfactory answer, I would agree to Niiniluoto's conclusions (pp. 349–50) that there are "objectively random" processes in nature and that the notion of a *complete* inductive-probabilistic explanation can be defined in non-epistemic terms without deterministic commitments.

All this being granted, it seems to me that one can still argue – on grounds which have nothing to do with questions of determinism – for the epistemic relativity of inductive-probabilistic explanations which claim to be complete. Let us assume that the RMS-principle is formulated in the 'ontic' mode, *i.e.*, without reference to a knowledge situation. Then in order to pronounce an explanation complete for a given case, we must believe that the principle is satisfied for this case. But unless we accept an interpretation of probability statements (or of statements of statistical relevance of partitions of a reference class) which makes it possible to verify conclusively such statements, we shall never know whether the principle is satisfied or not, and therefore never know whether the inductive explanation is complete or not. All we can do is to assume, believe, or think that this is so. We may, however, be mistaken – not only because unknown probabilistically (statistically) relevant partitions of the reference class may exist, but also because changes in the *knowledge situation* may cause changes in our opinion about the relevance of partitions, and thereby in our opinion about the completeness of the suggested explanation. For these reasons, inductive probabilistic explanations can be pronounced complete only relative to a mutable state of 'knowledge and ignorance' or better: a mutable state of beliefs, induced by a varying body of inductive evidence.

Deductive-nomological explanations are not *in this sense* 'epistemically relative'. They are complete when the individual case has been subsumed under a covering law. The covering-law *may*, of course, be false. We may come to know its falsehood later on, in which case the suggested explanation collapses, loses its validity. Or we may come to doubt the truth of the covering law, in which case we should no longer rely on it in explanations. So, in *a sense* covering-law explanations, too, are relative to a knowledge situation.

Niiniluoto argues for the indispensability, on ontological grounds, of inductive probabilistic explanations in science. I think he is right in holding that "explanation by means of probabilistic laws constitutes a mode of explanation *sui generis*" (p. 350). But I also think myself that the *sui generis* character of probabilistic explanations (in relation to nomological deductive explanations) can be established independently of questions of ontology and of the determinism-indeterminism issue. And though I sympathize with Niiniluoto's leanings to some form of *indeterminism*, I doubt whether these leanings can gain support from a clarification of the meaning of random distribution and completeness of probabilistic explanations.

(c) In the concluding section (pp. 358–64) of his paper Niiniluoto wants to argue that "the thesis of the dispositional character of explanation of action can be defended, if and only if *inductive* dispositional explanations are taken into consideration" (p. 359).

I do not wish to dispute this. I think it is true – and that Niiniluoto's argumentation is quite successful as it stands. But there are a number of points raised in connection with the argument which I find debatable or in need of further expansion. I shall here single out two for brief discussion.

The first is, in a way, terminological. It concerns the use – in my opinion misuse – of the word 'disposition' to cover a number of conceptually highly disparate things, many of which bear only a remote, if any, relation to what one would in non-philosophic discourse call a 'disposition'. I regret that Niiniluoto uncritically adopts this now fashionable but obscuring jargon.

What then would be an *un*objectionable use of disposition-terms in connection with human action? For example: when we speak of features of a man's temperament. We say "he is choleric and disposed to angry reactions, if something goes against his wishes". This gives a background for a 'dispositional explanation' of certain of his actions. (But it is noteworthy that we often prefer to speak of them as his *re*actions). Another type of example is when we refer to a man's habits. Knowing what his habits are, we predict that he will do something, or we explain something which he did. "O, this is just his habit" we may answer, when someone asks why he behaved in a certain way.

The above remarks are sketchy – but they should help us to see that there are many contexts in which it makes good sense to speak of *dispositional explanations of action* – although I think that even here caution in the use of the term 'disposition(al)' is in place. Niiniluoto hints at some such contexts (pp. 361–3). But if one speaks, as he does (p. 359) of the *dispositional character of the explanation of action* one intimates, or maintains as a thesis, that *all* explanation of action is, in some reasonable sense, 'dispositional'. Against this, however, I must protest.

I think there are, in fact, *several* important patterns or schemas for action explanation which should not be called dispositional – for the reason, among others, that they are interestingly and sharply *different* from types of explanation which uncontroversially can be called by that term. *One* such schema is the one for which I have used the name 'practical syllogism'. I should now much prefer to call it *intentionalist explanation* of action. (See above pp. 373–4.)

In an intentionalist explanation an individual action is seen as something to which an agent becomes committed by his intention and his opinion of how to achieve the object of this intention. "This is what under those circumstances he *had to* do" we say and thereby we explain (understand, make intelligible) why he did it. There is something 'deterministic' about this kind of explanation. It says *nothing* about "dispositions, inclinations, tendencies and propensities to act in a certain way in certain kinds of situation" (p. 359). It is as unlike an inductive-probabilistic explanation as could possibly be the case.

The determinism involved in intentionalist explanations of action must not be confused with the determinism at stake in the free-will controversy. An intentionalist explanation, one could say, illustrates the way in which the *will* (*i.e.*, intention to do this or that) determines our *action*. The question of what, if anything, determines the will, is another matter.

Niiniluoto says (p. 361) that "the situations to which practical syllogism applicable can be viewed as special, or limiting, cases of the range of a general explanatory schema with a covering law" of a certain form. This I would not deny. But I am anxious to stress that it is in the very (conceptual) nature of intentionalist explanations that they are *not* covering law explanations, neither of the deductive-nomological nor of the inductive-statistical type. Intentionalist explanation represents an *opposite point of view* in explanation theory from covering-law explanations. One should try to understand the specific nature, applicability and limitations of both kinds of explanation – and not try to assimilate the one kind to the other.

After having said this in disagreement with Niiniluoto, I should like to conclude my reply on a note of rather far-reaching agreement, as I see it, between us on the question of the relation between the practical syllogism and the inductive-probabilistic schema. The agreement concerns, however, *prediction* rather than *explanation* of action.

Though intentions are anything but dispositions, there is nevertheless a close connection between the two concepts. This connection is, so to speak, mediated by the idea of *deferred action.* For example: When I intend to do something to-morrow, or shortly, or at an indefinite future time, the intended action is 'deferred'. When action is deferred, there is, of course, no logical tie between the having of the intention and the execution of the action. All sorts of things may happen 'in the meantime'. The agent may become prevented from acting. Or he may, as we say, 'change his mind', *i.e.*, give up or modify the original intention. Contingencies of this kind may be subject to probabilistic prognostization based on inductive evidence about past cases. In the case of the second contingency, the 'change of mind', it is natural – or at least not *unnatural* – to speak of dispositions. The agent, we say, is more or less strongly *disposed* to (try to) put his intentions for future action into effect. We say such things as "if he intends to go to see his aunt to-morrow, he is likely to do so – he is a determined and reliable person". Or we may say "if he intends to do this next week, you can be sure he will never do it – he is an immensely capricious and inconsistent fellow". The opinions on which such locutions as these are based, can be used for a probabilistic prediction of behaviour. This I think is their most frequent use. But on occasion they can also be used, retrospectively, for explaining behaviour. We say perhaps "he probably did p with the intention to q, because he said, some time ago, that he was anxious to q and that he thought p'ing necessary for this, and we know from experience that he is strongly disposed not to change his mind, once he has got something in his head".

Such cases as those just indicated provide, I think, genuine examples of inductive-probabilistic explanation and prediction. The explanations can be called dispositional explanations of action. They are not instances of intentionalist explanations – nor are practical syllogisms limiting cases of them. But one could express the relation between the two types of explanation in a nutshell by saying that in dispositional explanations we evaluate the probability that it is correct to understand the action according to the intentionalist schema.

Although Professor Martin is explicitly concerned with criticizing some of my views, I think there is a fundamental agreement between our approaches to the problems of explanation in history and the social sciences. As far as I can see, both of us think the covering law model inappropriate as a master-model for those fields of inquiry. In both fields understanding plays a characteristic role, which distinguishes them from the natural sciences. Perhaps it is right to say that Martin is prepared to go farther than I am in the direction of Collingwood and Dilthey in the philosophy of history, whereas the reverse may be true of our positions with regard to the methodology of the social sciences.

Martin has a criticism against my *PI*-schema which I associate, rightly or wrongly, with his 'Collingwoodianism' and which I have been unable to understand quite well. A case from social anthropology, reported by Kroeber, exemplifies his criticism (pp. 310–2). I take the 'point' of the criticism to be that the *PI*-schema does not, by itself, establish an intelligible or perspicuous connection between the agent's intention and epistemic attitude on the one hand and his action on the other hand (pp. 313–5). It may fail to satisfy our standard of *understanding* (p. 315). Hence Martin's conclusion (p. 324) that "von Wright has failed to *integrate* understanding into his account of the teleological explanations of actions". This, he says (*ib.*) is "the foremost shortcoming" of my work in *E & U*.

I do not wish to rule out the possibility that there may be a serious shortcoming of this sort in my action theory. But I cannot see that Martin has shown this. In particular, I fail to see the force of the example from Kroeber. I shall make some comments on it here:

In order to make the case in question intelligible and perspicuous we want further information about it, a deeper insight into or understanding of the cultural background, one could say. (Cf. the quotation on p. 311.) According to Martin, what we want is not an explanation of the savage's strange belief. Such an explanation belongs in a 'different dimension'. This may be so. But it appears to me that the passage quoted on p. 311, which is supposed to provide the needed 'filling in' of the action-explanation, looks suspiciously like a belief-explanation. However this may be, my own diagnosis of the case would be somewhat different. If confronted with the case (in real life) I think we should ask ourselves: Does the man *really believe* that his treatment of the knife will have this influence on the wound? On what evidence is our interpretation of the facts as a *belief* based? We see him do certain things to the knife, and he is evidently worried about the wound. He *seems* to think that treating the knife as he does will help him get rid of the pain. Perhaps we ask him questions through an interpreter. Words will have to be translated from his language into ours. Some word or phrase the interpreter translates by the English 'belief'. But is it belief in our sense? In order to find out, we should like to know more about other occasions when the same words (in the savage's language) would be used to describe his attitude. We ought in fact to get to know a great deal more about him and the society to which he belongs in order to be sure that he *believes* that treating the knife will help him cure the wound. Getting to know this is precisely what is needed to make the case perspicuous and intelligible. Not until perspicuity is attained can we be sure that the premisses of our practical inference, *i.e.*, our assumptions about the savage's intentions and epistemic attitudes, are correct. But, I would suggest, if we have satisfied ourselves that the premisses are in order, then the explanation works. Therefore I still think it right to say: "Relevant is only what the agent intends and believes." (Cf. Martin p. 310 and *E & U* p. 97).

Thus the 'filling in' on the premisses, making the case perspicuous to our understanding, is really a verification procedure for establishing the intentions and beliefs of an agent. This verification procedure, I venture to suggest, is essentially the same as that which Martin refers to as "narrative understanding". It is the narration which gives us the context within which alone we can see which the premisses of the practical inferences really are. If I am right, this meets at least *part* of Martin's criticism of what he thought was the most serious shortcoming of my book.

As so often happens, one tends to exaggerate one's new insights. This was what happened to me when I first came to take interest in the *PI*-schema. I did not see then, as I think I do now, the existence of other, different explanatory patterns – particularly for explaining actions in a social setting. As Martin rightly points out at the end of his paper (p. 328) I ought to address "directly and with extensive argument", the question whether the schema really is "the principal model for explanations in the social sciences". I think the answer will be that in the social sciences

(a) explanations usually do *not* have the character of intentionalist *PI*-explanations – but that

(b) even here the role of the *PI*-schema is *pivotal* in the sense that the other explanatory mechanisms all seem to revolve round this schema as their core. How this happens I hope to have shown, or at least indicated, in my essay in the present volume 'Determinism and the Study of Man'.

NOTES

[1] On this I have tried to say something in my paper 'Determinism and the Study of Man', included in the present volume.

[2] For a fuller statement of my view of causal laws I must refer the reader to *Causality and Determinism*, Columbia University Press, New York 1974, pt. iii, Sections 3 and 4.

[3] It has sometimes been suggested that, if c and c' both have the generic power of causing e and if c and c' (and e) happen on the same particular occasion, then neither c, nor c' 'caused' e. This is correct if it means that neither c nor c' was *the* cause of e. But it is incorrect, if it means that neither of them was *a* (sufficient) cause of e. If one subscribes to the latter view, which I think is wrong, then it is indeed relevant to the subsistence of the causal relation in a particular case that the effect should not have occurred unless a particular intervention, causing it, took place. But there is no indication in the essay that MacIntyre would subscribe to this view.

[4] Cf. my *Norm and Action*, London 1963, p. 51.

[5] 'On so-called Practical Inference', *Acta Sociologica* **15** (1972), p. 45.

[6] Cf. my paper 'On Probability', *Mind* **49** (1940), p. 277 *seq.*

GEORG HENRIK VON WRIGHT

DETERMINISM AND THE STUDY OF MAN

I

By determinism I shall understand ideas according to which that which *is* also (somehow) *had to be*. Such ideas play an important role both in the natural and in the human sciences. Whatever the ultimate answer may be to the question as to how these two types of sciences are related, the following difference between them is striking on the surface:

In the natural sciences, deterministic ideas are connected with such other ideas as those of universal regularity, repeatability, and experimental control. In the human sciences the immediate connections are with ideas such as motivation and social pressure, goal-directedness and intentionality. In the natural sciences determinism serves in a large measure the forward looking aims of prediction; in the human sciences there is a relatively much stronger emphasis on retrospective explanation, or understanding, of what is already a *fait accompli*.

These differences between naturalistic and humanistic study in relation to determinism I would attribute to the following source – well aware that what I say may sound provocative: Natural science can be characterized as a study of phenomena under the 'reign' of natural law. Human science again is primarily a study of phenomena under the 'reign' of social institutions and rules. I shall argue for what may be termed a 'methodological parallelism' between laws of nature and rules of society. That is, I shall argue that deterministic ideas in the human sciences have a relation to societal rules which is *analogous* to the relation in the natural sciences between deterministic ideas and natural laws. I am not, of course, saying that laws of the state and other social rules are, in themselves, like laws of nature. They are, on the contrary, very different. The former are normative, the latter descriptive, as we say. And from this profound difference between the two types of law it follows that, if the 'methodological parallelism' I am making is at all correct, determinism in the study of man means something utterly different from determinism in the study of nature.

Manninen and Tuomela (eds.), Essays on Explanation and Understanding, 415–435. *All Rights Reserved.*

I can foresee immediate objections. The theoretical aim of the human sciences, it will be said, is to discover laws which state, not regulations and rules of conduct, but factual regularities of individual and collective behaviour and of institutional change. Such laws, if there are any, will 'determine' or 'govern' the life of men and of society in a similar way to that in which the laws of nature 'determine' or 'govern' natural phenomena. On this view, determinism in the study of man would mean essentially the same thing as determinism in natural science.

Deterministic claims of this character have often been made for the human sciences. Sometimes with the qualification that the laws of behaviour are not as 'rigid' as the laws, say, of classical mechanics, but are rather of a probabilistic than of a 'strictly' causal nature. Or it is said that these laws, or the conditions under which they apply, are so complex that it has not yet been possible to formulate them exactly. The social sciences are still young and we must not expect too much of them in their infancy.

To view matters in this light is, I think, to be guilty of a serious 'methodological misunderstanding'. It is symptomatic of an illegitimate transfer of conceptions and ideals from the natural to the human sciences. To say this is not to deny legitimacy to a study which ascertains factual regularities in human and social behaviour. Nor is it to exclude the possibility of relying on such regularities for successful predictions. But it is an invitation to view in a new light the study of man and all theorizing and philosophizing in the social sciences.

This shift in point of view will also have consequences for the age-old philosophic problem of 'the freedom of the will'. This is, roughly, the question of the relation between the actions of an individual and various forces working from within him: his will, his wants, his passions and sentiments, and his deliberations. There is a deep-rooted tendency to wed this question to the further question of the relation between neural processes and macroscopic reactions in the individual's body. This 'internalization' of the problem is another illegitimate transfer to a sphere where it does not belong of an attitude appropriate to the natural sciences, in this case physiology. The way to a solution is what I propose to call an 'externalization' of the problem. This consists in working one's way from considerations of individual action to considerations of the factors which influence ('determine') the individual's conduct as a member of a community, in the context of institutionalized human relationships.

It would be a slight, but still useful, exaggeration to say that the problem of the freedom of the will, like the problem of freedom generally, is essentially a question of social philosophy.

II

As a basis and starting point I shall present and briefly discuss two patterns of explanation of action.

The first explanatory pattern I shall call *intentionalist* explanation. It is related to a type of reasoning sometimes called 'the practical syllogism'. In its simplest form this reasoning goes as follows:

> *A* intends to *p* (*e.g.*, go to the theatre tomorrow).
> *A* thinks that unless he *q*'s (*e.g.*, reserves a ticket in advance), he will not be able to *p*.
> Therefore: *A* takes steps to *q*.

The inference remains valid, if for 'intends' we substitute 'has decided' or 'is resolved' or 'is determined', perhaps also 'is anxious'. For 'thinks' in the second premiss one can also put 'considers', 'realizes', 'knows', or 'believes'.

It is easy to see what practical reasoning has to do with action explanation. Assume *A*, as a matter of fact, *q*'s. We are curious to know *why*. It would be a satisfactory answer to this question to point out that he intends to *p* and considers *q*'ing necessary to this end. And the same answer would explain, why he tried to *q* in a situation when he failed.

It is quite natural to say here that *A*'s behaviour was *determined* by his intention and epistemic attitude. Given them, he had to do what in fact he did. We can speak of the intention and epistemic attitude as *determinants* of the agent's action and say that they jointly constitute a (sufficient) ground or *reason* for *q*'ing. The thing for the sake of which *A* undertook to *q* I shall call *A*'s *object of intention*, and the *q*'ing itself, I shall say, was part of the *requirements of the situation*, as *A* saw it, upon his action.

Suppose that *A* considers *q*'ing sufficient, though not necessary, for attaining the object of his intention. He intends to go to town and knows he will have to use a public conveyance to get there – say, either take a bus or a train. He takes steps to catch the bus. Should we here too say that his actual choice of the bus is fully explained, determined, by what he intends to do and knows about the means of making his intention effective?

Obviously we should *not* say this. The behaviour which we can explain on the basis of the facts, as I presented them, is the 'disjunctive action' which consists in the agent's taking a bus *or* taking the train. This action he can perform in one of two ways, *viz.* by taking a bus and, alternatively, by taking a train. So, if now he chooses the bus, he performs the disjunctive action. *This* action is then fully determined by the agent's intentions and beliefs – but not his actual choice of alternative.

But could not his choice be determined too? Certainly it could. Various *reasons* might have existed for his choosing to go by bus: perhaps it is safer or cheaper or quicker than going by train. If his choice of means of transportation can be attributed to some such reason, then it is also true to speak of the *choice* as determined. But it is important to note that, although a man's choices between alternative courses of action *can* be in this sense determined, they *need not* be. To insist that they must would be sheer deterministic dogmatism. Choice *can* be completely 'fortuitous'.

III

Many actions are performed in response to a verbal (or other symbolic) challenge. The challenge can be, for example, an order which I obey, or a request with which I comply, or a question which I answer – or a traffic light to which I respond. Why did I reach out for the salt on the table and hand it to my neighbour? Because he asked for it. This can be a complete explanation. His request determined my action, constituted for me a sufficient ground or reason for doing what I did.

Normally, when I respond to a challenge of this kind, I cannot rightly be said to *intend to* respond. I simply *respond.*

Assume, however, that the saltcellar had slipped from my grip and the salt poured out on the table. I hear people cry out: "What are you doing?" Then I could quite truthfully reply: "I intend*ed* ('I meant') to hand the salt to *X*, who asked for it". The fact that I can give this reply, when I fail in the performance, shows that complying with a request is intentional action – and not just a conditioned response to a stimulus.

In our example the object of intention was to hand the salt to my neighbour. This object was, so to speak, set by the request, and only reconstituted by me in retrospect. For this reason I shall call it externally set – and contrast it with the internally set objects of intention which are there when I can say, before proceeding to act, what I intend to achieve.

Similarly, we can divide the determinants of action into internal and external ones. Intentions and epistemic attitudes are of the former, symbolic challenges of the latter kind.

Response to verbal and other symbolic challenges is participation in various *institutionalized* forms of behaviour or *practices*. That the practices are 'institutionalized' means that they are shared by a community into which we are reared by being taught to participate.

Response to symbolic challenge is only one form of participation in an institutionalized practice. Another is behaviour in conformity with rules such as the laws of the state or the codes of morality and good manners or customs and traditions. "Why don't you park your car here?" (It would perhaps be convenient.) The answer might be "It is not allowed". Here a traffic regulating rule functions as a determinant of my behaviour. Obedience to it is an externally set object of my intentional acting.

The proportion of our actions which are determined internally and externally respectively, is not fixed. It varies from society to society, and it varies with the position of the individual agent in society. In a social order with many taboos and ritualized ways of life external determinants can steer the actions of men in the minutest details. In such societies the margins of individual freedom are very narrow.

IV

External determinants of our actions are given to us like stimuli to which we react. Such responses have to be learnt – as is the case with conditioned reflexes. Learning to participate in institutionalized forms of behaviour is connected with a characteristic motivation. I shall call this motivational mechanism *normative pressure*.

Non-conformity to legal and moral and other rules of conduct and good manners is likely to have unpleasant consequences for the agent. In the case of legal norms the 'administering' of these consequences is itself institutionalized and consists in various coercive measures against the agent. In the case of moral norms the consequences are disapproval, ostracism, loss of esteem or confidence – things which make a man ill at ease in society.

I think it is essential to the idea of participating in institutionalized patterns of behaviour that it should be surrounded by this 'aura' of normative pressure. This does not mean that the answer to the question why people participate or conform to rule, is always or even normally teleological. People do not usually conform *in order to* escape the unpleasant consequences of non-conformity. But sometimes they do this – *e.g.*, when participation or conformity is against their personal interests or connected with discomfort or when it seems pointless. Then the answer to the question: Why did you do *x*? could be: Had I not done *x*, *y* would have happened to me and this I am anxious to avoid. Here the determinant of my action is not the invitation to participate – but the internal determinant constituted by my intention to avoid a certain thing which I consider likely to happen to me, if I do not participate.

In educating people, particularly children, to participate in practices and obey rules, rewards also play a characteristic role. When reward is merely an alternative to punishment in making people conform, I shall call reward external. On the whole external reward seems to play a subordinate role in the institutionalization of patterns of behaviour. Philosophers have noted and tried to give an account of this fact. I think the explanation should be sought along the following lines:

Institutionalization of behaviour normally serves a purpose. Institutions have what we call a social function. Without traffic regulations there would be chaos on the roads. This nobody wants. Participation in the practice, by everybody concerned, is therefore supposed to be in the 'public interest', *i.e.*, something which will be in the interest of each individual participant, bring him some good. This good can be thought of as a 'reward' connected intrinsically with the practice, *i.e.*, with the idea of having it. Therefore it is a further feature essentially connected with the institutionalization of behaviour that the reason for conforming to the set patterns should, on the whole, *not* be the impact of normative pressure, but simply acceptance of the rule. When rules function in this way, they are also said to be *internalized* with the members of the society in question. The more often normative pressure determines behaviour, the more strongly is the coercive force of society felt and the less 'free', in a subjective sense, are the individual agents. But internalization is also a loss of freedom of a kind. For it means that externally given stimuli are allowed to determine the actions. It is on these two forms of non-freedom that social critics play. They question the fact of internalization, *e.g.*, raise and make people reflect on the question whether various institutions and practices *are* in the 'public interest' or whether they perhaps only serve to cement interests not at all 'public' but, say, those of a ruling class. Thus their criticism contributes to an increase in the normative pressure felt within the society. The society becomes more and more coercive and its institutions malfunctioning. Hereby the ground is prepared for institutional changes.

V

Assume that it were true that *A q*'ed *because* he intended to *p* and thought *q*'ing necessary for this. What sort of connexion does this 'because' establish between an intention and epistemic attitude on the one hand and action on the other hand? This is a question on which philosophers violently disagree.

Some hold that the connexion is causal. This position can be understood in two ways. I shall call them the trivial and the non-trivial.

The trivial understanding of the causalist position stresses the fact that intentions are quite commonly *called* 'causes of actions'. This is in order, and we should not attempt to reform language here. The only objection which one can have to this kind of talk is that it obscures the difference between the sense in which intentions can uncontroversially be called causes of actions and some other important senses in which things are said to be causally related. One of these other senses is often called 'Humean'. The existence of a Humean causal relation entails that there is a general law connecting instances of logically independent generic phenomena as cause and effect.

The non-trivial interpretation of the causalist position in action theory holds that a specific combination of intention and epistemic attitude is a Humean cause of a specific kind of action.

Defenders of this position sometimes think that it requires a reinterpretation of intentions and cognitive states in neurological terms. The causal relation is then in the first place between certain brain events and certain movements of limbs and other parts of the body. Of this view I shall here only say the following:

We need not doubt that there are causal relations of the kind just mentioned. But the neurological interpretation of volitional and epistemic attitudes is, at best, only a contingent correlation of them with equivalents in the brain and not a necessary connexion. And whether the movements caused by those brain events are actions or not is again a contingent matter, depending upon other facts about the agent than movements of his body and processes in his nervous system. These observations, when more fully substantiated, suffice, in my opinion, to wreck the nontrivial causalist thesis about the relation between intentions and actions. But I cannot argue the point at length here.

The contrasting opinion holds that there is a conceptual or logical connexion between an action and its grounds in an intention and epistemic attitude. This opinion is sometimes called the Logical Connexion Argument or the intentionalist view. I think it comes nearer the truth than the causalist view. But it is difficult to argue for it correctly.

Thus I think it is a mistake – of which I myself and others have been guilty – to understand the intentionalist view to mean that there is a relation of logical entailment between the premisses and the conclusion of a practical argument. (Cf. my book *Explanation and Understanding*, London 1971, pp. 97–118.)

Consider the following example. (*Ibid.*, p. 116f.) – A man is firmly resolved to assassinate a tyrant. He has access to his room, aims at him with a loaded revolver – but cannot bring himself to pull the trigger. Nothing which we later find out about him would make us think that he had changed his intention or come to a different opinion about the things required of him to make it effective. Is this conceivable, *i.e.*, logically possible?

It should be noted that we do not assume here that the assassin *in spe* forbore to pull the trigger. Had he done so, it would be, I should think, a contradiction to say that the man (still) intended to shoot the tyrant and knew what he had to do. (It would be a contradiction reminiscent of the so-called Moore paradox.) What is assumed is simply that he did not do anything at all just then which was relevantly related to his resolve to shoot the tyrant. He was 'paralyzed' – but neither physically nor mentally in a way which would make us revise the description of his intentions and cognitions.

In view of this example I think we should say the following about the nature of the relation between intentions and actions:

An intention and an opinion of what is required for it to become effective, constitute, as was already said, *a sufficient* ground or *reason* for acting accordingly. If the agent then acts accordingly, we understand completely why he is doing what he is doing, *e.g.*, trying to kill the tyrant by firing at him. No further information can help us understand *this* better. (We may, of course, wonder why he should have had the intention he had or how it was that he thought as he did – quite wrongly perhaps – about the requirements of making his intention effective. But these questions do not concern the determinants of his action but the determinants, if there are any, of these determinants.) If again the agent fails to 'act accordingly' we do not understand him at all. His behaviour is incomprehensible to us and in this sense *irrational* or, considering that he had sufficient grounds for acting in a certain way, anti-rational.

The relation between what I have called internal determinants of an action and the action itself is thus neither a relation of entailment nor a causal relation. We must resist the temptation to reduce it to something which it is *not*. But there is a sense in which we can call the relation *conceptual* – and something remains to be said about this before we have a full understanding of its nature.

VI

What is action? One could answer: Action is normally behaviour understood, 'seen', or described under the aspect of intentionality, *i.e.*, as meaning something or as goal-directed.

Intentionality can quite rightly be said to be in the behaviour. But not like a 'quality' inherent in the movements of limbs and other parts of the body. For these movements we can describe completely without mentioning intentionality. So what then is the intentionality of behaviour?

To understand behaviour as intentional, I shall say, is to fit it into a 'story' about an agent. We see a person walking in the street, carrying a parcel in his hand. He drops it and bends to pick it up. We should normally think of his picking it up as intentional. Why?

We may not know at all why he picked it up. But we can name hundreds of reasons why he might have done this, reasons which are such that had they been his at the time, they would explain his action completely. Perhaps the parcel contained something he was anxious not to lose; or a gift which he had bought for somebody. Or perhaps he took care not to litter the street or maybe was just following an existing regulation requesting one not to do this. We are, in other words, familiar with a number of possible, internal and external, determinants of his action here. We think it likely that some such determinant will be at work. This is what it is to 'see his behaviour as intentional'.

There are on the whole reliable ways of coming to know, of verifying, what a person intends to do, what he thinks are the requirements of the situation, and whether he acts accordingly. Verification may not always be conclusive, and ascertaining one of the three things mentioned may sometimes have to rely upon the accepted verification of one or both of the other two. But normally the verificational procedures here are applicable independently of one another.

A standard way of ascertaining a person's intentions and epistemic attitudes is by asking him. If we doubt whether his answers are reliable, there are usually other checks available. Ascertaining what a person does again, is usually a matter of simple observation. We literally *see* him do various actions, *i.e.*, we can describe his perceived behaviour under an appropriate aspect of intentionality. We may be mistaken, but normally we take what we see to be conclusive.

Relying on these verificational procedures we can establish a predictive correlation between the premisses and the conclusion of practical inferences. Having verified the premisses, we expect the conclusion to turn out to be true. The reliability of the prediction is in the following sense a function of *time* here:

When a person intends to do something, the object of his intention is in the future. This is so also when he intends to do something 'right now' – for 'now' then means the time immediately ahead of him. For this reason a person will normally have time to change his intention, and also his opinion of how he has to act, before he proceeds to making the intention effective. The more time he has to change his mind, the greater the risk that he will actually do so. If he does, the prediction may fail.

It is essential, however, that for short time intervals the reliability of the predictions should be high. This seems to be a feature of the way in which the *concepts* of intention, of the various epistemic attitudes, and of action are related to one another. If it were normally the case with a given person that he did not act in conformity with his shortly before professed intention and understanding of the situation, we should doubt either the veracity of his reports or doubt whether *he* knew what it *is* to intend and to believe something. And if this were the case with people generally, it would modify our view of what intentions and beliefs are. We could no longer be taught nor teach to others the present use of the words and therefore we should not *have* the concepts either, at least not in their present form. One could say that the language-games we now play with action-words and with epistemic and volitional terms rest on (presuppose) a high degree of correlation between intention and action in accordance with the understood requirements of a situation.

The prediction of actions from a background of intentions has a certain resemblance with the prediction of effects from knowledge of their causes. But there are also important differences to be noted.

One difference is this. The prediction of action is subject to a clause that no change occurs in the volitional and epistemic background before action is supposed to take place. The frequency of the failure of the prediction, with a given agent, – assuming that his intentions and epistemic attitudes are known for certain – is therefore a measure of the degree to which he is, as we say, 'capricious' or 'unreliable' or even 'irrational'. The only *hypothetical* element involved in the prediction is, in other words, that a certain volitional and epistemic attitude of the agent should remain constant in an individual case and that the agent should not act 'irrationally'. This is different from a typical causal context in the natural sciences. Here the failure of a prediction can always in principle recoil back also on a hypothetical *law* relating the cause to the effect.

In the case of predictions of actions there simply are no such 'covering laws' to be confirmed or refuted. To say that such and such intentions and beliefs, assuming they do not change, will normally result in such and such behaviour is not to state an empirical generalization based on observations or experiments. It is to state a necessary truth to which anybody familiar with the concepts involved will agree off-hand. And therefore this truth is very seldom stated – except, perhaps, in philosophic debates.

VII

What is the nature of the relation between actions and their external determinants?

Assume that it were true that *A* *q*'ed *because* he had been ordered to do so. It is quite obvious that the connexion between the determinant and the action cannot here be a relation of logical entailment. For, it would then be selfcontradictory ('unthinkable') that *A* had been given the order, had understood it and was able to carry it into effect, and yet did not do so. This, however, is far from unthinkable. – Is the connexion then causal? (Here it may be worth noting that it is much *less* natural to call an order a 'cause' of an action than to call an intention by this name.)

If the connexion were one of Humean causation, there should exist a *law* connecting cause and effect. This law cannot be that *A* always obeys when he is ordered to *q*. Perhaps this is true of *A*. But most probably it is not true of everybody who has learnt to obey orders. It may, for example, not be true of *B*. Still it could very well be the case that *B* too, on some occasion, *q*'ed *because* he had been ordered to do so. So whether the agent concerned happens to be one who always obeys orders to *q*, or not, is quite irrelevant to the question of the nature of the connexion, in an individual case, between the order and the action.

As far as I can see, the 'because' does not rest on any law at all here. And if this is so, then the relation between the determinant and the action is not one of Humean causation. What then is it? I propose to call it a relation of *justification.*

Let it be that *A* answers when asked why he *q*'ed: "Because I was ordered". May he not be lying or even mistaken about his own motives? When pressed with further questions, he may admit that really he *q*'ed because he feared the anger of the order-giver, *i.e.*, acted under the influence of normative pressure, and not just in response to the order. But if he does not admit any such other motive – not even 'to himself', 'in his heart' – then we must take him at his word and say that he *q*'ed because he was ordered. There is no 'external' way of deciding the truth of the 'because'-statement to which we could concede ultimate authority here. The connexion between the external determinant and the action is, as I have said, not intrinsic in the sense that it were a logical entailment. But it is in a characteristic sense an 'internal' relation, dependent upon the agent's judgement of why he acted as he did. Therefore it is not, in any good sense of the word, a 'causal' relation.

Just as one can, within limits, predict the actions of an agent from antecedent knowledge of internal determinants, one can make predictions on the basis of knowledge of external determinants. The degree of reliability of such predictions may vary with the agents concerned and also with the society under consideration. One can use this degree as a measure of the responsiveness of an individual to external determinants (of one sort or another, or generally). For example, one could use it as a measure of his obedience to the law or to his superiors. One can also use it to measure the degree of internalization and of normative 'cohesion' of a given society. The characteristics thus measured do not explain predictability. Predictability is their criterion. Nor is there any other general law besides a rough statistical correlation which connects the determinants with the actions.

It is of some interest to ponder why we do not willingly speak of such correlations as 'laws'. Is it because of their unprecise and statistical nature? Or because of their dependence upon individual agents and individual societies? An even weightier reason for not calling them 'laws' is, I think, their dependence upon factors, *viz.* norms and institutionalized patterns of behaviour, which are themselves susceptible to change in the course of history as a result of human action. 'Scientific laws', we tend to think, must not for their validity be dependent upon historical contingencies. They should hold true *semper et ubique*.

VIII

Determinants of action, I have said, are either internally set or externally given. By referring to their determinants we explain the actions, *i.e.*, answer questions why agents acted as they did.

Such answers, however, are only explanations 'in the short perspective'. They give rise to further questions. For example: Why do people have the intentions they have?

Sometimes the answer to that question is given in the terms of a further intention. Why did *A* intend to go to the concert tomorrow? The answer could be that he intends, is resolved, to acquire some education in music. Going to concerts here serves as a means to an end. But why should he be resolved to pursue this remoter end? The answer in terms of intentions will ultimately take us, I suggest, to one of two main types of determinants of intentions. I shall call them *wants* and *duties*.

Let us first consider wants. – Why did *A* intend to go to the concert? A frequent type of answer could be: Because he wanted to hear *B* perform or because he wanted to hear symphony *S* again or simply because he wanted to hear some music. Such answers are normally complete explanations. One wanted *just this*, and there is no further thing for the sake of which one wanted it, as a means to an end.

If the further question is raised as to why one wants what one wants, the answer sometimes is: Because it pleases or one likes it or because it is thought fine or nice or amusing. But such answers do not point to determinants of the want. They merely specify it by setting it against a background which is there independently of my present intention to reach out for the wanted thing. If, for example, I like to listen to music, I shall, given an opportunity and in the absence of any other determinant already at work on my intentions or actions, listen to music. On such an occasion I might say "I want to hear some music, I like music". My liking of music is, so to speak, a latent want which manifests itself in my intention, say, to hear some music now or to go to the concert tomorrow.

One cannot ask why people should want things they like or take pleasure in. It is, one could say, 'in the nature' of pleasant and liked things that they should be wanted – as it is 'in the nature' of unpleasant and hurtful things, such as illness or punishment, that they should be shunned. Shunning things is wanting not to have them or wanting to get rid of them. When we intend (decide) to follow a rule or order because we are anxious not to risk punishment, it is our shunning of something intrinsically unwanted that determines our intention.

A person says: "I intend (have decided) to go to the Canary Islands during my vacation". "Why do you want to go there?", we ask. We are anxious to know the *want* behind his intention. He answers: "Oh, I just want to see the place, it is supposed to be nice." This could be his sole motive force (want) here. But assume he answers: "I think it will do me good, I have been very tired and run down lately". Then the planned action is seen as a means to an end, the end being one's health or well-being generally. This is what one wants to promote.

Health, well-being, and happiness are 'natural' objects of want. Other things being equal we pursue them of necessity, for their own sake. In this they resemble the things we like or which give us pleasure. It makes no sense to ask "Why do you want to be healthy?" But it does not follow that a person will necessarily care for his health, or pursue his pleasures or happiness. There can be overriding considerations. He may have 'no time' for his pleasures or 'be forced' to neglect or even to ruin his health. The determinants which can override the influence on our intentions and actions even of things which are 'by nature' objects of wants, have the character of *duties*.

IX

The word 'duty' should here be taken in a broad and somewhat loose sense. The range of things which I have in mind can be roughly characterized as follows:

As a member of a society any man usually holds one or several positions in which he is expected, or sometimes even obliged, to do various things. Some such position a man can be said to hold 'by nature', such as the position of a parent; others he holds, *e.g.*, by appointment or by election. But in either case the actions or types of action expected of him are defined by the explicit or implicit rules (laws, customs, conventions) of the society to which he belongs. I shall call such positions *roles* and the things expected of a *role*-holder *duties*. (The etymology of the word then suggests that they are things which he 'owes' to the rest of the society by virtue of his position in it.)

Thus a head of state is expected to care for his country's prestige, its power and prosperity. This will make him form intentions and take decisions which, as a 'private citizen', he neither could nor would contemplate. The objects of these intentions form part of what he and others consider his duties. Failure to perform need not have legal implications for him, but will surely have consequences which it is in *his* ('personal') interest to shun, such as loss of popularity or an unfavourable 'verdict of history'. So, failing a motive 'from duty', there will be a motive 'from want' ('self-interest') to make him have the action-guiding intentions which are appropriate to his role.

A policeman is seen jumping into a car and speeding away. Why this behaviour? We are told that he intends to catch the thief who was seen running in the street. Why should he intend this? As a private citizen he may even have felt inclined to let the poor man escape. But his role as policeman 'imposes' this intention on him with all the actions following upon considerations about the means of making the intention effective. If he does not realize this and act accordingly, he runs the risk of being fired or even punished.

Similar considerations apply, *mutatis mutandis*, to all holders of roles. It is an essential part of the picture here that roles should be surrounded by an aura of normative pressure which, when needed, makes people perform in their roles – perhaps somewhat 'against their will' but still in agreement with what they want, lest something worse happens to them. In this regard there is a parallel between roles as external determinants of intentions and rules and symbolic challenges as external determinants of actions. But as with rules it is also the case with roles that it is essential to our notion of a functioning society that role-performance should on the whole *not* be motivated by normative pressure, but should be 'internalized', that is: the duties unquestioningly accepted as an ultimate determinant of what we intend.

It is perhaps right to say that duties implicit in various roles more than any other determinants mould men's intentions and therefore indirectly guide their actions. But the extent to which this happens is different within different societies and with different roles. Therefore questions of role-distribution within a society are inseparably connected with the problems of individual freedom.

So-called 'free-time' is that part of a man's life when he can do what he *wants* and temporarily forget the demands on him of at least some of his assigned roles. When a man has no time for his wants, only for his duties, he is a slave to his roles. This he can be both as servant in the meanest position and as a master endowed with the greatest power.

X

In addition to wants and duties there is also a third type of factor which determines a man's intentions – and through the intentions his actions: his *abilities*.

Unlike wants and duties which 'prompt' people to action, abilities determine actions negatively, restrictingly. They delimit the 'horizon' or 'domain' or 'range' of a man's freedom to act. This range will then wax and wane with variations in ability.

To have an intention to do something presupposes that the agent thinks, rightly or wrongly, that he can achieve the object of his intention. What he does not think he *may* accomplish, he will not attempt either. To say this is to make a *conceptual* observation on the relation between the volitional and epistemic attitudes involved in a practical argument. There is not, be it observed, a corresponding relation between 'want' and 'can'. A man may want to do things he knows he cannot do. But if his want is not to remain merely an 'idle wish', he will have to form an intention to acquire the ability. He may, for example, be resolved to learn to do it. This again presupposes, logically, that in his subjective estimate, he *can learn* to do the thing in question.

Abilities are either 'innate' or else determined by biological and physical factors, or they are acquired. Intelligence and memory, health and bodily strength are gifts of nature – and nature endows people unequally and also makes the abilities which depend upon these endowments vary within the lifespan of the same individual. But within rather broad limits people are *roughly* equally endowed in these regards, and the differences which exist between them can to some extent be equalled out as a result of care or training.

Acquisition of abilities happens through learning, instruction, and education. These are largely socially institutionalized processes. But new abilities are also acquired thanks to individual inquiry into the possibilities of doing things and thanks to creative efforts. Of abilities, thus acquired, those which have the character of *technological innovations* occupy an important and peculiar place, chiefly because of their consequences on the social level.

The fact that learning and education are parts of the social fabric can be responsible for great inequalities in the possibilities which men have of acquiring abilities. A man may not be able to afford or his social position may not allow him to avail himself of facilities for education which some other men enjoy. The range of things he is able to do may on this account remain very restricted. Then it is also likely that within this range his duties, 'the pressing necessities of life', much more than his wants, will determine his intentional actions.

There are thus a great many abilities which a man 'by nature' could acquire but which for deontic reasons, *i.e.*, reasons built into the normative structure of the community of which he is a member, he cannot acquire.

Many abilities, moreover, require for their exercise *equipment* in the form of instruments or machinery. This is true of all abilities which are conditioned by technological innovations, but particularly of the ability to use technology for the production of commodities. This fact is the greatest source of inequalities in the freedom of men and also the greatest urge to make men equal by changing the institutionalized restrictions on what each one of them *can do*.

The prospects of social philosophizing which open up here are easily recognized – but their further exploration must remain outside the scope of this paper.

XI

What a man can do in a given situation is, however, only partly conditioned by his abilities. An equally important condition is formed by the *opportunities*. A child may have learnt how to open a window, but if the windows in his surroundings are already open, it *cannot*, in that situation, open a window. The ability is a generic feature of an agent; the opportunity, again, an individual feature of a concrete situation.

Every action by any man creates and destroys opportunities for actions – by the agent himself and by other agents. By shutting a door I create an opportunity for opening it; by leaving the room I may destroy the opportunity for another man to request my help in an important job.

The opportunities are thus in a constant flux as compared with the *relative* stability of abilities and wants and duties – not to speak of their background in the institutions of society. Intentions fall in a middle zone here. As the situations change, creating new opportunities for action, intentions articulate under the already existing wants and duties and within the frame of given abilities. This interplay between situational change, intentionality, ability, and a motivational and normative background I shall call the *logic of events*. It constitutes the cogwheels of the 'machinery' which keeps history moving.

Sometimes the changed situation which makes new actions possible, or imperative, results from the working of natural forces alone. This is the case when, for example, an earthquake or a flood upsets human conditions. The intentions for acting formed under the impact of such changes are often the outlet for wants (and shunnings) shared by practically all men at all times and which might also be called a 'will to survival'. People seek refuge or migrate to new abodes – or they join hands to take various countermeasures such as building walls against floods or protecting the environment against industrial pollution. Such measures may in their turn require (in the sense of the second premiss of a practical argument) changes on the institutional level of society.

A very different type of logic of events is exemplified by changes in situation which primarily result from people acting in roles, for example when the actions are those of a government, or a corporation, or the army, *etc.* A country conquers a province from another to safeguard its borders. "Unless we do this, they might invade us" is now the reason-giving premiss of a practical syllogism. In the new situation, created by the conquest, the neighbouring country, *i.e.*, its government, thinks it necessary to safeguard its independence by entering into an alliance with a third power. The two now jointly constitute a threat to the first and further actions become imperative for its rulers. And so forth. Each new action by one party 'triggers off' the conclusion of a 'latent' practical argument by another party – the 'latency' of the argument consisting in the fact that the goal-structures, both the duties of the role-holders and the wants ('hopes and fears') of those who vest their expectations in the role-holders, are fixed, and the requirements of the situation appear univocal in the light of past experience or of traditional standards of assessing them.

Such chains of successively created sufficient reasons necessitating action are particularly impressive when events are heading for disaster or towards decay. The origins of imperialistic wars, the fall of empires, the decay of an economy often follows this pattern and thereby assumes an air of 'historical inevitability'.

Again a somewhat different type of 'logic' is presented by the great creative innovations, particularly in the realm of technology, which open new possibilities of action and thereby become an outlet for latent wants rather than for pre-existing institutionally determined duties. The chain-reactions 'let loose' by such changes often create a tension between the duties of role-holders and external objects of intention set by the rules on the one hand, and the direction given to the action-opportunities by the flux of situational change on the other hand. Internalization of the institutional forms becomes more difficult and more dubious, normative pressure increases and the 'internal contradictions' of the community crystallize into grounds for changes in institutions.

XII

The description which I have tried to give of the motivational mechanisms and the working of chains of necessitation of action should help us answer the question whether there are 'laws' in history.

In fairness to the question it should be said that if there existed law-like connexions between concrete historical events, we should rather think of them as instantiations of general laws of sociology, and perhaps of economics, than as 'laws of history' proper.

A number of so-called laws of economics easily come to the layman's mind: Say's or Gresham's laws, the law of supply and demand or of diminishing marginal utility, *etc.* In sociology there is much less unanimity about what deserves the name of a law – but candidates are not lacking, for example the several Marxian principles concerning the dependence of social structure upon productive forces and relations.

My suggestion now is that such laws are applications to specified types of activity and types of historical situation of the very general conceptual patterns which I have outlined in my paper. Even the most elementary laws of economics presuppose some institutionalized forms of exchanging commodities on a market and of rough standards of measuring the value of goods to a producer and a consumer. It is usually not too difficult to see under which assumptions concerning the institutional frame these laws are conceptual necessities about the way in which wants and abilities regulate behaviour. Within different frames different laws are valid. This means that different frames require different laws, if the logic of events is to be correctly described. The complexities of theoretical economic analysis largely consist in devising conceptual instruments appropriate to the description of economic behaviour within the institutional structure of a historically given situation. Thus, for example, in the strongly 'manipulated' market of late capitalist societies the laws of 'classical' market economy cannot be expected to hold good. For this reason it is sometimes said that the laws of economics and sociology are themselves subject to historical change – unlike the laws of nature which are valid *semper et ubique*.

Social 'laws' are not generalizations from experience but conceptual schemata for the interpretation of concrete historical situations. Their discovery, or rather, invention, is a matter of the analysis of concepts and their application a matter of analysis of situations. On this account one can say that social study occupies an intermediate position between philosophy and history. It can move in the direction of the one or the other of the two poles, but it cannot live a self-contained life divorced from either of them.

History, when it is 'scientific' and not mere chronicle or narration, is an inquiry into the logic of events in a fragment of the past with named actors and institutions. It is a study of history from a deterministic point of view in as much as it studies the interplay between historical change and the determinants of human action. As we have seen, these determinants have to a great, not to say overwhelming, extent their roots in the structure of the social fabric: in the distribution of roles and the institutionalization of behaviour-patterns. With changes in these societal determinants of actions, actions too will be different. But changes in the determinants are in their turn the results of action – except for the cases when they are man-independent changes in nature. Thus the actions of men are determined by their historical situation, but the historical situation is itself the result of the actions of men. There is no circularity of a logically vicious kind in this fact that mankind is both slave and master of its own destiny.

The determinants of natural change are causal laws – and them man cannot change. But he can use his knowledge of the laws to steer natural change by producing and suppressing opportunities for causes to work. Man's foresight, however, is limited and what further causal consequences his manipulations of nature will call forth may be humanly impossible to foresee. We are reminded of this by the eroded landscapes in lands of ancient cultures – but also by the ecological problems facing modern industrial society. That man has made himself master of nature to the extent he has is one of his greatest achievements as a species. To exercise the restraint and skill needed in order not to be dethroned is the most serious challenge facing him today. It is unlikely that it can be successfully met without profound changes also in that law-regulated realm in which man's mastery can never be challenged and where he is for ever sovereign, *viz.* his societies.

Academy of Finland
and Cornell University

INDEX OF NAMES

SYNTHESE LIBRARY

Monographs on Epistemology, Logic, Methodology,
Philosophy of Science, Sociology of Science and of Knowledge, and on the
Mathematical Methods of Social and Behavioral Sciences

1. J. M. BOCHEŃSKI, *A Precis of Mathematical Logic*. 1959, X + 100 pp.
2. P. L. GUIRAUD, *Problèmes et méthodes de la statistique linguistique*. 1960, VI + 146 pp.
3. HANS FREUDENTHAL (ed.), *The Concept and the Role of the Model in Mathematics and Natural and Social Sciences, Proceedings of a Colloquium held at Utrecht, The Netherlands, January 1960*. 1961, VI + 194 pp.
4. EVERT W. BETH, *Formal Methods. An Introduction to Symbolic Logic and the Study of Effective Operations in Arithmetic and Logic*. 1962, XIV + 170 pp.
5. B. H. KAZEMIER and D. VUYSJE (eds.), *Logic and Language. Studies Dedicated to Professor Rudolf Carnap on the Occasion of his Seventieth Birthday*. 1962, VI + 256 pp.
6. MARX W. WARTOFSKY (ed.), *Proceedings of the Boston Colloquium for the Philosophy of Science, 1961–1962*, Boston Studies in the Philosophy of Science (ed. by Robert S. Cohen and Marx W. Wartofsky), Volume I. 1973, VIII + 212 pp.
7. A. A. ZINOV'EV, *Philosophical Problems of Many-Valued Logic*. 1963, XIV + 155 pp.
8. GEORGES GURVITCH, *The Spectrum of Social Time*. 1964, XXVI + 152 pp.
9. PAUL LORENZEN, *Formal Logic*. 1965, VIII + 123 pp.
10. ROBERT S. COHEN and MARX W. WARTOFSKY (eds.), *In Honor of Philipp Frank*, Boston Studies in the Philosophy of Science (ed. by Robert S. Cohen and Marx W. Wartofsky), Volume II. 1965, XXXIV + 475 pp.
11. EVERT W. BETH, *Mathematical Thought. An Introduction to the Philosophy of Mathematics*. 1965, XII + 208 pp.
12. EVERT W. BETH and JEAN PIAGET, *Mathematical Epistemology and Psychology*. 1966, XII + 326 pp.
13. GUIDO KÜNG, *Ontology and the Logistic Analysis of Language. An Enquiry into the Contemporary Views on Universals*. 1967, XI + 210 pp.
14. ROBERT S. COHEN and MARX W. WARTOFSKY (eds.), *Proceedings of the Boston Colloquium for the Philosophy of Science 1964–1966, in Memory of Norwood Russell Hanson*, Boston Studies in the Philosophy of Science (ed. by Robert S. Cohen and Marx W. Wartofsky), Volume III. 1967, XLIX + 489 pp.

15. C. D. BROAD, *Induction, Probability, and Causation. Selected Papers.* 1968, XI + 296 pp.
16. GÜNTHER PATZIG, *Aristotle's Theory of the Syllogism. A Logical-Philosophical Study of Book A of the Prior Analytics.* 1968, XVII + 215 pp.
17. NICHOLAS RESCHER, *Topics in Philosophical Logic.* 1968, XIV + 347 pp.
18. ROBERT S. COHEN and MARX W. WARTOFSKY (eds.), *Proceedings of the Boston Colloquium for the Philosophy of Science 1966–1968*, Boston Studies in the Philosophy of Science (ed. by Robert S. Cohen and Marx W. Wartofsky), Volume IV. 1969, VIII + 537 pp.
19. ROBERT S. COHEN and MARX W. WARTOFSKY (eds.), *Proceedings of the Boston Colloquium for the Philosophy of Science 1966–1968*, Boston Studies in the Philosophy of Science (ed. by Robert S. Cohen and Marx W. Wartofsky), Volume V. 1969, VIII + 482 pp.
20. J. W. DAVIS, D. J. HOCKNEY, and W. K. WILSON (eds.), *Philosophical Logic.* 1969, VIII + 277 pp.
21. D. DAVIDSON and J. HINTIKKA (eds.), *Words and Objections: Essays on the Work of W. V. Quine.* 1969, VIII + 366 pp.
22. PATRICK SUPPES, *Studies in the Methodology and Foundations of Science. Selected Papers from 1911 to 1969.* 1969, XII + 473 pp.
23. JAAKKO HINTIKKA, *Models for Modalities. Selected Essays.* 1969, IX + 220 pp.
24. NICHOLAS RESCHER *et al.* (eds.). *Essay in Honor of Carl G. Hempel. A Tribute on the Occasion of his Sixty-Fifth Birthday.* 1969, VII + 272 pp.
25. P. V. TAVANEC (ed.), *Problems of the Logic of Scientific Knowledge.* 1969, XII + 429 pp.
26. MARSHALL SWAIN (ed.), *Induction, Acceptance, and Rational Belief.* 1970. VII + 232 pp.
27. ROBERT S. COHEN and RAYMOND J. SEEGER (eds.), *Ernst Mach; Physicist and Philosopher*, Boston Studies in the Philosophy of Science (ed. by Robert S. Cohen and Marx W. Wartofsky), Volume VI. 1970, VIII + 295 pp.
28. JAAKKO HINTIKKA and PATRICK SUPPES, *Information and Inference.* 1970, X + 366 pp.
29. KAREL LAMBERT, *Philosophical Problems in Logic. Some Recent Developments.* 1970, VII + 176 pp.
30. ROLF A. EBERLE, *Nominalistic Systems.* 1970, IX + 217 pp.
31. PAUL WEINGARTNER and GERHARD ZECHA (eds.), *Induction, Physics, and Ethics, Proceedings and Discussions of the 1968 Salzburg Colloquium in the Philosophy of Science.* 1970, X + 382 pp.
32. EVERT W. BETH, *Aspects of Modern Logic.* 1970, XI + 176 pp.
33. RISTO HILPINEN (ed.), *Deontic Logic: Introductory and Systematic Readings.* 1971, VII + 182 pp.
34. JEAN-LOUIS KRIVINE, *Introduction to Axiomatic Set Theory.* 1971, VII + 98 pp.
35. JOSEPH D. SNEED, *The Logical Structure of Mathematical Physics.* 1971, XV + 311 pp.
36. CARL R. KORDIG, *The Justification of Scientific Change.* 1971, XIV + 119 pp.
37. MILIČ ČAPEK, *Bergson and Modern Physics*, Boston Studies in the Philosophy of Science (ed. by Robert S. Cohen and Marx W. Wartofsky), Volume VII, 1971, XV + 414 pp.
38. NORWOOD RUSSELL HANSON, *What I do not Believe, and other Essays*, ed. by Stephen Toulmin and Harry Woolf, 1971, XII + 390 pp.

39. ROGER C. BUCK and ROBERT S. COHEN (eds.), *PSA 1970. In Memory of Rudolf Carnap*, Boston Studies in the Philosophy of Science (ed. by Robert S. Cohen and Marx W. Wartofsky), Volume VIII. 1971, LXVI + 615 pp. Also available as a paperback.
40. DONALD DAVIDSON and GILBERT HARMAN (eds.), *Semantics of Natural Language*. 1972, X + 769 pp. Also available as a paperback.
41. YEHOSUA BAR-HILLEL (ed)., *Pragmatics of Natural Languages*. 1971, VII + 231 pp.
42. SÖREN STENLUND, *Combinators, λ-Terms and Proof Theory*. 1972, 184 pp.
43. MARTIN STRAUSS, *Modern Physics and Its Philosophy. Selected Papers in the Logic, History, and Philosophy of Science*. 1972, X + 297 pp.
44. MARIO BUNGE, *Method, Model and Matter*. 1973, VII + 196 pp.
45. MARIO BUNGE, *Philosophy of Physics*. 1973, IX + 248 pp.
46. A. A. ZINOV'EV, *Foundations of the Logical Theory of Scientific Knowledge (Complex Logic)*, Boston Studies in the Philosophy of Science (ed. by Robert S. Cohen and Marx W. Wartofsky), Volume IX. Revised and enlarged English edition with an appendix, by G. A. Smirnov, E. A. Sidorenko, A. M. Fedina, and L. A. Bobrova. 1973, XXII + 301 pp. Also available as a paperback.
47. LADISLAV TONDL, *Scientific Procedures*, Boston Studies in the Philosophy of Science (ed. by Robert S. Cohen and Marx W. Wartofsky), Volume X. 1973, XII + 268 pp. Also available as a paperback.
48. NORWOOD RUSSELL HANSON, *Constellations and Conjectures*, ed. by Willard C. Humphreys, Jr. 1973, X + 282 pp.
49. K. J. J. HINTIKKA, J. M. E. MORAVCSIK, and P. SUPPES (eds.), *Approaches to Natural Language. Proceedings of the 1970 Stanford Workshop on Grammar and Semantics*. 1973, VIII + 526 pp. Also available as a paperback.
50. MARIO BUNGE (ed.), *Exact Philosophy – Problems, Tools, and Goals*. 1973, X + 214 pp.
51. RADU J. BOGDAN and ILKKA NIINILUOTO (eds.), *Logic, Language, and Probability*. A selection of papers contributed to Sections IV, VI, and XI of the Fourth International Congress for Logic, Methodology, and Philosophy of Science, Bucharest, September 1971. 1973, X + 323 pp.
52. GLENN PEARCE and PATRICK MAYNARD (eds.), *Conceptual Chance*. 1973, XII + 282 pp.
53. ILKKA NIINILUOTO and RAIMO TUOMELA, *Theoretical Concepts and Hypothetico-Inductive Inference*. 1973, VII + 264 pp.
54. ROLAND FRAÏSSÉ, *Course of Mathematical Logic* – Volume I: *Relation and Logical Formula*. 1973, XVI + 186 pp. Also available as a paperback.
55. ADOLF GRÜNBAUM, *Philosophical Problems of Space and Time*. Second, enlarged edition, Boston Studies in the Philosophy of Science (ed. by Robert S. Cohen and Marx W. Wartofsky), Volume XII. 1973, XXIII + 884 pp. Also available as a paperback.
56. PATRICK SUPPES (ed.), *Space, Time, and Geometry*. 1973, XI + 424 pp.
57. HANS KELSEN, *Essays in Legal and Moral Philosophy*, selected and introduced by Ota Weinberger. 1973, XXVIII + 300 pp.
58. R. J. SEEGER and ROBERT S. COHEN (eds.), *Philosophical Foundations of Science. Proceedings of an AAAS Program, 1969*. Boston Studies in the Philosophy of Science (ed. by Robert S. Cohen and Marx W. Wartofsky), Volume XI. 1974, X + 545 pp. Also available as a paperback.
59. ROBERT S. COHEN and MARX W. WARTOFSKY (eds.), *Logical and Epistemological*

Studies in Contemporary Physics, Boston Studies in the Philosophy of Science (ed. by Robert S. Cohen and Marx W. Wartofsky), Volume XIII. 1973, VIII + 462 pp. Also available as paperback.

60. Robert S. Cohen and Marx W. Wartofsky (eds.), *Methodological and Historical Essays in the Natural and Social Sciences. Proceedings of the Boston Colloquium for the Philosophy of Science, 1969–1972*, Boston Studies in the Philosophy of Science (ed. by Robert S. Cohen and Marx W. Wartofsky), Volume XIV. 1974, VIII + 405 pp. Also available as paperback.
61. Robert S. Cohen, J. J. Stachel, and Marx W. Wartofsky (eds.), *For Dirk Struik. Scientific, Historical and Political Essays in Honor of Dirk J. Struik*, Boston Studies in the Philosophy of Science (ed. by Robert S. Cohen and Marx W. Wartofsky), Volume XV. 1974, XXVII + 652 pp. Also available as paperback.
62. Kazimierz Ajdukiewicz, *Pragmatic Logic*, transl. from the Polish by Olgierd Wojtasiewicz. 1974, XV + 460 pp.
63. Sören Stenlund (ed.), *Logical Theory and Semantic Analysis. Essays Dedicated to Stig Kanger on His Fiftieth Birthday*. 1974, V + 217 pp.
64. Kenneth F. Schaffner and Robert S. Cohen (eds.), *Proceedings of the 1972 Biennial Meeting, Philosophy of Science Association*, Boston Studies in the Philosophy of Science (ed. by Robert S. Cohen and Marx W. Wartofsky), Volume XX. 1974, IX + 444 pp. Also available as paperback.
65. Henry E. Kyburg, Jr., *The Logical Foundations of Statistical Inference*. 1974, IX + 421 pp.
66. Marjorie Grene, *The Understanding of Nature: Essays in the Philosophy of Biology*, Boston Studies in the Philosophy of Science (ed. by Robert S. Cohen and Marx W. Wartofsky), Volume XXIII. 1974, XII + 360 pp. Also available as paperback.
67. Jan M. Broekman, *Structuralism: Moscow, Prague, Paris*. 1974, IX + 117 pp.
68. Norman Geschwind, *Selected Papers on Language and the Brain*, Boston Studies in the Philosophy of Science (ed. by Robert S. Cohen and Marx W. Wartofsky), Volume XVI. 1974, XII + 549 pp. Also available as paperback.
69. Roland Fraïssé. *Course of Mathematical Logic* – Volume II: *Model Theory*. 1974, XIX + 192 pp.
70. Andrzej Grzegorczyk, *An Outline of Mathematical Logic. Fundamental Results and Notions Explained with all Details*. 1974, X + 596 pp.

SYNTHESE HISTORICAL LIBRARY

Texts and Studies
in the History of Logic and Philosophy

Editors:

1. M. T. BEONIO-BROCCHIERI FUMAGALLI, *The Logic of Abelard.* Translated from the Italian. 1969, IX + 101 pp.
2. GOTTFRIED WILHELM LEIBNITZ, *Philosophical Papers and Letters.* A selection translated and edited, with an introduction, by Leroy E. Loemker. 1969, XII + 736 pp.
3. ERNST MALLY, *Logische Schriften*, ed. by Karl Wolf and Paul Weingartner. 1971, X + 340 pp.
4. LEWIS WHITE BECK (ed.), *Proceedings of the Third International Kant Congress.* 1972, XI + 718 pp.
5. BERNARD BOLZANO, *Theory of Science*, ed. by Jan Berg. 1973, XV + 398 pp.
6. J. M. E. MORAVCSIK (ed.), *Patterns in Plato's Thought. Papers arising out of the 1971 West Coast Greek Philosophy Conference.* 1973, VIII + 212 pp.
7. NABIL SHEHABY, *The Propositional Logic of Avicenna: A Translation from al-Shifā':al-Qiyās*, with Introduction, Commentary and Glossary. 1973, XIII + 296 pp.
8. DESMOND PAUL HENRY, *Commentary on De Grammatico: The Historical-Logical Dimensions of a Dialogue of St. Anselm's.* 1974, IX + 345 pp.
9. JOHN CORCORAN, *Ancient Logic and Its Modern Interpretations.* 1974. X + 208 pp.
10. E. M. BARTH, *The Logic of the Articles in Traditional Philosophy.* 1974, XXVII + 533 pp.
11. JAAKKO HINTIKKA, *Knowledge and the Known. Historical Perspectives in Epistemology.* 1974, XII + 243 pp.
12. E. J. ASHWORTH, *Language and Logic in the Post-Medieval Period.* 1974, XIII + 304 pp.
13. ARISTOTLE, *The Nicomachean Ethics.* Translated with Commentaries and Glossary by Hypocrates G. Apostle. 1975, XXI + 372 pp.
14. R. M. DANCY, *Sense and Contradiction: A Study in Aristotle.* 1975, XII + 184 pp.
15. WILBUR RICHARD KNORR, *The Evolution of the Euclidean Elements. A Study of the Theory of Incommensurable Magnitudes and Its Significance for Early Greek Geometry.* 1975, IX + 374 pp.
16. AUGUSTINE, *De Dialectica.* Translated with the Introduction and Notes by B. Darrell Jackson. 1975, XI + 151 pp.